D0989674

AMERICAN ABYSS

AMERICAN ABYSS

*Savagery and Civilization
in the Age of Industry*

DANIEL E. BENDER

CORNELL UNIVERSITY PRESS
ITHACA AND LONDON

HC
105.7
.B38
2009

Copyright © 2009 by Cornell University

All rights reserved. Except for brief quotations in a review, this
book, or parts thereof, must not be reproduced in any form without
permission in writing from the publisher. For information, address
Cornell University Press, Sage House, 512 East State Street, Ithaca,
New York 14850.

First published 2009 by Cornell University Press

Printed in the United States of America

Library of Congress Cataloging-in-Publication Data

Bender, Daniel E.
    American abyss : savagery and civilization in the age of industry /
Daniel E. Bender.
        p. cm.
    Includes bibliographical references and index.
    ISBN 978-0-8014-4598-9 (cloth : alk. paper)
    1. Industries—Social aspects—United States—
History.   2. Eugenics—Social aspects—United States—
History.   3. Alien labor—United States—History.
4. Women—Employment—United States—History.
5. Minorities—Employment—United States—History.
6. United States—Economic conditions—1865–1918.
7. United States—Social conditions—1865–1918.
8. United States—Civilization—1865–1918.
I. Title.

    HC105.7.B38   2009
    338.0973—dc22

                        2009013039

Cornell University Press strives to use environmentally responsible
suppliers and materials to the fullest extent possible in the publishing
of its books. Such materials include vegetable-based, low-VOC inks
and acid-free papers that are recycled, totally chlorine-free, or partly
composed of nonwood fibers. For further information, visit our website
at www.cornellpress.cornell.edu.

Cloth printing   10   9   8   7   6   5   4   3   2   1

CONTENTS

University Libraries
Carnegie Mellon University
Pittsburgh, PA 15213-3890

ACKNOWLEDGMENTS

"And beneath will be the abyss, wherein will fester and starve
and rot, and ever renew itself, the common people, the great bulk of the
population. And in the end, who knows in what day, the common people will
rise up out of the abyss; the labor castes and the Oligarchy will crumble away;
and then, at last, after the travail of the centuries, will it be the day of the
common man. I had thought to see that day; but now I know that
I shall never see it." He paused and looked at me, and added: "Social
evolution is exasperatingly slow, isn't it, sweetheart?"

Jack London, *The Iron Heel*

Slow, indeed. Historians take a long time to state their thoughts. We depend on friends, colleagues, students, and family to listen to us, comment on our ideas and our writing, and love us during the years it takes to complete a project. This particular set of thoughts took me about half a decade. From the moment I formulated my questions, then reveled in fascinating evidence, put pen to paper, and finally relinquished my files to storage boxes, I undertook my own evolution. I am a different scholar and, in many ways, a different person at the end of this project. This book was a journey, from question to question, library to library, and archive to archive. It also took me through a set of jobs and two countries. I am grateful to all those who helped and supported me and listened to me while I formulated all these thoughts. And, I am indebted to those who provided distraction, friendship, and love along the way.

I was lucky to have a wide and diverse group of scholars who read all or part of this manuscript. All provided indispensable friendship. I believe that this is a better book because of the efforts of Eileen Boris, Elspeth Brown, Rosanne Currarino, Donna Gabaccia, Cindy Hahamovitch, Rick Halpern, Franca Iacovetta, Eric Jennings, Russ Kazal, Molly Ladd-Taylor, Scott Nelson, David Offenhall, Steve Penfold, Kimberly Phillips, David Roediger, Natalie Rothman, Jayeeta Sharma, and Michael Wayne. I owe Rick Halpern a special thanks as a terrific friend, supporter, and constant partner in planning the most challenging of academic events.

Audiences at conferences in Canada and the United States and at invited talks at the University of Tokyo, University of Delhi, V. V. Giri National Labor Institute (New Delhi), Jadavpur University (Calcutta), Carnegie Mellon University, York University, and the University of Toronto offered wonderful questions and thoughtful critiques. I hope I have been able to answer them. I am especially grateful for the commentaries of Himani Banerjee, Sabyasachi Bhattacharya, Alan Kraut, Samita Sen, and Yujin Yaguchi. Here in Toronto, Jimmi S. B. Khargosh digested the manuscript in his own way.

My undergraduate students at the University of Toronto Scarborough, have been a constant source of inspiration. Progressive and exuberant, they remind me of the urgent need for accessible historical writing that critiques the past in order to change the present. Two brilliant undergraduates, Michelle Baigorri and Leah Ko, read the entire manuscript at short notice.

In the University of Toronto graduate department of history, my MA and PhD students have reshaped how I understand American cultural and intellectual history. Holly Karibo and Paul Lawrie proved to be indispensable research assistants.

I received needed research support at all the right moments. At Princeton University and the University of Waterloo, fellowships supported archival research. At the University of Toronto, the Connaught new faculty grant and an award from the Canada Fund for Innovation helped me finish my research and writing. As a Canada Research Chair, I have enjoyed the rare ability to travel, research, and support the research of students and scholars; it is a privilege.

The University of Toronto Scarborough, and the University of Toronto, as a whole, has proved an extraordinary place to teach and research. At the University of Toronto Scarborough, my chair, William Bowen, vice-president for research John Coleman, and dean Ragnar-Olaf Buchweitz provided essential resources and research time. They have created an atmosphere of support for humanities research. I am grateful as well to the efforts of Monica Hretsina and Laurel Wheeler. I am indebted to the tenaciousness of Elizabeth Seres and the interlibrary loan staff at the University of Toronto Scarborough library. They saved me uncountable hours and trips by finding the rarest of volumes.

I wrote this book while taking part in an extraordinary series of interdisciplinary events on transnational labor. A three-year project on global labor history entitled "Transnational Labor, Transnational Methods" was an odyssey of meetings in Hyderabad, India; Campinas, Brazil; and Toronto.

At each event, surrounded by wonderful scholars from around the world, I felt like I had returned to graduate school. "Global Economies; Working Communities," a yearlong Sawyer Seminar, amply funded by the Andrew Mellon Foundation, helped frame my understandings of the relationship between the domestic and the transnational realms. I am grateful to the many students and scholars who took part in these events, especially Christine Berkowitz, Sidney Chalhoub, Betsy Esch, Paulo Fontes, Chitra Joshi, Jeremy Krikler, Caroline Levander, Alex Lichtenstein, Minelle Mahtani, Ken Mills, Prabhu Mohapatra, Nico Pizzolato, Susie Rosenbaum, Nisha Shah, Marcel van der Linden, Barbara Weinstein, and Derek Williams.

Beyond Toronto, staff at the American Philosophical Society, Brown University, Cold Spring Harbor Laboratory, Columbia University, Radcliffe Institute, Truman State University, University of Minnesota (Social Welfare Archives), Wisconsin Historical Society, and Yale University provided friendly advice and help.

At Cornell University Press, Michael McGandy inherited this project when he arrived and was immediately supportive. He has a fine and intuitive touch as an editor. I am proud to have worked with him.

Scholarship without friendship is merely words on paper. Family and friends across the globe have truly made this book possible. In South Asia, I am especially grateful to Shahana Bhattacharya and the Sharma, Baruah, and Bordoloi families—my Indian kin. I am lucky to have brilliant and warm friends and colleagues in the departments of humanities and history (many mentioned above). The friendship of Raja and Pooni B. Babu (and family), Beth Landau-Halpern, Geoffrey Leonardelli and Kenya Leonardelli-Thompson (and family), Alejandro Paz, and Diane Swartz has made Toronto a permanent home. In the University of Toronto department of history and the University of Toronto Scarborough department of Humanities, Katherine Blouin, Katie Larson, Tong Lam, Michelle Murphy, Ian Radforth, Steve Rockel, and Alison Smith and many others may not have read the manuscript, but they created the perfect warm, friendly, and intellectual atmosphere that nurtured it.

One of the pleasures of writing a book is the chance to thank family. Mine is a family of love, nurture, support, bad jokes, politics, and conversation. If a book is a labor of love, then much of that love is for my brother, Michael Bender, my sister-in-law, Tamara Koss, and their new son, Stephen Bender; my cousins Joan Blumenfeld, Marjorie and Bob Boyar, Jeremy and Ingrid Feldman, and Alice Schumacher; and my aunts, Ellen Bender and Leslie Waldbaum, and my uncle, Larry Waldbaum.

A book is personal; it is political; and it is dedicated to loved ones: my grandmother, my parents, and my partner and wife. Mildred Waldbaum passed away just as I began writing my first draft. She was my grandmother, a confidant, a supporter, and a model for a life beautifully lived. My parents, Carl and Jessica Bender, have created a family of love, pleasure, commitment, and intellect. And, to Jo Sharma, a fellow historian, a beautiful friend, my love, and my life partner: together, we share thoughts about the past and love in the present and in the future. You are my greatest joy. Finally, the already beloved Piya Rose's arrival is, as I write, just days—or minutes—away. If it was completed before her birth, it is written for her future.

Toronto, Ontario
March 2009

AMERICAN ABYSS

INTRODUCTION

One of the great books of our century will be some day written on
the selection of men, the screening of human life through the actions of man
and the operation of the institutions men have built up. It will be a survey of
the stream of social history...and the effect of each and all of its conditions
on the heredity of men. The survival of the fit and the unfit in all degrees
and conditions will be its subject-matter. This book will be written,
not roughly and hastily, like the present fragmentary essay.

David Starr Jordan, 1907

In 1912, the sociologist Edward A. Ross set out to explore the industrial urban landscape of the United States. In the mines of Duluth, as in
the garment shops of New York City, Ross found workplaces filled with
immigrants. These "foreign" workers, strangers and laborers from distant
shores, as much as their communities, workplaces, schools, and places of
worship, emerged as the objects of his study. What started as an industrial
survey became industrial ethnography. Ross used every tool in the social
scientists' box. Anthropology, sociology, biology, economics, politics, and
history merged as Ross meandered from metropolises to mill towns to pit-
head villages.

He filled four notebooks with ethnographic research. He measured the
skulls and features of immigrant workers, both men and women. He described their living and working conditions and commented on their habits
and character. He interviewed the religious leaders, teachers, social workers,
politicians, bosses, and policemen charged with guiding, reforming, con-
trolling, and managing their labor. In his scrawling shorthand, these notebooks provided the evidence for his 1914 book *The Old World in the New:
The Significance of the Past and Present Immigration to the American People.*

The warnings Ross offered in *The Old World in the New* were dire. The
book and his notebooks reveal Ross's apocalyptic fears about the arrival of
immigrant workers that he regarded as part of distinct and inferior races.

Across the country, in big metropolises and small industrial cities, Ross found immigrants whose bodies he saw as twisted and ugly and whose skulls, lips, foreheads, and brows showed the telltale racial and medical signs of inferiority. He described dusty houses and tenements devoid of flowers or any other signs of pride or cleanliness. He detailed the proliferation of "fifty cent" brothels, rising crime, rampant alcoholism, and a growing army of unemployed tramps.

The strong American race that had conquered the frontier, he worried, was being replaced by immigrant races with low standards, expectations, vitality, and potential. The typical Jewish immigrant, for example, was hardly the "stoical type who blithely fares forth into the wilderness, portaging his canoe, poling it against the current, wading in the torrents, living on bacon and beans, and sleeping on the ground...'to keep hard.'"[1] New immigrants, like these flaccid, feeble, bacon-eschewing Jews, were taking jobs from superior races because they would accept lower wages. At the same time, such inferior immigrant races had a higher birthrate than the native born. Immigration, as Ross would argue elsewhere, was a foreign invasion leading to "race suicide." According to Ross, where once life on the American frontier had led to the selection of the fittest and strongest, natural selection was now taking place in the industrial city. And this new kind of social and biological competition seemed to favor inferior races.

Investigating Industry

Why was Ross so frightened? He was exploring and writing at a time when the United States had emerged as the world's preeminent industrial power and had pacified a two-ocean empire. The nation had outstripped its nearest competitors in the production of essentially all industrial products. It was nearing completion of the Panama Canal, which would link the two halves of America's imperial possessions and dependencies. The decades leading up to World War I were an era of industrial and territorial expansion punctuated by economic depressions and panics, working-class agitation, and resistance by colonized peoples. Ross surveyed an industrial landscape and saw on the horizon inferior races eager to arrive on American shores. Like many of his contemporaries, Ross came to view industry as the highest incarnation of civilization. Yet he believed that industrial civilization was threatened from within and from without by savagery.

Like Ross, historians have long placed industry at the center of turn-of-the-last-century American history. They have explored its significance, in particular, as a catalyst for waves of migration, urban growth, and class formation. Industrialization has been understood by historians as fitful efforts at mechanization and innovation that began in the young nation.[2] It has been reduced to a set of domestic processes and structural transformations contested by working people and their political allies. As a focus of cultural interrogation and intellectual explanation, the history of industrialization is comparatively impoverished.[3] Historians—even as they cite their work and evidence—find themselves at odds with earlier observers, who placed industry at the center of their efforts to understand simultaneously the origins of civilization, the persistence of savagery, the conflict of races, the emergence of class hierarchy, and the origins of gender divisions.

This book explores turn-of-the-twentieth-century notions of industry as the highest stage of civilization. Industry and industrialization referred, most obviously, to the emergence of factories and other locations of machine and mass production, including mines, mills, sweatshops, even homes. It also referenced a transformation in society and morals, accented by the growth of cities and by the rise of a distinct laboring class. Industry also suggested a way of life, characterized by diligence, energy, and innovation. For observers of American industry, from official government investigators to social scientists to socialist organizers, industrialization was part of a long process of human evolution that stretched back far beyond the Early Republic to cavemen and the travails of the lower animals and insects.[4] Industrial and labor history, they argued, began with cannibals and cavemen, not with the invention of the waterwheel or the emergence of the first textile mills. To write the cultural and intellectual history of industrialization, we must follow their debates to the most unlikely of places. Texts about contemporary labor problems or factory organization frequently included discussions of cavemen or bees; texts about bees or cavemen often ended with material on modern, human industry. The history of industrial evolution was, as the eugenicist and Stanford president David Starr Jordon so aptly put it, "the survey of the stream of social history." Factories and mechanized production, the highest achievement of mankind, they argued, were evidence of the evolutionary development of certain races.

The development of industry was racial progress from savagery to civilization. This perspective helps suggest why an essentially domestic development—the growth of factories, mines, and mills—was studied in the context of transnational processes, including imperialism and immigration. American historians in recent years have sought to place American history in a transnational context. They have traced global connections and the rise of an American empire. Too often, the domestic and the transnational have been deemed separate spheres. Industry has generally been examined within the domestic sphere (with the notable exception of a number of recent studies that examine production on a global scale).[5] The assigning of different topics to distinct spheres has led to an unfortunate disconnect between histories of industry, immigration, and empire.[6]

Yet turn-of-the-twentieth century observers considered these topics in interconnected ways and employed the same vocabulary of civilization and savagery. The omnipresence of conversations about civilization, in particular, has occupied cultural historians in recent years, especially as it related to the expression of manhood and the articulation of racial hierarchy.[7] The era of industry also produced efforts to define the savage and the primitive. Liberated from older ideas about the noble savage, observers engaged with evolutionary science to describe a continuum of racial development.[8] Especially for anthropologists and novelists, faraway tribes played the role of living fossils. They illustrated the evolutionary past of industrialized peoples. Such a sweeping formulation of primitivism naturally had its critics, most notably Franz Boas and his students. Yet Boas's significance is easy to inflate, partly because his cultural explanations of race seem more benign than the biologically inflected understandings he denigrated.[9] In fact, Boas's influence, at least until the 1920s, was limited primarily to academic circles. His best-known and most cited work remained his curious study of the head shape of Jewish and Italian migrants. Even Boas, despite his cultural focus, continued to work within the binary of civilization and primitivism.

The equation of industry with civilization forged a set of enduring connections. The study of a factory in Chicago, for example, demanded knowledge of colonial Africa because the opposite of the industrial factory was the nonindustrialized world, deemed savage, and nothing represented savagery in the American imagination more than Africa. The lives, conditions, and character of workers entered into the same conversations as the fate of Philippine tribes. The framework for understanding industry was expansive—far broader than historians have realized. It encompassed jungles, caves, tribes, immigration, and empire. To contemporary readers, these ties might seem bizarre. Factory management appears far removed from jungle survival. What can musings about the lives of cavemen—whether anthropological or fictional—really offer to the study of industrial development? Other connections, for example, between industrial labor and the arrival of immigrants, were forged at the same moment. Yet they have become virtually self-evident. The study of migration has become central to the study of industry.

The domestic was transnational and debates about industry, empire, and immigration merged. At the most practical level, as American industrialists worried about domestic overproduction, untapped or captive foreign markets seemed a possible solution. In almost the mirror image of such sentiment, especially after 1904, with the startling military victory of Japan over Russia, Americans worried about industrial competition from the nonwhite world. These intersections ran even deeper to the way American observers explained how immigration was transforming industrial cities, and how such cities should be investigated and portrayed, and how some races had achieved industry while others remained mired in savagery. When Ross and his contemporaries sought to understand the effects of recent immigration, they looked first to prehistory. Drawing on anthropological interpretations of migration as the process through which races confronted each other and battled for superiority, critics of immigration described the arrival of newcomers from Asia and the periphery of Europe as a form of negative colonization.[10] Given their advantages in the industrial struggle, these migrants would, effectively, conquer superior races. The stature of Americans would decrease and their physical appearance would deteriorate—everything that Edward Ross thought he discovered on his exploration of industrial America.[11]

American observers worried about the ability of civilized native racial stock to withstand the immigrant "invasion." Immigration represented the colonization of the metropole. By the turn of the century, American readers had embraced a European enthusiasm for books and magazines that described adventures and peoples in faraway imperial hinterlands. The narrative style of imperial travel writing was replicated in descriptions of visits to immigrant neighborhoods and workplaces written by journalists, reformers, and even government inspectors. Slum tourism was a kind of urban safari in which visitors, however well meaning, tested their courage. Even as social reformers and settlement workers emerged as some of the most vocal opponents of imperial expansion, they cast themselves as explorers of foreign colonies.[12]

Methods of investigating, exploring, and describing foreign and immigrant colonies merged. Through their study of primitive peoples, American observers also explained why only certain races had advanced toward industry. They looked to Europe for methodological inspiration, but to the nonwhite world for evidence. Through their adaptation of European imperial science developed in response to fears that white settlers would degenerate in the tropics, American geographers, racial theorists, and policymakers came to see climate as either the engine of or a hindrance to evolution.[13] In temperate zones, nature was fickle; survival demanded adaptation and innovation. Nature in the tropics, by contrast, promoted indolence. Observers laced industry with racial meanings, as the achievement of white races. They distinguished between colonizer and colonized, efficient and indolent. In worrying about the fate of white colonizers, especially in the Philippines, observers also defined African Americans—descendents of tropical races—as unfit for American industrial civilization. They believed that the African American, like the American Indian, was destined for extinction.[14]

Survival of the Unfit

This book is based on the premise that when turn-of-the-twentieth-century social observers, such as the leading economist Richard Ely, claimed to write the history of American "industrial evolution," they believed that they were documenting the most advanced stage of human development. Though too many historians have assumed that these observers had merely dropped the "r" from industrial revolution, the essence of how industrialization was understood, described, and studied lay in the term evolution. When economists, social reformers, sociologists, labor leaders, and popular journalists and writers used the term "industrial evolution," they truly meant evolution. For these turn-of-the-century observers, the industrial revolution was just a single step in the broader span of industrial evolution. If present-day historians are most interested in the specific transformations of the industrial revolution, turn-of-the-twentieth-century observers were concerned primarily with the larger processes of human development within industrial evolution. In applying theories about natural selection to the emergence of factories, mass production, manners of efficiency, and a distinct working class, they blurred the boundaries between economics, sociology, paleontology, anthropology, and biology. Industry, for such observers, included the development of new sites of production, urban areas where factories were located and workers lived, and a standard of hard work and behavior. Industry, in short, was a racial accomplishment.

The story of the dissemination of evolutionary thought from Europe to the United States is not new, but it is worth summarizing. Charles Darwin was part of wave of nineteenth-century social thinkers—by no means limited to biologists—who revisited the ideas of evolution voiced by Jean-Baptiste Lamarck and others.[15] By the 1860s, the debate about evolution had spread to the United States where it met the resistance of both clerics and biologists.[16] However, evolutionary thought gained adherents in the United States, especially as the nation tried to come to terms with a shifting racial order after Emancipation. Even more than in Europe, evolution became integrated into the language of public political debate. Herbert Spencer's triumphant 1882 tour of the United States marked the growing importance of evolutionary thought within and beyond the academy. Hailed as a hero, Spencer was feted with banquets, including a feast on the last night of his tour at Delmonico's in New York, the city's toniest restaurant.

The reception of Spencer as a popular hero in the United States highlights the way evolutionary theory in turn-of-the-twentieth-century United States was from the start applied widely to the study of human societies as much as to the origin of species. At a time when, as historian Dorothy Ross reminds us, the walls between the university and the larger public remained remarkably permeable, evolution was a public science in America. It could be used as much by the expanding generation of academics who defined themselves as sociologists, political scientists, and anthropologists as it could by biologists.[17] Moreover, evolutionary theory was employed equally by those applying social science to the questions and problems of society, including a host of social reformers and socialist activists, and by a range of artists and writers, loosely labeled naturalists.[18]

Evolutionary theory was a unifying theory that linked the study of the natural world to the study of the city, the exotic jungle to the urban jungle, economics to biology, sociology to social reform. How work was divided within families, how many children women bore, the prevalence of disease, alcoholism, and feeblemindedness (a term coined by evolutionists in the early 1900s)—all became questions as integral to economics as inflation, prices, supply, or demand. The ability of immigrants to withstand the struggle for survival within industrial civilization emerged as a critical economic variable. Biological conflict in the age of industry, evolutionists worried, differed from earlier forms of selection; losers did not simply perish and disappear. As Yale's Albert Galloway Keller argued, the progress of civilization mitigated the primordial violence and brutality of natural selection. "Societal selection," he wrote, "is logically certain to preserve those who would perish under nature."[19]

What exactly was the measure of fitness in an industrial civilization? Fitness in the economic struggle for jobs did not necessarily mean biological superiority or long-term social desirability. The fit, in the short term, paradoxically, might be the unfit in the long term. Industrial competition, especially in the face of immigration, favored those who would accept lower wages and declining working and living conditions. The poor conditions that Ross noted among European immigrant workers were ironically evidence of their short-term advantage in the economic struggle.[20] Economic competition within the working class did not preserve the best and most desirable races. Rather, it favored those more able to get and keep industrial jobs. Even those observers most sympathetic to immigrants worried about a process of race substitution in which the most advanced races lost their jobs to new and less civilized immigrants. Economic competition might not produce race progress. Savagery persisted within the borders of civilization.

Social investigators and governments officials alike connected the study of industrial competition to the identification of the character of races. Government committees, such as the Industrial Commission on Immigration (1901) and the Immigration Commission (1911), explored the traits of immigrant races to determine the most desirable races of workers. Similarly, Ross, in his exhaustive study of the effects of immigration, sought to explain the essential character of different immigrant races, and the economist John Commons, in his *Races and Immigrants in America,* included details on the ability of different races to withstand industrialization.[21]

If biological economic competition was racial, so too was it gendered. Observers agreed that industrial evolution led to a differentiation between the sexes. In primitive societies, as in lower races, men and women had similar roles in families. These families, moreover, were not necessarily monogamous pairings. Even those deeply hostile to women's wage work admitted that industry had its primitive origins in cavewomen's child rearing, not in men's hunting and warfare. Feminist economists and socialist feminists alike argued that although women's retreat to the home at the time of the emergence of savage tribes was an adaptation that allowed them to protect offspring from the violent male world, male superiority was merely a subversion of the natural order. Yet in an age of factories, the adaptation of domesticity was no longer beneficial. Women, instead, had been excluded from what the University of Chicago's W. I. Thomas called the "white man's world of practical and scientific activity."[22] For critics such as Thomas, the movement of women into the workforce accelerated a degenerative race to the bottom not simply in wages but in racial quality. The sexual division of labor collapsed as male workers were replaced in this race to the bottom. Their fall from the ranks of respectable workers to the hordes of paupers was more than impoverishment; it was a retreat to primitiveness.

Male and female losers in the industrial struggle for survival led a primal, primitive life. A host of poverty experts, settlement workers, reformers, socialists, and eugenicists cast paupers, tramps, and prostitutes as "savage survivals." Immigrant children in street gangs as well as working women found themselves compared to men of the lower races. Ominously, like the lowest animals, racially inferior humans seemed to be breeding faster than their moral and economic betters. The fear of the fecundity of immigrants, the lower classes, and paupers set off a panic about "race suicide," a term coined by Ross and echoed even by Theodore Roosevelt, who criticized elites for allowing the lower classes to breed faster. Typical of concerns about race suicide, Frank Fetter of Cornell University warned that "the ignorant, the improvident, the feeble-minded, are contributing far more than their quota to the next generation."[23]

The idea of industrial evolution raised the specter of degeneration, the moral and physical reversion to primitivism. American observers understood domestic industrialization as part of global processes of race competition. Some races would advance. Others would remain mired in primitivism. Still others, at the heart of civilization, would decline. American observers considered workers within the United States alongside not only prehistoric man but also foreign peoples. Sometimes they looked toward colonized peoples. They also examined the experience of industrialization in Europe, especially in Britain. British theorists of poverty, such as Charles Booth, as well as advocates of reform, such as settlement house proponents, found receptive audiences in the United States. Americans regularly cited British evidence and statistics to warn that British problems could reappear in the United States. They borrowed evolutionary theories and their close cousin, the idea of racial degeneration. The idea of degeneration was framed in Europe and connected in the United States to concerns about immigration, poverty, and urban life.[24] By the 1880s, the outcasts of society, the feebleminded, vagrants, prostitutes, criminals, drunkards, and paupers, were linked as degenerates.

The idea of degeneration was obviously hostile to those who occupied the lowest rungs of the working-class hierarchy. Paradoxically, notions of degeneration advanced between the 1880s and the early 1910s were wielded in support of Progressive reform and, remarkably, also socialism.[25] The new movements of the age, indeed the very origins of the Progressive Era activist impulse, were grounded in anxiety over the future of industry and the looming perils of degeneration. In intersecting ways, socialism, reform, and eugenics—three of the major political strands of the day—sought to achieve what socialist author Jack London termed "a new law of development." All engaged with different interpretations of scientific theory and each cast their politics as a manifestation of biological advance. Socialists described themselves as the first incarnation of a new type of evolved man, inclined toward mutual aid rather than individualistic competition. They traced an evolutionary history from cannibals to tribal combatants to warlike savages to capitalists and proletarians and, finally, to socialists. Reformers and settlement workers, meanwhile, described their work as the result of an evolution of sympathy.[26] Primitive brutality gave way to religious charity which, in turn, was superseded by the scientific philanthropy of progressivism. Eugenicists, finally, would boast that they depended on the most advanced science to reverse the racial effects of ill-advised charity.

My goal here is to shift analytical focus from histories that have provided a patchwork vision of turn-of-the-century notables such as Florence Kelly and institutions such as Hull House. This book examines the motivations and tensions behind efforts to confront the problems of degeneration and the survival of the unfit. It reveals the organic connections between different movements as they struggled to adapt their politics not only to the social reality of industrial America but also to changing scientific understandings, in particular, of heredity. Such an approach illuminates the painful transition from triumphant reform to ascendant eugenics, a process that included the marginalization of socialism. The continuing development of the science of evolution challenged how evolution could be wielded in the social realm. Changing ideas of heredity that denied the significance of environmental factors undercut the rationale for reform and socialism in favor of eugenics. By the 1910s, eugenicists suggested that degeneracy was caused not by poor environment, but by defective "germ plasm."[27]

Historians have attempted to trace the passage from reform to eugenics, perhaps the most dramatic political shift of the years around World War I. The historian of American nativism John Higham noted that by the 1910s there were more mentions of eugenics in the popular press than about all the favorite topics of reform, from gymnasiums and playgrounds to tenement regulation.[28] In fact, the transformation was part of a critique of charity and philanthropy that gathered force over the course of the twentieth century. More than historians have ever noted, reform and, to a lesser extent, socialism never evinced an ecumenical belief in the fitness of the urban impoverished. Rather, they depended on a careful balance between campaigns for the segregation and elimination of the unfit and efforts at the preservation and protection of the unfortunate. For eugenicists, reform and socialism were out of step with scientific orthodoxy and simply preserved the unfit. Yet eugenics inherited much from reform. Not only did numerous reformers move increasingly to eugenics by the late 1910s and 1920s but the same contingent of college-educated women who had once served as the backbone of reform had become the field-workers of eugenics. Eugenicists' efforts constituted a direct attack on reform insofar as they expressly reinterpreted the very same evidence reformers had often cited to prove that heredity was related to environment and, therefore, could be altered by human benevolence. If progressive environmentalists and eugenicists disagreed strenuously on the causes of degeneracy, they agreed fully on its symptoms.

In short, the emergence of organized eugenics—what historian Daniel Kevles calls "mainline eugenics"—represented a scientific and political rupture. It ripped asunder the political consensus of the Progressive Era that had coalesced around concerns about the racial effects of industrialization. This consensus had loosely united reformers, liberals, socialists, and trade unionists in a shared belief in the environmental roots of degeneration. Eugenics threatened the scientific justification for reform and placed the blame for degeneracy squarely on the frail shoulders of workers and immigrants. Immigrants' racial inferiority was not a problem exacerbated by poor conditions, but a permanent degeneracy. Degeneration was not an ongoing process; it was a fait accompli. By the 1920s, eugenicists helped rewrite the nation's immigration laws and articulated new racialized fears of communism as the ultimate triumph of the biologically unfit. Yet they regarded the passage of the landmark 1924 Johnson-Reed immigration restriction act with its stringent system of national quotas as only the beginning of a process of racial salvation.[29] Even as the bill inched its way through Congress, they were already bemoaning its loopholes, especially those that permitted the immigration of Mexicans. This was a new "conquest." Observers of the new industrial reality shifted their focus from the racial character of European migrants and warned of a "rising tide of color." They feared a conflict that would pit colored races against white races.[30] The age of white supremacy had passed, first, with the defeat of Russia in 1904 and, later, with the racial "civil war" of World War I, a cataclysm, they argued, that had served only to carry off the fittest of the white races. The process of "becoming white" that has been the focus of recent immigration histories was, in fact, part of a retreat from empire, in the aftermath of World War I and in the face of the growing industrial power of Asia.[31]

Evolving America

When the popular race theorist and immigration critic Lothrop Stoddard warned of a post–World War I "revolt against civilization" led by Bolsheviks and colored races, he mourned the passing of an era of white supremacy in industry and empire. In the coming decades, questions of whiteness and blackness—as Stoddard predicted—would increasingly move to the center of the nation's political, cultural, and economic debates. Many of the ideas of an earlier age reappeared in new guises. Eugenics, as historians have often noted, left its mark, even as its original proponents faced criticism as shrill racists. Theories of industrial evolution bequeathed a legacy in a new language of "development." Immigration, as well, is still frequently cast in racial terms as invasion and conquest.

Stoddard's warnings, like Ross's earlier concerns, suggest anxiousness about the future of an age of industry. This book draws on that anxiety to suggest a historical synthesis of the turn of the century and challenges distinctions between the domestic and the transnational, metropole and colony.[32] This synthesis retains some familiar events, movements, and individuals. The analysis of the obsession with the fate and direction of civilization helps identifies new key moments, such as three Race Betterment conferences called to assess methods of racial improvement; crucial debates, such as those about the effects of tropical climates; startling ideas, such as the notion of immigration as colonization; and key texts, including popular fiction about life in the prehistoric world.

We begin in prehistory—or, at least, with the turn-of-the-twentieth-century fascination with prehistoric man. Chapter 1 examines the roots of the idea of industrial evolution in the shadowy and largely imagined world of cavemen and savages. Like the theory of evolution itself, the concept of industrial evolution was expansive and synthetic. It linked ideas of labor, empire, race, gender, family, and migration and divided the world into civilized and savage, industrialized and primitive. Chapter 2 explores the shared conversation about industry and empire. American observers studied the effects of climate to explain why temperate races had advanced, while tropical races remained stagnant. Although this racial science questioned the ability of American colonizers to maintain their high racial standard, especially in the Philippines, it also cast domestic industrialization as a racial achievement. Theories of climate may have placed civilization in the temperate zone, but modern migration made it simple for inferior races to travel to the heart of civilization. Critics compared ancient and contemporary immigration, the subject of Chapter 3. Where once only the bravest and strongest would attempt migration, steamships had reversed the selective processes of immigration. Less civilized races confronted their racial betters in the struggle for industrial jobs. Critics warned of a process of race substitution that favored the fit. Critics feared that while industrial competition might lead to the extinction of African Americans, old-stock white Americans might not be able to withstand the invasion of European and, especially, Asian immigrants. The competition for jobs not only pitted race against race but also women and children against men. Chapter 4 analyzes the intersection of theories of industrial evolution and the study of the origins of the family. In novels and social-scientific tomes, feminist economists and socialist feminists argued that the origins of industry in the first use of tools for production lay in women's primitive domestic labors. Yet arguments for the feminine origins of industry did not translate into sympathy for women and child workers. By the first decade of the new century, observers across the political spectrum were warning that industrialization was leading to race suicide and racial degeneration.

Chapter 5 begins a discussion of the way evolutionary understandings of industrialization shaped strategies of social reform, socialism, eugenics, and, finally, working-class radicalism. Worried critics traveled to slum neighborhoods—immigrant "colonies"—in search of evidence of degeneration, the overarching racial problem caused by industrialization. They employed the same methods as explorers visiting imperial hinterlands. Reformers and socialists outlined strategies to transform the degenerative environment of the slum. Their success, however, depended on the identification, segregation, and elimination of tramps, prostitutes, and the feeble-minded. Socialists and settlement workers, the subject of Chapter 6, argued that philanthropy was itself an evolutionary development. They defined their politics as the product of advancing civilization and as necessary for continued progress. Their plans for countering degeneration depended on a belief in the importance of environment in determining racial progress or decline. They believed that improvements in environment could be passed on to progeny. Yet by the second decade of the twentieth century, it became clear that their understanding of evolutionary science ran counter to an emerging scientific orthodoxy that focused on the significance of ingrained heredity, not environment. Chapter 7 traces the rise of a generation of eugenicists who sought to revitalize links between biology and social science. They accused reformers and socialists of preserving the unfit. Chapter 8 examines the fractured politics of the late 1910s and 1920s. Radicals, increasingly at the margins of American politics, advanced a limited counternarrative to the notions of fitness advanced by eugenics. Whereas earlier socialists had condemned the tramp, these radicals, especially those sympathetic to the Industrial Workers of the World (IWW) union, celebrated the tramp as a proletarian hero. World War I not only silenced such radical voices but also emboldened critics of immigration. The war heightened concern about degeneration and the threats to civilization from Asia and from Bolshevism. As immigration restrictions were extended in the 1920s, notions of racial difference shifted. Critics described a white race besieged by Mexican immigrants and African American migrants. Fears of invasion and degeneration persisted into the interwar years, as Chapter 9 shows.

Finally—the "abyss." In 1902, when Jack London set out to write about city slums with "an attitude of mind which I may best liken to that of the explorer," he plunged into the "abyss." Using a common, but dramatic word so often favored by social investigators, he found an "under-world" life that was "dwarfed and distorted." Settlement workers in New York were appalled, but not surprised, by his discoveries. "One almost refuses to believe that such poverty and misery exist," they wrote. But "our own knowledge of conditions in New York, dispel our doubts." When socialists, settlement workers, or eugenicists looked out over the racial landscape of industrial civilization, they peered into the abyss.

Equally, when the imaginary revolutionary hero of London's 1908 novel *The Iron Heel,* Ernest Everhard—a name which evoked visions of the fit, manly proletarian hero—led the unsuccessful "second revolution" in "the spring of 1932 A.D." he envisioned the emerging civilization of the "common man." London's novel is a musing on violent "social evolution." It lavished vivid prose on the savagery of class conflict, the violence of revolution, and the degeneration of the working class. "The condition of the people of the abyss," Everhard declared, was pitiable. "They lived like beasts in great squalid labor-ghettos, festering in misery and degradation." London's socialist fantasies of working-class degeneration, in fact, hauntingly evoked Ross's own real-life observations. Urban denizens had returned to the jungle, a brutish, primitive world of apes and tigers:

> Men, women, and children, in rags and tatters, dim ferocious intelligences with all the godlike blotted from their features and all the fiendlike stamped in, apes and tigers, anaemic consumptives and great hairy beasts of burden, wan faces from which vampire society had sucked the juice of life, bloated forms swollen with physical grossness and corruption, withered hags and death's-heads bearded like patriarchs, festering youth and festering age, faces of fiends, crooked, twisted, misshapen monsters blasted with the ravages of disease and all the horrors of chronic innutrition—the refuse and the scum of life, a raging, screaming, screeching, demoniacal horde.

London had turned turn-of-the-century anxiety on its head, glorying in the perils of degeneration. In the end, after his failed revolt Everhard turned back the evolutionary clock and retreated to a cave, before his capture and execution. After all, as Everhard himself warned, "'social evolution is exasperatingly slow, isn't it, sweetheart?'"[33]

1

CAVEMEN IN THE PROGRESSIVE ERA

From Savagery to Industry

Thus is civilization menaced by dangers perhaps as grave as those that overshadowed it at the beginning. It was threatened then by barbarism beyond its
walls. To-day it is threatened by the savagery within its gates.

Franklin Henry Giddings, 1896

Old Long Beard took another bite of the bear carcass, smacking his lips.
His grandchildren sat around the fire and, in between their grandfather's
animalistic gnawing on the bones of the bear, listened to his history of
industrialization.

This is the opening scene of Jack London's short story, "Strength of the
Strong," the lead story in his 1912 collection of short stories that stretched
from prehistoric bear feasts to modern general strikes to conflicts over immigration. The combination, eclectic to the modern scholar, made sense
for the turn-of-the-century reader. For London and his contemporaries,
through the bear-muffled voice of Old Long Beard, industrialization was a
long process of natural struggle that began in prehistory, in an age of cavemen. But, it was not the simple triumph of the most fit envisioned by social
Darwinists such as William Graham Sumner. For London, strength, fitness,
and the ability to endure the struggle for survival were not synonymous.
Sometimes the weakest survived.[1]

Long Beard originally dwelt in the treetops and stole his arboreal neighbors' mates. As he descended from the trees, he recognized that he and
his neighbors shared the bonds of tribe. Soon, the communal Eden of the
tribe was disrupted by Little-Belly who discovered that he could overcome
the fickle generosity of nature by building a fish trap. Industry was part of
mankind's ongoing conflict with nature—in short, a biological adaptation.

The Fish-Eaters tribe now had a steady stream of fish, but Little-Belly alone owned the means of production. Long Beard was the caveman common worker. He was paid wages for the savage task of catching fish.[2]

As the tribe divided between those who worked and those who owned, a system of seashell currency emerged. Concurrently, mate stealing had given way to a more stable world of families. At the same time that the Fish-Eaters recognized the commonality of the tribe and developed the first families, they discovered an enemy in the Meat-Eater tribe that lived in the next valley. The Meat-Eaters, undergoing a similar process of industrialization and social stratification, enjoyed a rising population. Freed from the vicissitudes of nature, this primitive industrialization, that is, the first forms of production that used rudimentary tools, permitted population growth, which, in turn, encouraged mass migration. The Meat-Eaters conquered the Fish-Eaters and cemented their superiority by seizing their wives. Long Beard, vanquished in the social struggle, underwent a process of biological degeneration. He fled to the hills—long considered by social scientists the refuge of the least-fit races—and returned to hunting and gathering. Thus, this racial degenerate was sitting on his haunches, eating the carcass of a bear, and narrating the story of industrialization.

The stuff of thinly veiled allegory, the "Strength of the Strong" reflected turn-of-the-century understandings of industrialization as a process that could be traced to humankind's descent from the trees, the recognition of kinship or enmity with fellow humans, and the first hints of family relations and stable marriage. The primeval setting presented by London was intentional, as was the division of his characters into Meat-Eater and Fish-Eater tribes. London captured in fiction the dominant social-scientific representation of industrialization as an evolutionary saga that pitted group against group. Turn-of-the-twentieth-century observers, from social scientists to socialists to novelists, examined industry as a stage in a process of human evolution that stretched back to prehistory. Such an interpretation cast industrialization not merely as technological advance or consumer abundance but as a state of civilization itself.[3] Industry, of course, referred specifically to factories, sweatshops, workshops, and mines as well as to an ethos of hard work and efficiency. Less obviously, it stood for a stage of racial development, the high-water mark of civilization.

Linking industry to civilization broadened its history. Turn-of-the-twentieth-century observers studied industrialization by searching for the engines of human evolution, rather than simply by evaluating new work processes.[4] Most looked beyond the development of the individual; modern man was no stronger or smarter than the ancient Greek. Instead, they found explanations for what they termed "industrial evolution" in the global origins of families, tribes, races, castes, and classes.

These human divisions, in turn, were fostered by the currents of ancient migrations. Such migrations long predated the modern context of steerage, restriction, or medical inspection.[5] Yet such observers viewed modern migrations through the lens of the primitive. They were writing from a perspective in which they saw contemporary migrants as the backbone of the modern factory workforce. They were also intertwined in their own world of international migration, this time of ideas and scholarship. As American social scientists struggled to create the definitions and institutions of their emerging fields, they found themselves deep in conversation with their European counterparts. In the process, they intertwined migration, evolution, civilization, savagery, and industry. Their conclusions laid the foundation for concern about the future of industry: toward higher civilization or backward to savagery.

Industrializing the Social Organism

The history of industrialization began with troglodytes. Industry emerged not from evolving human intelligence, but from the evolution of the "social organism."[6] Social scientists' understanding of industrialization as part of a long process of human development was indebted to recent conceptions of human evolution that lay at the heart of the development of American social sciences. The emergence of American social science—whether located formally in established universities such as Yale or in newer centers of learning such as the University of Wisconsin and Johns Hopkins University or informally in the pages of popular magazines and literature—was profoundly influenced by European developments. As historian Dorothy Ross has noted, by the turn of the century, a new generation of American social scientists, devoted to disciplines such as economics or sociology, rejected the strictures of the American academy. Many sought training in European universities and returned to reinvent a social science directly engaged with the problems of industry and informed by recent advances in biological theory. As the economist Henry Carter Adams would declare, the development of social science depended on "observing new truths to emerge from the growth of the social organism."[7]

Among the social sciences, sociology and economics were especially concerned with the rise of industry and often drew upon anthropology for evidence. Sociology was a new field. Indeed, the word was coined (in French) by Auguste Comte only in 1838. Comte's positivist ideas about the stages of human life were reinterpreted into a biological frame by the Englishman Herbert Spencer in his 1874 *The Study of Sociology*. American sociologists, most notably Lester Frank Ward, joined Spencer in linking Comte's original sociology to evolutionary biology. Ward's 1883 *Dynamic Sociology* placed sociology squarely in the realm of science, not religion. The development of man—the rise of civilization itself—was a biological, organic development: "His progress has been the progress of nature, not...the result of foresight and intelligent direction....It is natural selection that has developed intellect."[8] Ward's passionate belief in the place of sociology within the biological constellation was no surprise. He had first forged his public reputation as a biologist and botanist. Likewise, Columbia's Franklin Giddings, another towering figure of early American sociology, readily acknowledged his debt to evolutionary theory: "The evolutionist explanation of the natural world has made its way into every department of knowledge....It was inevitable that the evolutionary philosophy should be extended to embrace the social phenomena of human life."[9]

From its incarnation, sociology was concerned with the question of labor and industry. Ward placed innovation and invention at the center of his story of human development but, just like bird nests or beaver dams, described them as "the products of natural selection." Others, like Albion Small, the founder of the pioneering University of Chicago sociology program, described industry as a cause of increasingly complex social ties, "as industries become diversified, as division of labor and competition become territorial and international." The Montana rancher was connected to coal miners, machinists, and carpenters whose labors made the shipping of his herds by railway possible. The silver miner in Nevada was affected by lawmakers in faraway Washington. For the socialist economist J. Howard Moore, the industry of the "higher races" depended on complex and organic linkages: "They are grouped into great cities and states and maintain vast industries....Life is *co-operative*."[10]

In comparison to sociology, the emergence of a biologically inflected economics was characterized by rupture and rebellion. By the 1880s, a younger generation, led especially by Richard Ely, began critiquing an American economics that remained in the cocoon of classical economics. He accused economists of being uninterested in historical context or in the social effects of their study. The German-trained Ely and his allies proposed a more socially aware economical science. They attacked laissez-faire as dangerous and immoral and advocated an active role for the state and for reform. In 1885, they founded the American Economic Association. The AEA quickly rose to prominence, despite the dramatic 1894 University of Wisconsin disciplinary hearing of Ely, its preeminent member, on charges of radicalism and socialism. Triumphantly acquitted, Ely helped Wisconsin become a leading center for the study of industrial evolution.[11] By 1906 he had brought Edward A. Ross, himself fleeing charges at Stanford, and John R. Commons to Madison. All three would extend the study of industry by exploring the effects of immigration and the details of American industrial history.

By the turn of the century, the AEA had established itself as the leading economics association in the United States. Key to the success of the AEA and to its attack on the older practice of economics was its dedication to evolutionary science. This placed it squarely at the center of the biological turn in the social sciences. Illustrative of the new evolutionary focus in American economics was Thorstein Veblen's 1898 article "Why Is Economics Not a Biological Science?" Veblen critiqued the older economics for its classical bent and disconnect from contemporary social relations. Instead, he promoted a new engagement with evolutionary science. He urged economists to engage, like biologists, with a "theory of process, of an unfolding sequence."[12] Though written in impenetrable prose ("All the talk about cytoplasm, centrosomes, and karyokinetic process") many of Veblen's critiques were increasingly incorporated into American economics. Ely articulated the new reality of biologically inflected economics in 1901 when he would confidently write that "nothing could well be more unscientific in the present age of science than to leave evolution out of account in our examination of anything so fundamental in society as competition." After all, he insisted, "competition is the chief selective process in modern economic society, and through it we have the survival of the fit."[13]

Ely and Commons may have repudiated socialism, but socialists remained engaged with social science. As historian Mark Pittenger has noted, the development of American socialism, its rebuff of utopianism, and the rise of the American Socialist Party all depended on a foundation in evolutionary theory. Socialist intellectuals relied on evolutionary theory to proclaim the coming of socialism as an inevitable and natural step in human development. The social linkages fostered by industry were part of "race progress" toward socialism. As Murray King declared, in a typical effort to naturalize socialism as biological development, "*Socialism has co-existed with the race; it is but the expression of the growth of society.* From the primordial brute world...to the present time and onward, the vast social processes of integration work to the making of man."[14]

In their shared focus on human and natural evolution, socialists and nonsocialists alike drew on multiple and often contradictory biological theories of human development. American scholars drew freely from the theories of Lamarck, Darwin, Spencer, T. H. Huxley, and others, even when these theories were in obvious contradiction.[15] The world of evolutionary science, especially in North America, blurred boundaries between animal and human evolution. A study of bees might include a comparison to human industry. A text on the rise of the factory system might equally reference the experience of insects or the adaptations of animals. The socialist Frankenthal Weissenburg for instance, looked to the "ways of the ant" as examples of "a more perfect communalistic society."[16] Among nonsocialists, Veblen included details of the coral polyp in his survey of human economic evolution and the noted explorer and geologist John Wesley Powell (known, among other things, for his mentoring of Ward) intertwined his study of competition with examples from human, plant, insect, and animal life. This mingling of the various categories of the natural world helped Powell reach a conclusion similar to that of Ely, that economic competition among humans was natural struggle: "Competition among plants and animals is fierce, merciless, and deadly." The adaptations of the bees and the birds and the innovations and inventions of humans emerged through this competition.[17]

If innovation had its origins with the lowest animals, human civilization represented a biological rupture. As Powell argued, civilized humans no longer competed with other animals or plants in the struggle for survival. Moreover, human development was not centered on the evolution of intelligence. Biologists and sociologists, regarding the Greeks enviously, admitted that human intelligence had not expanded significantly. Ward noted that one need only look to the desperate moral and intellectual conditions of the working classes to realize that evolution had not fostered the growth of intelligence.[18] Human evolution, instead, was the product of increasing specialization and cooperation combined with the selective forces of race conflict emerging from ancient migrations.

Scholars imagined society as a "social organism" that went through its own process of evolution. In the chain of development of "protoplasm to man," the first stage in human evolution may have been the emergence of intelligence. The rise of "his empire," however, depended on the evolution of social instincts. Among humans, the fittest were the most sociable and "sociability," for Giddings, was the prime factor in human evolution.[19] Ward, likewise, argued that "society is simply a compound organism whose acts exhibit the resultant of all the individual forces which its members exert." Society was "a natural object," as complex as any animal.[20]

The notion of the social organism was critical to theories of "social evolution" that emerged as early as the 1880s. The term social evolution was the

brainchild of Spencer and, like Spencer himself, received an enthusiastic audience in the United States. By the turn of the twentieth century, the study
of social evolution had moved beyond a dependence on the work of Spencer.[21] A generation of American scholars from Albert Galloway Keller, the
intellectual protégé of the conservative William Graham Sumner, to Ward,
Giddings, and Simon Patten, all opponents of laissez-faire, placed "social
evolution" at the center of their sociological treatises. Regardless of their
politics, these American observers understood social evolution as leading
to the emergence of industry. They cast industrialization as the most developed incarnation of the social organism. It demonstrated the most complex
specialization that divided mankind into workers and managers, each analogous to the body parts and organs of the higher animals. If workers were
the blood cells of the social organism, managers and industrialists were its
vital organs. "The individual is no longer industrially independent," concluded one sociologist. Instead, "social differentiation has made him part of
a living structure, an organism."[22]

The notion of "industrial evolution" emerged in the context of the larger
study of social evolution. The term "industrial evolution" was first used in
a title in 1893 by the German economist and classicist Carl Bücher, who
cast the modern factory economy as the height of civilization. He traced
its ancient origins in prehistory and savagery. He found his evidence in anthropological studies of living primitives and in images of the prehistoric
past when "little groups like herds of animals…roam about in search of
food, find a resting-place for the night in caves."[23] His method captured
the attention of American observers. In 1901 his text was translated into English and published in the United States. Soon, drawing heavily on the work
of their European counterparts, American scholars presented their own
theories of industrial evolution. Their engagement with European theories
helped American observers understand domestic industrialization as part
of a global process of human development. Like Bücher, Ely's *Evolution of
Industrial Society* (1903) began when "our ancestors were living mere animal existences." Using evidence from the tribes of northern Canada and the
Australian outback, Ely traced the mental, moral, and industrial characteristics of the earliest hunting and fishing stage. The Chicago factory owner
was linked to the Australian Aborigine as extremes in the process of human
evolution.

For Ely, industrial society—not an imagined bucolic agrarian republic—
was the result of a long process of development that stretched back to "the
beginning of human existence."[24] The transformation of American social
science from a nostalgic interest in the independent farm and rural life to a

deep engagement with industry was a fundamental product of the idea of industrial evolution. As Ely cemented his position in Madison, in 1904 he launched an ambitious effort to document American industrial evolution. He placed his Bureau of Industrial Research in the hands of Commons. Under Commons's direction, the Bureau produced multivolume documentary and narrative histories of American labor. Commons sought to place American industry in the context of global human development. In 1907, for example, he initiated a correspondence with leading anarchists in an effort to compare different biological understandings of competition. They found the roots of competition in the age of industry in the distant age of "brutes and savages." Similarly, in 1909, in one of the earliest publications to come out of the Bureau's work, Commons examined the case of labor organizing among shoemakers as a case of "industrial evolution." He described the changing nature of organizing as symptomatic of the progress from primitivism: "Each of these organizations stands for a definite stage in industrial evolution, from the primitive...to the modern factory."[25]

Commons and his associates began their monumental two-volume history of the American labor movement by placing the American experience within the context of global "industrial evolution." They looked to socialist and nonsocialist theorists of industrial evolution from Bücher to Karl Marx (and his interpretation of American anthropologist Lewis Henry Morgan) to the British John Hobson to the Belgian Emile Vandervelde. Even their belief in the exceptionality of American class conflict was grounded in their global comparison of civilization and savagery. The peculiarity of the American trade union movement could be traced partly to the fact that American colonialists, even though they constructed a new nation in a primitive boreal environment, had come from an advanced race that had already evolved through the early stages of industrial evolution. Colonists had already passed through the primitive manufacturing stage, and thus "industrial evolution in colonial times was not the evolution of tools and processes, but the evolution of markets fitted to utilize the tools and processes already evolved." For Commons, therefore, contestations over this rise of markets, not over the means of production, would determine the contours of American class conflict.[26]

American socialists eagerly engaged in scholarship that defined industrial evolution as progress from savagery to industry, ultimately culminating in the socialist state. The socialist Vandervelde's landmark *Collectivism and Industrial Evolution* was translated and published in the United States in 1904 by the leading American socialist publisher, Charles Kerr. For many socialists, society advanced in stages—"social metamorphoses" as King called them—from primitivism to warfare to capitalism and, eventually, to the "co-operative commonwealth."[27] Even the trust—an industrial organization most prominent in the United States—came to be understood as the most developed example of the specialization of the social organism. Historians have often noted the ambivalent relationship of turn-of-the-twentieth-century social scientists and socialists to the trusts. Such observers criticized the trusts for their exploitative labor policies and disregard for the concern of consumers yet praised their rational industrial organization. Still, for socialists, the trust stood as a prime example of a social organism moving toward perfection. Socialist intellectuals seized upon the trust as evidence for the scientific inevitability of socialism. The trust, they argued, mirrored the kind of vast and rational social organization of industry that would result in the socialist state.[28]

Cavemen into Workers

For turn-of-the-century observers, the history of American industrialization began long before the development of the first factories in New England in the first decades of the nineteenth century.[29] Moreover, its study depended on evidence drawn from far beyond the borders of the United States. Bücher's landmark text advanced the argument that industrialization had its roots in prehistory. His study began with the "economic life of primitive people" and traced the origin of the sexual division of labor, the earliest technologies of food procurement, and the first "embraces [of] what we designate industry" in the construction of simple tools. He found his evidence in a vast array of anthropological studies of living primitive tribes and in imperial travel narratives. In a similar fashion, Veblen's plea for an evolutionary economics was grounded in following the progression from "primitive conditions" to modern civilized industry. Under primitivism, Veblen believed, mankind was under the sway of immediate environmental conditions "much as beasts are" and evolved "modern mechanical processes" to cope with a harsh nature.[30]

The theoretical pleas of Bücher, Veblen, and Ely, among others, for the centrality of the Stone Age were reflected in popular representations of industrial evolution. In the early years of the twentieth century, Ely was a frequent figure on the lecture circuit. Significantly, his lectures on "industrial evolution" rarely even ventured past primitives. One 1912 lecture, typically, was rich with details drawn from the lives of primitives within the American empire: North American Indians and Filipino tribes.[31] Otis T. Mason, the first curator of the Smithsonian Institute, followed suit, arranging the museum's primitive artifacts collection to illustrate technological evolution. Much to the chagrin of later observers, notably Franz Boas, Mason gave pride of place to the story of industrial evolution over the specifics of individual tribal cultures.[32]

The celebration of American industrial achievement in publicly accessible books and, even more obviously, in world's fairs depended on contrasts with non-American primitivism. In her broad survey of American industrial evolution, Katherine Coman, a Wellesley historian and economist, praised the racial success of the United States in converting "virgin" territory into bustling industry. This success became meaningful only in contrast with the savages who had given way to the first European migrants. The key to the triumph of a nation was the racial character of its inhabitants, more than its natural resources: "The most propitious physical endowment can avail little if the inhabitants of the land are so ignorant or so sluggish as to leave its resources unexploited. The Aborigines of North America, so far as history knows them, were lacking in these essential economic traits....The most civilized of all, the Aztecs of Old Mexico, had hardly passed beyond the barbarous stage of evolution."[33] For those unwilling to read Coman's lengthy narrative, the incessant comparison between American industry and foreign savagery appeared dramatically at the Chicago World's Columbian Exposition, a celebration of American industrial evolution as the progress of civilization. Most obviously, the fair's architecture depended on contrasts between civilization (represented by glorious temples of industry) and savagery (represented by the ribald entertainments and displays of native villages on the Midway). Even within the hallowed halls of the Manufactures building, visitors were encouraged to contrast American industrial production and the primitive industries of the world. In neoclassical grandeur the building itself depicted the various stages of industrial evolution, from hunting to artisan crafts to modern metalworking. The exhibits, meanwhile, showcased American products through contrasts with products of countries "whose manufacturing industries are yet in their infancy, such...as Zanzibar and the Orange Free State."[34]

Cavemen mattered in a modern world obsessed with measuring the distance between civilized and savage.[35] The University of Chicago sociologist W. I. Thomas, for example, warned against neglecting the study of the "institutional life of savage society." This was a lesson the United States commissioner of labor, Carroll Wright, took to heart in a series of public lectures in 1906 that explored the causes of contemporary labor unrest. He turned first to the experience of prehistory and then to the ancient civilizations of

Figure 1.1. This simian caveman earned pride of place at the front of a synthetic history of social and industrial evolution. Frontispiece of F. Stuart Chapin, *An Introduction to the Study of Social Evolution* (1921).

Egypt, Greece, and Rome. To understand contemporary labor problems, he reasoned, twentieth-century observers needed to look back to prehistory and ancient eras. Labor conflict was tied to an evolutionary transition from tribal conflict to armed militancy to industrialism.[36] In similar fashion, F. Stuart Chapin's *An Introduction to the Study of Social Evolution* assigned a central place to the caveman. Even his frontispiece featured an imaginatively primitive caveman with appropriately simian features and a heavy club.[37]

Popular books about the Stone Age, such as *The Later Cave Men* by Katherine Dopp, kept their focus firmly on the development of industrial life. *The Later Cave-Men*, anticipating "the discovery of the fact that steam is a force that can do work," was part of a school series on "Industrial and Social History." The series began with *The Tree-Dwellers* and ended with *The Early Sea People*. In between, *The Later Cave-Men* was divided into short stories that described incremental innovation, food gathering, and tribal ritual as part of the daily life of a tribe of cavemen. One story, for example, traced the effort of the cavemen to discover new ways of making new spears: "Perhaps they did not want to do it. But they had to do it or die." Another story described the discovery of snowshoes in the midst of a brutal snowstorm.[38] Similarly, Margaret McIntyre's novel *The Cave Boy of the Age of Stone* built its adventures around the process of invention and evolutionary progress. From its drawings of cavemen inventing stone tools to anecdotes of hunting and making needles from bones and thread from tendons cut from their prey, the novel turned industrial evolution into popular entertainment.[39] Even as serious a sociologist as Giddings tried his hand at primitive fiction with his *Pagan Poems*. His poems ranged from marauding cave bears to the "belching smoke and deafening din" of "industry."[40]

Darwinism and evolutionary science in general helped open up to public fascination prehistoric millennia previously shrouded by biblical chronologies. Americans were enthralled as biologists revealed past ages for public imagination and archeologists and anthropologists pieced together lost primitive worlds. Cavemen, ancient societies, and living primitive tribes were on the minds of early twentieth-century Americans. Popular newspapers were filled with the news of major discoveries in faraway tropical lands, as well as closer to home in Europe and North America. A survey of the leading American archeology journals reveals as well a changing shift in focus. Although excavations of classical ruins remained important, they were increasingly joined by expeditions to uncover a prehistoric American past. In 1885, the inaugural issue of the *American Journal of Archeology* was limited to articles on the Parthenon and sepulchral vases from Alexandria. By 1900, however, the archeologist Henry Haynes noted a change in American archeology. Supported by growing university museums as well as the Smithsonian Institution and the Bureau of Ethnology, American scholars turned their attention especially to North American Indians and the primitive American past. They had begun to search for "the existence of Paleolithic man in North America."[41]

Attention to the Stone Age coincided not only with the scientific challenging of biblical understandings of time but also with the redefinition of non-European and nonwhite societies as living fossils, that is, as modern primitives. This was a process that historical anthropologist Adam Kuper calls the "re-invention of the primitive." The conditions of primitive life could be read from the lives of the living. As early as the 1860s, Kuper argues, scholars working in Europe and later in the United States, most notably Lewis Henry Morgan, assigned traits to primitive peoples regardless of place or time. Morgan used evidence of living Indians to describe the life of ancient or extinct primitives. Primitives, he reasoned, followed similar paths of development.

Even as a new generation of anthropologists in the 1910s, led by Boas, challenged this rigid evolutionary framework to argue for different racial responses to environmental stimuli and for unique patterns of development, they still relied on the evidence of living primitives. Boas's 1911 *The Mind of Primitive Man* and books by his students, such as Robert Lowie's 1920 *Primitive Society,* sought to realize an anthropology of primitivism that moved beyond Morgan's thesis.[42] Rather than envisioning a single path for development, Boas and Lowie advocated the close study of individual cultures, in Boas's case of Northwest Coast Indians and Lowie's of Plains and Southwest Indians.[43] Focused narrowly on the history of anthropology, such a position, indeed, represented an intellectual rupture. However, in its broader significance, especially for the study of industrial society, even Boas's revision still contrasted primitive tribes and modern industrial civilization. For Morgan, as for Boas and Lowie, history did not progress in a linear fashion. For some peoples, the Stone Age was very much in the present. Moreover, Morgan, though critiqued harshly by Boas, would remain popular well into the twentieth century, with observers using his findings to construct arguments about the origins of industry. From socialist activists to progressive economists, observers of industry continued to rely on Morgan's thesis. Even novelists such as McIntyre sought to ground imaginative adventures of "thousands of years ago" in evidence gathered from living primitives. Her novel abruptly concluded by moving from storytelling to popular anthropology. She described similarities between the ancient tribe of Strongarm with more contemporary Eskimos, "the Red Men of our own country," and "the people of our time who were most like the cave men." The cavemen that she portrayed were much like Tasmanian aborigines who made stone tools and "drew rude pictures on bark."[44]

Based on the idea of a living Stone Age, anthropological evidence gathered from living primitives illustrated the industrial past of the civilized. As Giddings argued, "The oldest remains of human workmanship show that Paleolithic and Neolithic man had the same arts that savage man have at the present time." Thus, when the Yahi Indian, Ishi, walked out of the wild in California in 1911, he was whisked first to a vaudeville theater and then to the San Francisco Museum of Anthropology. As a human exhibit, Ishi lived under the protection of Alfred Kroeber, an anthropologist and student of Boas, and provided demonstrations of archery and other Stone Age arts. Ishi was more than a spectacle; he was a living fossil for the history of the human race.[45] The examination of primitive life whether living or extinct helped associate industry with the highest form of social evolution.[46] Yet even those who lived within industrial societies but found themselves at the bottom of the class hierarchy could feature alongside the nonindustrial races as modern primitives. The Chicago World's Fair constructed displays of primitive life along the Midway to contrast with the White City's palaces of industry and machinery. Yet among the attractions on the midway, alongside a Dahomey village, was a workingman's home.

Public fascination with cavemen and living primitives helped popularize the complex concepts of social evolution. London's "Strength of the Strong" was just one of a number of novels and short stories set in an imagined prehistoric past that offered parables about human development. Even the *International Socialist Review,* a journal that imagined itself as a serious intellectual outlet for discussion about the science of socialism, offered fictional and comic accounts of industrial evolution. One cartoon in the journal highlighted its fascination with cavemen. A typically simian caveman armed with a club stands next to the modern scientist, joining prehistoric savagery and modern science. Beginning in 1909 and continuing nearly every month until 1916, the journal published "Stories of the Cave People." The stories were advertised on the journal's normally staid cover with a sensual image of a cave woman. These stories traced the industrial evolution of

Figure 1.2. This erotic image introduced Mary Marcy's stories of "The Cave People" for a socialist audience. *International Socialist Review* (1909).

the tribe of Cave People and appeared alongside advertisements for reprints of Morgan's *Ancient Societies*. Authored by Mary Marcy, the managing editor of the journal, the series continued even at the height of World War I, when socialists in the United States were facing repression and censorship.

Marcy's stories described a tribe's migrations, confrontations with enemy tribes, the development of mutual aid, and innovation. Innovation was linked to the struggles of the tribe to survive in a fickle environment and in conflict with competing tribes. During the short period of summer abundance, the tribe regressed: "These were not the days of progress or discovery, and the minds of the Cave People grew torpid and they forgot many things they had learned in times of hunger and activity." Warfare with the neighboring Arrow Throwers tribe, by contrast, led to the invention of the bow and arrow. Marcy's stories were largely the product of a fertile imagination about the life of Stone Age savages as well as contemporary anthropological studies. Laughing Boy scrambled up trees, just as "among many of the savages living today, great skill and agility prevails. We are told of tribes whose members are able ... to mount trees in a sort of walk." She cited, in particular, Jack London's observations of "the natives of the South Sea Islands." Based on evidence about contemporary primitive rituals, the death of the tribal leader Strong Arm was mourned in a cannibalistic feast in which Laughing Boy was given the leader's hands so that he might inherit his strength.[47]

Migrating from Tribe to Race

"Had no ape ever revolted against economic conditions there would be no race of men upon the planet," declared the *International Socialist Review*.[48] Such revolts were the first animations of the social organism. According to observers such as Bücher and his American counterparts, the earliest primitive society consisted of animal-like herds of humans. As pure hunting gave way to gathering, pastoralism, and agriculture, herds became differentiated tribes. Tribes became more complex, with different economic roles assigned to men, women, elites, and slaves.[49]

For sociologists and biologists, the advance of man and the transition from individual to group life emerged from what Giddings called "like-mindedness" or what Ward termed "consciousness of kind."[50] Through these processes, humans identified commonality with others, first as members of a family, then as members of a tribe, and later as members of a race and a nation. As London's story suggests, consciousness of kind facilitated the development of first family and later more abstract, nongenetic (what

Giddings called "demotic") ties. George Winkler put this idea into poetic form for his socialist readers. His poem "The Survival of the Fittest" featured "Neolithic man" scrambling up a tree to flee a cave bear. He called out to "his tribesmen." Saved by their arrival and rescued from the top of the tree, he learned an important lesson in mutual aid: "So, although the fit survive, / In a world where all must strive / Is it true the fit would win out all alone?"[51] The first associations beyond undifferentiated animal-like herds were families. Giddings looked to the "lowest savages"—Brazilian Amazonians or northern Greenland Eskimos—for examples of the earliest families. He described these as short informal pairings that lacked modern rituals of formal marriage. The evolution of the family could be traced by examining its many existing and varying forms. Thus, Giddings examined the form of the family in tribes from Hawaii to Turkey, from China to India. He described the eventual emergence of a monogamous family, a mating for life.[52]

For Giddings, these genetic associations were the crucial initial phase of social evolution, soon followed by demotic associations of families linked, first, into "hordes" and then later into tribes and races. Like the Veddahs of Ceylon, the Mincopis of the Andaman Islands, or the Bushmen of South Africa, which Giddings described as living examples of hordes, families were brought together by the need for food and for protection. More advanced races, such as the Kasias of Bengal, the Iroquois of North America, or the Tongans of Polynesia, organized themselves into more complex tribes spread over greater space and governed by councils, complex rules, and rituals.[53] As tribes spread and consolidated their control over more lands, the consciousness of kind expanded to a larger grouping, that of race. Bound up in the very process of evolutionary development, such groupings were biological as well.

Consciousness of kind demanded mutual aid in addition to the articulation of a common enemy. Alone among animals, humans engaged in group conflicts, as the anarchist C. L. James noted in his correspondence with the Bureau of Industrial Research. Like bees, ants, beavers, or wolves, man engaged in mutual aid, but "man is the only animal which competes against his own species."[54] Such conflict, however destructive, was also the engine of progress and industrial development. Tribes had first emerged in isolation from others because they were limited in their movements by natural features such as mountains and waterways. However, rising populations, facilitated by the first steps of social evolution, punctured this splendid isolation. The earliest industrial innovations—Ely cited the example of "corn-cribs" developed by North Carolina Indians—had the effect of further increasing population, which in turn forced tribes to consider migration. Migration, inevitably, led to competition and warfare with neighboring tribes. As the Yale sociologist Thomas Nixon Carver noted, "inherent antagonisms" were the natural by-product of the processes of innovation, rising population, and migration. Yet conflict also spurred further innovation. Migration became an engine of industrial evolution. It turned the caveman into an economic actor. Man "became an economic being, an adapter of means to ends....In short, the process of industrial civilization, of social evolution, had made its first faint beginning."[55]

From its first prehistoric incarnation, human migration was central to industrial history. As the geographer Ellen Churchill Semple argued, the earliest migrations were "the outward expression of ... [a] complex of economic wants [and] intellectual needs." Migration, she insisted, was the norm for an evolving society. The most successful tribes sought new territory and learned to overcome natural barriers. "The earth's surface is at once factor and basis in these movements," Semple argued. "In an active way it directs them; but they in turn clothe the passive earth with a mantle of humanity. This mantle is of varied weave and thickness, showing here the simple pattern of a primitive society, there the intricate design of advanced civilization."[56]

Migration was necessarily accompanied by violence. Social scientists such as Giddings portrayed migration as conquest, first of tribe over tribe and then of race over race. "A migrating race," asserted Giddings, "is a conquering race." Nomad tribes overran native Egyptians to launch their empire. The Assyrians conquered the Akkadians and the Latins overran the Etruscans. "The evidences are inexhaustible," he concluded, "that the great historical peoples were created by the superposition of races."[57] Ross found in the process of these ancient migrations the very causes of "race superiority." "Repeated migrations," he declared, "tend to the creation of energetic races of men." "Successive waves of conquest" revitalized civilization by replacing quiescent with progressive races. The Asian cities conquered by invading Greeks represented high points of classical culture. The Arabs and Moors achieved the climax of Saracenic culture in faraway Spain. From the Hebrews to the Rajputs, migrating races were civilizing and progressive. Even in modern America, the Anglo-Saxons on the West Coast showed none of the "listlessness and social decay" of the old communities in the East.[58]

Migration caused by population pressures became racial conquest, enacted not simply through the force of weaponry but also through sex, rape, and murder. As one race conquered another through the warfare that emerged out of migration, the men of the vanquished race were slaughtered or reduced in status. The women of the race, by contrast, were kept as brides and mates. In the British Isles, the range of types featuring different head shapes, hair, and eye color revealed an ancient past of migration, colonization, rape, and slaughter. This diversity suggested a process of intense selection through conquest. For many, this was an explanation for British racial superiority.[59] Similarly, the rape of the Sabine women, in which early Romans seized the women of the competing Sabine tribe, a popular theme of paintings and of ancient mythology, gained new meaning at the turn of the century as an oft-cited example of race conquest through migration, rape, and slaughter.[60]

Given their significance for social evolution, migration, rape, invention, and racial conquest were common themes in popular fiction about prehistoric life. Stanley Waterloo in his *The Story of Ab* described the ascent of his hero Ab to tribal ruler. Using his mastery of fire, Ab's tribe grew in size and eventually came into conflict with other migrating tribes. They defeated the Eastern Cave Men led by Ab's nemesis, Boarface. The "wives and children" of the defeated tribe became part of Ab's people. London's *Before Adam*, meanwhile, recounted the story of a defeated tribe. The hordelike Folk of the narrator, Big Tooth, were defeated by the superior tribe of Fire-Men. The men were immediately slaughtered. Only Big Tooth managed to escape with his wife Swift One, but only to a bleak seaside cave. The last survivors of a defeated tribe, "we never made another migration." The Sun-Men of Ashton Hilliers's *The Master-Girl* used their invention of the bow and arrow to defeat neighboring tribes. They slaughtered the men, but incorporated the women and young children.[61]

Harvard political economist William Z. Ripley celebrated this concept of migration and sexual amalgamation as true "colonization" because it was a form of complete race conquest. As Ripley argued, true colonization must be accomplished domestically. The forces of armies could not match the ability of rape to submerge a vanquished race into the ranks of their conquerors. The Vandals in Africa left "neither hide nor hair, in a literal sense." The English armies occupied Aquitaine for three hundred years, but left little residue in the local racial composition and the Romans left little mark on Britain. The Teutonic invasion of Britain was more successful as a migration of hordes who mated and settled.[62] This valorization of sexual brutality was important in understanding the emergence of industrial civilization, as well as the character of contemporary imperialism. Ripley cast modern empires as incomplete colonization because they lacked the all-important process of race amalgamation. Modern imperialism was merely the pacification of inferior races. It involved neither rape (or enough of it) nor the slaughter of adult men. Inferior races could persist and continue to breed

under the gaze of rulers that some, such as Charles Woodruff, an American
army surgeon stationed in the Philippines, warned were too benevolent.
Imperialism could only serve a racial purpose if it could become a form of
race conquest. It was dangerous, conversely, if it protected the inferior from
extinction, allowing them to reproduce with impunity.[63]

In this violent celebration of rape and murder as the engines of civilization, scholars also highlighted preindustrial migration as a selective force
that favored the fittest races and individuals. Only the strongest could survive such arduous travel. Through the sexual violence of race conquest, the
weakest were eliminated. Ironically, celebrants of these ancient migrations,
such as Ripley, Ross, or Francis Amasa Walker, an early president of the
American Economic Association, would also become some of the nation's
fiercest critics of modern immigration. In fact, by looking back to ancient
pasts and to common savage ancestors, they offered a scientific basis for
their antipathies toward recent immigrants. If ancient migrations actively
selected the fittest races, modern migrations had become too easy, as Walker
warned the AEA in his keynote address. Ripley similarly worried that "modern industrial life with its incident migration of population does more to
upset racial purity than a hundred military campaigns."[64]

Mapping Migrations: Class and Race

If migration was an engine of evolution, then retracing the ancient routes of
human movement could illuminate the biological roots of racial and class
hierarchies. Migration pitted race against race and class was the aftermath
of race conflict. The most far-reaching attempt to trace past migrations was
presented by Ripley at a series of public lectures in 1896. Ripley stands as
one of the luminary figures of the late nineteenth century who united the
biological and social sciences for the study of race, class, and industry. He
held university positions in both sociology and anthropology and was frequently called to serve on government committees. An expert in railroad
economics, he also focused on the biological origins of racial and class differences. Drawing on the science of measuring heads and imagined distinctions between long-headed and short-headed races, Ripley rejected
phrenology with its study of character through the examination of bumps
and brain capacity, but remained convinced that head shape revealed the
history of ancient migrations. He relied on the "cephalic index," that is, the
width of the head above the ears as a percentage of length from forehead
to back. If the length is held as a constant number of 100, the larger the

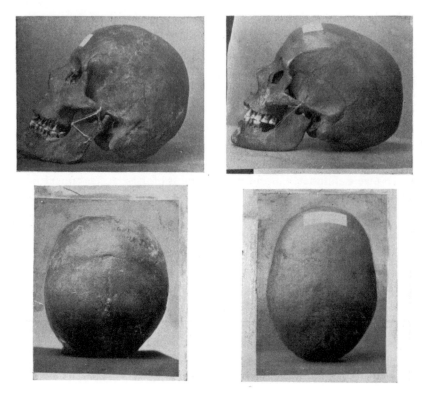

Figure 1.3. William Z. Ripley used the cephalic index to divide races as brachycephlic (rounded head) or dolichocephalic (long head). William Z. Ripley, *Races of Europe* (1899).

width of the head, the higher the cephalic index. Above a width of 80, he called the rounded head "brachycephlic"; below 75, the long head became "dolichocephalic."[65]

Ripley argued that skull shape was more revealing than other physical cues, especially skin color, for what it had to suggest about race. Skull shape, he insisted, linked race to the histories of ancient and prehistoric migrations. In fact, many of his contemporaries studying the complexity of race also relegated skin color to the level of a general division that could say little about racial origins or history. With pigmentation too difficult to connect to single or significant causes, only a closer inspection of the body could reveal the detailed history of race and of human migration. The notion of physical traits as the key to ingrained racial character endured well into the twentieth century. In fact, when later critics of ideas of racial difference led by Boas sought to deny ideas of ingrained racial ability, they returned to ideas of racial division by skin color, not bodily form, skull shape, or migratory history. Boas's *The Primitive Mind* talked about race by privileging skin color as the determinant of racial category. This turn to skin color allowed Boas to circumvent older, powerful ideas of racial history and to focus on mere visible difference. Ironically, he rejected earlier dismissals of skin color, for example, by Ripley, as comparatively weak clues to racial hierarchy.[66]

Ripley insisted that his global map of cephalic indexes revealed ancient histories of migrations. He focused centrally on Europe and pushed the rest of the continents to the periphery. This was based on the presumption that Europe had been the site of the most frequent migrations and was the site of origin for the races of modern America. Drawing on the measurements of many European investigators, Ripley pieced together a dizzyingly complex map of head shapes. Patterns of head shapes helped Ripley map the migrations of different European races, from the Teuton to the Mediterranean to the Slav to the Alpine races. Where skull shapes were most mixed, he identified locations where races had struggled for supremacy. Patterns also showed the battlegrounds of race conflict, highlighting how vanquished races were forced into the hills or the infertile plains.[67]

Moreover, Ripley insisted that the study of head shapes could also reveal the origins of class. Class was the outcome of ancient race conquest, not merely the recent product of the factory system of wage employment. In race conflict, male losers—if they were not immediately slaughtered—were reduced to the ranks of servants or peasants. Even centuries later, Ripley argued, the lower classes had different head shapes than their social betters. Class was a biological difference rooted in race difference. Conquerors kept for themselves the privilege of ordering society, claiming political and religious leadership. In fictional illustration, Waterloo's Ab realized that the surviving and defeated Eastern cavemen could be transformed into workers: "'Why not let them live and work for us?'...And Ab saw the reason of all this and the hungry, imprisoned men were given the alternative of death or obedient companionship."[68] The racially and physically distinct conquered people composed the "industrial society." As the biological product of race conflict, classes likewise developed their own "consciousness of kind." For Giddings, the very origins of the class division of labor emerged as a result of ancient migrations. The "division of labour" could not emerge in "an ethnically homogeneous group," he insisted.[69] A diversity of head shapes revealed a well-formed class system.

As Europe, especially northwestern Europe and the British Isles, had the most complex patchwork of head shapes, scholars concluded that this was the location of the most brutal and selective migrations. Ripley and others

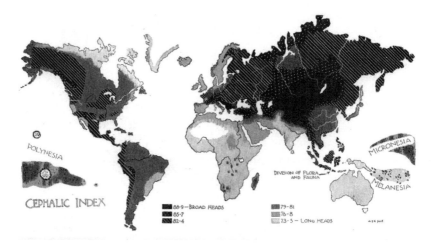

Figure 1.4. Using measurements of head shape, William Z. Ripley traced the migration of races. His study helped define migration as race conflict and conquest. William Z. Ripley, *Races of Europe* (1899).

argued that the continents beyond Europe featured more consistent race types. Confirming older prejudices about racial hierarchies, Africa featured the most homogeneous racial type. Some, such as Ripley and Giddings, insisted that Africa remained racially consistent because migration had been unidirectional. Migrants only left Africa, leaving behind the indolent and the inferior. Such a theory followed an older thesis, confirmed by archeological finds, that the origin of human life was in the benign climate of Africa. Social evolution, however, occurred outside of Africa where energetic migrant races clashed. The trials of migrations across the mountains and harsh climates of Europe had taken their selective toll.

Such complex measurements capture the way conversations about migrations were inherently about racial origin. Thus, nineteenth and twentieth-century discussions of immigration turned inevitably to race and subdivided modern immigrants into complex racial hierarchies. Place of origin and ancient histories of migration, more than skin color, provided signs of racial character. In contrast to definitions of race advanced later in the twentieth century, this understanding of race proposed that European immigrants were part of distinct races. The 1911 *Dictionary of Races or Peoples,* the most far-reaching American effort to catalog racial difference and racial hierarchy, therefore included details not only on contemporary racial character but also on past migrations. Albanians, for example, were cast as tall, muscular, and blond. By character, they were "turbulent in spirit—warriors rather than workers." Their broad face and head contrasted with the pure Greek. They were, therefore, likely part of the Celtic or Alpine races, their past shaped by ancient Slavic migrations. Bohemians, similarly, were affected by ancient migrations of Slavs and Mongols. Still, their broad heads revealed more of a "Teutonic admixture" than most Slavic peoples. This Teutonic racial element had made Bohemia the "brightest jewel in the Austrian crown."[70]

The *Dictionary* also assigned characteristics to different races, for example, as impulsive or disciplined, superstitious or scientific, nervous or staid. Such traits revealed the place of a race in the history of development. The tribal type, as shown by the evidence gathered from living savages, was "barbarous, and exceedingly superstitious." Later innovations, such as the bow and arrow, led to the selection of the "warrior" or "military" type. The ideal warrior displayed a spirit of adventure, prone to imagination and impulse. Still, the warrior, in turn, would be replaced by the "economic" or "commercial" type. "In the commercial type," declared Murray King, "is restored much of the keenness, cunning, resourcefulness, and initiative of the savage," but arrayed in the new context of the industrial economy. Socialists such as King foretold the coming of a new "social type" that would emerge with the evolution of the socialist, co-operative state.[71]

The Stone Age in the Progressive Era

Understandings of industry as a racial accomplishment helped intellectually to set the stage for a generation of progressive reformers, as well as for a further generation of eugenicists. In identifying the varying engines of racial progress, observers worried that invention was a haphazard method of progress. Instead, the processes of social evolution needed to be rationalized and directed in a process of "dynamic" evolution. The state could play a critical role in directing, regulating, and ordering evolution. Giddings warned that, as society developed, evolution necessarily became artificial. Society "can volitionally shape its own destiny. It can become teleologically progressive."[72] Suggestive of the common terrain around ideas of social evolution, a socialist commentator, similarly, celebrated the growing role of the state in an increasingly "complex and highly developed" social organism.[73]

In critiquing laissez-faire, the social Darwinist faith in the efficacy of unimpeded natural selection, socialist and nonsocialist commentators alike saw the need for the modern direction of evolution. The social linkages that Albion Small described as fostered by industry and empire made new laws and an active state necessary. If not, Small worried, the inevitable result was degeneration. Evolution could never be static; it either advanced or regressed. Indeed, the binary of civilization and savagery hinted at the fragility of evolution. Some modern primitives, such as the Fuegians of Tierra del Fuego, were set aside from their brethren and regarded by many anthropologists as degenerate types.[74] Perhaps they had once advanced but had regressed or, perhaps, they had simply reached a static level and faced a future of indolence or extinction.

The study of the ancient world suggested that the principal social divisions of the age were biological products. Moreover, the processes that began in prehistory were still active at the turn of the twentieth century in an age where the fish trap had become the factory and the cave had become the tenement. Yet Jack London's short story offered a note of pessimism that was echoed widely by a range of observers, from socialists on the left to ardent social Darwinists on the right. His story contested the narrative of progressive development. His narrator reveals himself subtly as a loser, but still a survivor, in the social struggle. He was still alive to describe his story of loss. In some cases, the vanquished were slaughtered and disappeared from history. In other cases, they survived as the lower classes and as inferior races. London's narrative structure introduced a complexity of tone in a story that otherwise read as a straightforward parable. The possibility of reverse lingered and was echoed, among others, by Ward: "Let no one, however, be deluded by the thought that this cosmical progress, even in its own slow way, can continue for ever....It will hereafter be shown that this swarming planet will soon see the conditions of human advancement exhausted, and the night of reaction and degeneracy ushered in, never to be again succeeded by the daylight of progress, unless something swifter and more certain than natural selection be brought to bear upon the development of the psychic faculty."[75] Although Ward saluted the progressive triumphs of the nineteenth century and optimistically awaited the twentieth—he would even dedicate his *Pure Sociology* (1903) to the twentieth century—he still noted that the distance between civilization and savagery remained frighteningly short. Industry had advanced and produced unheard of goods and tools. But the regulation of society through progressive politics lagged: "in the art of politics we are still savages."[76]

The goal of a new politics that emerged out of the idea of industrial evolution, whether socialist, progressive, or, eventually, eugenicist, was to control the very means of racial development. For some, this demanded scientific philanthropy, for others the co-operative state, and, for others, seizing the means of reproduction. In the process, each group came to see their politics as an evolutionary development, the expression of new and more civilized types. Progressives, such as Ward or Ely, cast themselves as guardians of progress and champions of a philanthropy that had emerged out of science and out of the very growth of the social instinct. Similarly, socialists described themselves as the new "social type," midwifing a society based on mutual aid. Both saw themselves as combating the tendency of evolution to stall. Eugenicists likewise cast themselves as combating degeneracy. They protected civilization from the unhealthy effects of the growth of social sympathy. Each of these groups—however disparate—found in ancient and prehistoric pasts lessons that made contemporary problems even more alarming.

Notions of industrial evolution posited that certain races had achieved civilization, while others remained mired in savagery. Progress became firmly linked to innovation, industry, and association, and migration became the crucial form of race conquest that helped spur innovation. Industry was, indeed, an emerging system of complex relations, demotic association, and large-scale production. It was also evidence of racial superiority and inferiority. Industrial evolution had raised certain races to civilization, but left others in poverty. Why had certain races, but not others, advanced? To understand how industrial evolution had left races scattered along a complex scale of progress, observers looked to mankind's relation to nature. They developed a racialized worldview that divided the world into temperate and tropical, energetic and indolent, progressive and stagnant, and—ultimately—civilized and savage zones. In the process, they inherently linked industry and empire.

2

MAPPING CIVILIZATION

Race, Industry, and Climate in the American Empire

> The influence of climate upon men may be likened to
> that of a driver upon his horse. Some drivers let their horses
> go as they please....Such drivers are like an unstimulating climate.
> Others whip their horses and urge them to the limit all the time. They
> make rapid progress for a while, but in the end they exhaust their animals.
> They resemble climates which are always stimulating....Still a...driver
> may whip his horse sometimes and sometimes let him walk....He
> conserves the animal's strength, and in the long run can cover more
> distance and do it more rapidly than any of the others. Such a
> driver resembles a climate which has enough contrast of seasons
> to be stimulating but not to create nervous tension.
>
> Ellsworth Huntington, 1915

In 1913 the Yale geographer Ellsworth Huntington began a project that he believed would be the climax of his career. He was going to construct a map of civilization.

Born in 1876 in Galesburg, Illinois, Huntington followed a traditional route into the comfortable world of Ivy League social science. He took his master's degree at Harvard in 1902 and completed his PhD in geography at Yale seven years later. In the break between his time at Harvard and Yale, he would accompany several expeditions to Central Asia, the accounts of which he related in accessible prose in his first two books, *Explorations in Turkestan* (1905) and *The Pulse of Asia* (1907).[1]

After completing his degree, Huntington settled into a chair at Yale. With a few breaks for expeditions to Palestine, Asia, and Latin America, Huntington would spend his time in New Haven building a reputation as the leading North American scholar of the influence of climate on human civilization and racial development. Huntington imagined the map of civilization as the centerpiece for one of his earliest books on climatic theory, *Civilization and Climate*. First printed in 1915, *Civilization and Climate* proved a foundational work in racial geography and enjoyed brisk sales and multiple printings.

He drew upon all of his contacts in the world of anthropology, sociology, and geography and upon the power of the Yale letterhead to get letters opened and answered by even the busiest and most powerful men. Huntington sent out 214 complicated surveys. The surveys asked men—no women—around the world, from different racial, though not class, backgrounds to arrange a series of cards representing different regions into a hierarchy of civilization. Huntington posted sixty-four letters to Americans; forty-three to Englishmen, forty-two to "Teutonic Europeans," and thirty-one to "Asiatics."[2] They were asked to consider "the distribution of those characteristics which are generally recognized as of the highest value...the power of initiative, the capacity for formulating new ideas and for carrying them into effect, the power of self-control, high standards of honesty and morality, the power to lead and to control other races...and other similar qualities."[3] In his map, Huntington articulated a definition of civilization linked to industrial evolution. The energy to labor and the capability to innovate and lead were the basis for civilization.

Huntington's surveys converted the world's most distinguished social scientists, missionaries, policymakers, and politicians into the jury of civilization. Collectively, they judged the success or failure of the world's races. Huntington's initial list of experts was impressive, although his monumental, synthetic effort confronted some resistance. Some, such as Robert H. Lowie of the Museum of Natural History in New York, politely begged out of the project. Lowie admitted that while he "would like to cooperate," he studied "undeveloped peoples and therefore...could not contribute."[4] Franz Boas criticized the basis of the project. Alone among Huntington's correspondents, Boas doubted that civilization could be neatly mapped.[5]

Yet Huntington persisted. Other chosen experts responded with enthusiasm. "Permit me to say how heartily I thank you for engaging in this enterprise," one participant wrote. "[S]uch a scheme will be invaluable to all students of human progress."[6] From every corner of the world the cards, neatly organized, returned to New Haven. Carefully using their responses, Huntington charted a series of maps, in preparation for his final, definitive map of civilization. He used the cards of Americans to trace maps of civilization globally and in North America. The cards of experts of Asian descent produced another set of maps. Not surprisingly, each regional group rated their place of origin more highly than outsiders. Americans located the highest zone of civilization in North America. It stretched all the way to the edge of the Rocky Mountains. Taken together, the average of responses described civilization as diminishing near the shores of the Great Lakes. Asia showed the largest discrepancy between native and outsider ranking. Asian experts argued that their civilization had reached much higher levels than Teutonic Europeans or Americans were willing to admit.[7] Finally, Huntington compiled all the responses to reveal the final contours of civilization.

To the surprise of no one, including Huntington himself, the map confirmed cherished racial beliefs: the center of civilization lay at the heart of the Anglo-Saxon North Atlantic. It stretched from Germany in the east to New England and the mid-Atlantic American states in the west. The rest of the racial hierarchy splayed out in nearly concentric circles surrounding this heartland of civilization. The map seemed to confirm the very precepts of migration theory: the most advanced races had fled their place of origin, probably in Africa, and drifted northward. Through generations of racial conflict, propagated by the effects of migration, the concentric circles of civilization emerged. Racial superiors occupied the heart of the world map of civilization. The losers in the racial struggle for survival occupied the hinterlands. The farther from the center and the closer to the presumed African origin of human life, generally, the lower the standard of civilization. The effects of climate, Huntington argued, led some races to evolve rapidly toward industrial civilization; they kept other races mired in savagery.

The map linked industry, whiteness, empire, and racial superiority within the unifying notion of "civilization." As comforting as the map of civilization must have been to Huntington and to his respondents, it seemed to challenge the justification of imperial colonization. The map suggested that civilization had drifted northward, "coldward" as one scholar put it, away from the tropical site of human origin. But, now the advanced races, in the form of conquering colonial armies, had returned to the tropics. Could they maintain their racial standards?[8] For Huntington, the map was the pinnacle of a career. For the imperial administrator, it was a warning of the perils of colonial work. For the factory manager, it was a key to new methods to optimize production. For the contemporary historian, the map and the science on which it depended help link previously distinct histories of labor, industrialization, race, and empire.

The American debate about the advisability of empire, in particular the colonization of the Philippines, were part of the same conversation as the domestic world of industry.[9] Just as American colonial administrators were seeking to solve the labor problems in the Philippines—the largest and least known of the nation's new possessions—factory managers at home were seeking new methods of workplace control and efficiency,

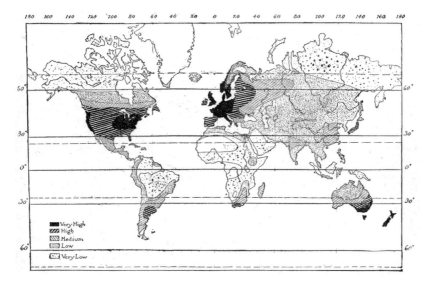

Figure 2.1. Ellsworth Huntington's Map of Civilization used surveys he distributed around the world to ground in science hierarchies of race and civilization. He would argue that climate played a crucial role in helping some races evolve toward civilization, while keeping others mired in savagery and indolence. Ellsworth Huntington, *Civilization and Climate* (1915).

culminating in a science of management.[10] In domestic mills, as in foreign colonies, notions of civilization applied racial terms to labor. Racial thought could explain equally the perils faced by white colonial administrators, the failures of native workers, and the success of domestic industry. Huntington's map confirmed many of the central theories of studies of industrial evolution. It seemed to prove, indeed, that ancient migrations had led to the triumph of fitter races over those destined for extinction or amalgamation. Anglo-Saxon races sat at the top of the racial industrial hierarchy. Huntington's effort to map and to explain civilization grounded understandings of this term in a specific racialized, imperial science. This was a science that was as useful in understanding domestic industry as in explaining overseas empire.

The theories of industrial and social evolution that helped Huntington and his collaborators produce their definitive map of civilization linked conversations about industrialization to those about empire. Too often historians of labor, foreign relations, and empire have explored issues of empire and industry in isolation. Yet for the turn-of-the-twentieth-century observer, labor and empire together raised critical questions about the origin of racial difference, the roots of the racial hierarchy, as well as the meaning and distribution of civilization. As labor was the root of civilization, empire was the political subordination of inferior races by civilized races.[11] Scholars of industrial evolution drew from and helped articulate an imperial science that, while rooted in the European colonial experience, spoke to the particularities of the new American empire and the new American factories. European imperial science had emerged slowly over the course of the nineteenth century as colonists confronted the obvious health and hygienic challenges posed by new regions and sought, at the same time, to justify imperial expansion through racial science. American imperial science, like the American empire itself, developed more quickly. Like political models of imperial governance, American imperial science was in constant dialogue with its European counterpart. As the United States inherited the bulk of the far-flung Spanish empire, most notably the Philippines, after 1898, American scientists and social scientists focused new attention on race, science, and empire. They quickly sought solutions to the challenges of governing tropical colonies. At the same time, they attempted to justify the idea that the Constitution did not, in fact, "follow the flag" though a complex racial science that cast the inhabitants of tropical colonies as indolent and incapable of progress and their rulers from temperate climates as vigorous, efficient, innovative, and progressive.

This imperial racial science, so crucial in the early years of the American empire, defined a geography of energy and lethargy, progress and degeneration. It divided the globe into regions, graded by their levels of civilization, which mirrored industrialized regions. Observers of race and industrialization all but ignored national boundaries. When they defined industry as civilization, they proclaimed ancient and biological commonalities between Europe and North America and privileged racial background over place of national origin as determining the ability to withstand the rigors of industry. Such commonality was contrasted with savagery, especially in such colonial hinterlands as the Philippines.

Climate, Race, and the American Empire

Huntington's map of civilization confidently proclaimed western Europe and the eastern United States as the heart of civilization. It was in this cradle of temperate climate that human energy was at its highest. Here, man could and must work hardest to survive; labor was dignified, innovation was essential, and civilization ascendant. Huntington's map was grounded in a science of climatology that looked to climate as the catalyst or hindrance to racial evolution and civilization. This climate science was part of a larger imperial science that sought to explain the history of racial difference and to rationalize the contemporary reality of the spread of European and American empires.[12] Yet this science also raised troubling questions about the feasibility of empires. Imperialism, after all, was leading Europeans and Americans out of the hale, temperate cradle of civilization to the heart of the primitive, indolent tropics. Given the processes of human evolution, European and, later, American scientists came to wonder about the possible fate of white settlers in tropical colonies. Could they maintain industry outside of the borders of the United States? Or, would they degenerate in the tropics to the lethargic levels of primitive natives? Later American observers, concerned about the ongoing growth of domestic industry and often wary of empire, came to wonder why civilization and industry seemingly flourished in temperate climates, but not in the polar or tropical climes.[13] Racial superiority and racial accomplishment seemed linked to location.

As historians of European medicine and empire have noted, theories of climate and geography had a long relationship to empire. Over the course of the nineteenth century and culminating in 1899 with the opening of the London School of Tropical Medicine, European scientists sought explanations for the poor health of colonial soldiers.[14] Like so much else that was tinged with empire, debates about health in the colonies came to be framed by ideas about race. Doctors, such as Patrick Manson, the founder of the London School of Tropical Medicine, and other tropical medicine specialists came to wonder aloud whether the white races could survive in tropical climates. They argued that at first white settlers benefited from their racial superiority over native nonwhites. They could work harder and longer. In short order, however, white settlers suffered from the enervating effects of the climate and began to degenerate.[15] Scientists urged colonial administrators and army commanders to construct hill stations that would reproduce the hale climates of the temperate zone. Throughout tropical colonies, hill stations provided a respite from oppressive heat and humidity. Most notably, the British constructed Simla as a summer capital in the Himalayas. The colonial government retreated to the cool of Simla in the heat of summer and reveled in its lavish social life.[16]

Turn-of-the-twentieth-century American scholars and colonial officials drew heavily on European theory and practice as they sought to understand both industry and empire. Some, like Ellen Churchill Semple, popularized European theories in publicly accessible texts. An affluent Vassar graduate, Semple traveled to Leipzig, Germany, where she received special permission to attend the lectures of the geographer Friedrich Ratzel, as women were not allowed to enroll at the University of Leipzig. Ratzel's notions about the effects of physical geography on human development would remain the guiding theory of her lengthy American career in geography that included the presidency of the Association of American Geographers. Her 1911 book *Influences of Geographic Environment,* for example, introduced Ratzel's theories of migration, climate, and colonization to American readers.[17]

Other American scholars confronted European conversations during their training in German or English universities. As historian Daniel Rodgers notes, many leading American social theorists had completed their education in Europe. They brought with them upon their return to American universities the theories and debates of the Old World. Among these debates was the question of the fate of white races in the tropics. At the precise moment that the United States replaced its aging Civil War–era fleet, confronted European powers over their expansion into South and Central America, and sought colonies and coaling stations of its own throughout the Caribbean and Pacific, theories about race, climate, and civilization took on a new public significance.[18]

European theories about the racial effects of life in the tropics could be used to explain domestic questions, especially the contrast between the vibrant industrial North and the agricultural, underdeveloped South. Observers pointed to the American South as an example of the fate of civilized whites in warm climates. Southern whites, Huntington argued, had degenerated in the hot climate of the South. He even suggested that the persistence of black slavery in the South and its early abolition in the North could be traced to the unwillingness and inability of whites to maintain the energy for productive labor. By contrast, the Northern white Puritan worker was simply too efficient and the African slave simply could not compete: "Slavery failed to flourish in the North not because of any moral objection to it…but because the climate made it unprofitable."[19] The fact that factories and free labor were concentrated initially in the North was proof that climate was the source of industrial civilization: "That climate is the original force which sets the wheel in motion seems to be evident because it is only in averse climates that we find the 'cracker' type of 'poor white trash.'"[20]

Not surprisingly, such a thesis provoked an angry response from Southerners. "We of the South are so accustomed to being misrepresented by New Englanders," editorialized the *William and Mary Quarterly.* Yet even this paean to Southern progress confirmed Huntington's notion that "the climate of the South…is too constant to afford the necessary stimulus for high civilization, while in the changeful weather of the North there is an incentive to ceaseless motion and activity." Instead of claiming equal human energy and human innovation in the South, the editors looked forward to the day, perhaps "a thousand years hence," when "the Northern weather may change to constancy, and the climate of the South became as terrible as that of New Haven."[21]

Issues of race, labor, and climate would move to the center of debates about American imperial policy after 1898. Some American scholars employed theories of race and climate to discourage imperial expansion, particularly in the Philippines. At the height of the debate over the colonization of the Philippines, geographer Walter S. Tower warned about the torrid climate of the archipelago. It had "an enervating effect on a Northern constitution." He shared the concerns of an army surgeon who served in the islands: "the great majority of white men in the tropics suffer a gradual deterioration...and year by year become less and less fit for active service."[22]

The experience of Europeans, cautioned Harvard's Robert DeCourcey Ward, should demonstrate to Americans the looming dangers of degeneration in the tropics.[23] Even as the Philippine insurrection against American rule was just beginning, the American papers were filled with details about the declining physical and mental health of American soldiers. The *New York Times,* for example, reported the case of a veteran admitted to a mental hospital. His breakdown was blamed on the Philippine climate. Even Adm. George Dewey, the hero of the Battle of Manila Bay, was rumored to have suffered a dramatic physical decline: "It is said by persons familiar with the Philippine climate that the second year is hardest to bear for a Caucasian, and Dewey will soon begin his second year there."[24] So alarmed was one congressmen by reports of casualties blamed on climate that he introduced a congressional resolution that demanded a report from the surgeon general on the effects of the Philippine climate.[25]

As historian Kristin Hoganson has demonstrated, when it became clear that American rule in the Philippines could only be enforced through some version of colonial administration, observers began to worry that life in the colonized tropics, rather than confirming racial superiority and virile manhood, might produce dissolute and feeble men.[26] Huntington, meanwhile, worried that Americans in new tropical colonies would suffer the same fate of English settlers in the Bahamas. Conditions among white residents of the island were so bad that Huntington recorded seeing white children being taught by black teachers in a sordid reversal of racial hierarchies. Whites seemed to have degenerated to the conditions of the surrounding blacks. Their homes were no cleaner, their work habits no more efficient, and their literacy no more advanced. In sum, these white tropical residents seemed to have "largely forgotten those accomplishments" of their race.[27] Strength gave way to weakness, education to ignorance, and industry to lassitude. As one resident reported to Huntington, "Until I came to the Bahamas I never appreciated posts. Now I want to lean against every one that I see."[28]

The experiences of the English—Anglo-Saxon racial cousins—were often cited to question the impending annexation of the Philippines. The English had been most successful when they had colonized the temperate zone. Americans, thus, could not hope to incorporate a tropical colony. One critic warned by paraphrasing the British social Darwinist Benjamin Kidd, "the Anglo-Saxon race cannot…live, work, and rear healthy offspring in the tropics." They could, at best, hope to "live there as divers live under water."[29] Even Amos K. Fiske, a journalist and advocate of annexation of the "sub-tropical" Hawaii and more generally of expansion, worried that the Philippines, unlike other new possessions, was "wholly within the torrid zone."[30]

Others even dismissed imperialism as true, biological colonization. True colonization, after all, demanded race annihilation or amalgamation, not merely subjugation. One antiannexationist, for example, warned that England's tropical outposts were merely "imperial possessions," not "true colonies." They had to be governed through the force of arms. In contrast, New Zealand, Canada, and Australia were successful examples of colonization, effectively carried out through race conquest.[31] William Z. Ripley similarly warned that "a colony can, however, never approximate even to the civilization of Europe until it can abolish or assimilate the native servile population; and yet one of the many things which are expressly forbidden to all colonists is agricultural labor." The lack of race conquest necessarily meant a crisis of labor. As historian Greg Bankoff has noted, American rulers immediately faced an acute shortage of native labor and fears of warm climates prohibited intense physical labor for white settlers.[32] Thus, the American occupation of the Philippines could only support a few officials and soldiers. Observers worried especially about the perceived indolence of natives. The islands appeared ill-suited to development or civilization.[33]

Colonies in the tropics, therefore, might only lead to short-term economic benefit at the cost of the degeneration of the most biologically fit men—those eager for adventure and conquest. Tropical lands were valuable for the products they could provide. Tropical "'spheres of influence' or 'colonies'"—Ward was not willing to call them true colonies and described them instead in quotes—were some of man's most "coveted possessions." The lure of the products of the tropics, from sago and coconuts to rubber and lumber, might draw the most adventuresome and innovative of white men. However, the effects of the tropical "debilitating and enervating climate" were simply too powerful. In the warmth and abundance of the tropics, "the will to develop both the man who inhabits the tropics, and also the resources of the tropics, is generally lacking." Ward concluded, "progress towards higher civilisation is not reasonably to be expected." Even worse, the most industrious of white men were likely to succumb to the climate—or perish in its unforgiving heat.[34] Charles Woodruff, a leading surgeon working in the Philippines and a major influence on Huntington, went so far as to place the blame on tropical light itself. The failure of whites to colonize the tropics, that is, to replace native primitive races, to labor effectively, and resist the pull of degeneration, could be blamed on the intensity of the tropical sun.[35]

Like Europeans, American imperialists sought a solution in higher altitudes. They urged the rapid construction of hill stations.[36] Not surprisingly, one of the earliest American imperial projects in the Philippines was the construction of a summer capital in the mountains near Manila. As American officials contemplated their new colonial possessions in the Philippines, they were confronted with the critical question of how to survive in the tropics. The Philippine Commission, charged with delineating the economic potential and practical challenges of imperial control, asked its many witnesses about the difficulties and dangers of the climate. The Commission focused especially on the need for a hill retreat that could double as a sanatorium and as a summer temperate retreat for the government in tropical Manila: "It becomes a matter of great practical importance to ascertain whether or not there exists…any accessible region presenting suitable climates…for the speedy recuperation of sufferers…from the injurious effects of long-continued residence in a hot climate."[37]

In fact, the first American colonial administrators continued a Spanish search for the ideal location for a hill station. Eventually, the Spanish—and following their lead, the Americans—focused on the hilly province of Benguet and, especially, on the meadow town of Baguio, forty-five hundred feet above sea level. Benguet was only about 250 kilometers from Manila; a railroad could be built. The Spanish conducted numerous detailed climatological and meteorological studies of Baguio. They recorded temperatures, humidity, fogginess, cloudiness, and storminess. American officials relied on such statistics as they began the process of setting up the hill station at Baguio. Americans described themselves as the inheritors of a Spanish science of *climatoterapia*—the science that reduced the effects of the tropics through a change of environment.[38]

Even as the rebellion against American rule raged, the Philippine Commission organized an expedition to test the suitability of Baguio. Indicative of its importance, two commissioners—Luke E. Wright and Dean C.

Figure 2.2. "Under the Pines at Baguio": The American expedition and its native guide relax in a temperate climate. Their reclining posture among the temperate vegetation highlighted the suitability of Baguio as a hill station and summer capital for the Philippines. *Reports of the Taft Philippine Commission* (1901).

Worcester—were its leaders. As the expedition climbed, they passed from the exotic and threatening tropical lowlands to the comfortably familiar temperate hills. Their joy and relief at discovering raspberries and pine trees was palpable. They imagined themselves "in the pine woods of northern New England," just like Worcester's childhood Vermont home. This was an "extensive highland region...free from tropical vegetation affording pasturage in plenty and suited to the production of many of the fruits, vegetables, and grains characteristic of the temperate zone." In their lengthy hearing about Baguio and its suitability as a hill station, the commissioners lingered on the presence of temperate vegetation and fruits. The success of the new empire hinged on the raspberry; vegetation and fruits could represent the safe and familiar, or the tropical and the degenerative.[39] To highlight the suitability of the climate at Baguio, the expedition included in its report a photograph of the expedition lounging in the cool shade of familiar pine trees. Their reclining posture, enhanced by careful photographic attention to the vegetation, attested to the comfort of the hills. Even before the construction of a hill station, these soldiers had found a temperate home in the tropics.

All, that is, except one figure—an unidentified native guide on the left side of the photo, set slightly apart from the white soldiers. Alone among the group, his face is partially obscured. His very presence, like his uncomfortable position, contradicted the familiar comfort implied by the soldiers' posture and disrupted their daydreams of northern New England. Baguio was not, as the expedition might have hoped, an empty temperate plateau. It was populated by several small towns of the local Igorot "tribe," which included this native guide. In the presence of a tribe described as savage, the expedition had to reconcile contradictory understandings of hills, as a retreat from the tropical plains and as a last refuge for those races beaten in the struggle for survival.

On the one hand, American observers placed the Igorot near the bottom of what historian Paul Kramer has described as a hardening Philippine racial hierarchy.[40] As a witness to the Commission insisted, the Igorot were poor and, like mountainous people elsewhere (for example, in the Alps), they were tightfisted and cunning. They displayed as well the mark of lower races, occasionally running off to join "headhunters." On the other hand, these same critics could not ignore the effects of living in temperate climes. The Igorot became the rare example of the industrious savage. The same witness who bemoaned the tendency of tribesmen to become headhunters also praised them as more honest and more moral than plains natives. Other witnesses cited the Igorot for their remarkable health and their industriousness as porters. A photograph amply illustrated this representation of the Igorot, caught between contradictory theories of climate and geography. The picture shows a tribesman and, like images of Filipinos that would become familiar to many Americans through displays at the 1904 St. Louis World's Fair, offered the visual cues to savagery. The tribesman was nearly naked, adorned only with exotic decorations and a loincloth. Yet he is stooped under the weight of a heavy pack. Unlike so many images of Filipinos engaged in strange rituals, here was the rare image of the working savage.[41]

Baguio seemed a "substitute prescription" for soldiers who needed to return to the temperate "motherland." The Commission moved swiftly. Shortly after the expedition's triumphant return, they passed an act to conduct a survey for the construction of a railroad from Manila.[42] In 1903, the chief architect of the Chicago World's Fair's White City, Daniel H. Burnham, was commissioned to redesign the malarial and anarchic capital at Manila and to turn a "great mountain meadow" into a safe summer seat of government. The designer of an architecture that once celebrated American industry and empire was now given the task of constructing the tropical seat

Figure 2.3. "Igorrote Packer": Like so many images of Filipino natives, this picture provides a voyeuristic view of the "savage" body. Yet, in the confrontation between different theories of race, climate, and hills, this muscle-bound native comes to represent both industriousness and savagery. This ends up as an untypical image of the native at hard work. *Reports of the Taft Philippine Commission* (1901).

of the new American empire. Given the harshness of the tropical climate, Burnham argued that his primary duty was the construction of Baguio. As early as 1904, the American government retreated to the hills in summer.[43]

Climate, Coolies, and the Crisis of Labor in the Philippines

The same science that demonstrated the enervating effects of tropical climates could be reversed to prove the salutary effects of the temperate zone. If the tropical climes were enervating and degenerating, then the temperate zones were progressive. In temperate zones, climate forced its inhabitants to adapt, evolve, and advance—or perish. For scholars as diverse as Semple, Ward, and Huntington, climate could be the engine or barrier of evolution. All agreed that the globe was divided into climatic zones: the tropical or "torrid zones," the temperate zones, and the polar zones. Characteristic of an age in which race, more than nation, shaped significant population borders, their notion of the world as divided by zones envisioned a world map that privileged racial over political boundaries.

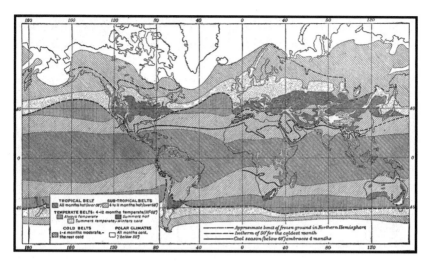

Figure 2.4. "Temperature Zones After Köppen": Robert DeCourcey Ward presented this world map of climatic zones. For Ward, climatic zones closely replicated hierarchies of race. Robert DeCourcey Ward, *Climate Considered Especially in Relation to Man* (1908).

The division of the globe into climatic regions that corresponded with racial, and not national, divisions had a long history in European and American climatic thought. The German Alexander Supan divided the globe into five zones: two cold caps, two temperate belts, and an expansive hot belt. Wladimir Köppen similarly divided the world into belts. For Köppen, the belts were determined by the amount of time that temperatures remained with fixed limits.[44] This conception of the division of the world based on climatic regions proved popular with American observers such as Ward. World maps like these grounded in science the idea that the world's races had evolved in different environments.

Climatic zones could be further subdivided into what Ward and Supan labeled "provinces" determined by a range of factors, including altitude, proximity to water, and storminess. Political boundaries had little consequence. Just as the division of the world into climatic zone matched the division of mankind into racial groupings, so too did the subdivision of zones reflect the proliferation of racial categories. Ward endorsed Supan's division of the world into thirty-five provinces, based on common climatic and topographical conditions. In fact, the provinces consciously reflected racial divisions. Ward understood the West European Province as similar to the Atlantic (eastern North American) Province, but as dramatically different from the temperate Sino-Japanese or Asiatic Mountain and Plateau provinces. Moreover, the conditions in the European and American provinces were far superior to the Asian provinces.[45] If Ward was the academic champion of zonal climatic theory, Huntington was its popularizer. When the Underwood Travel System company sought to create a series of stereographs to illustrate "zone life" for American school children, they turned to Huntington to write the captions. He responded with enthusiasm about the proposed slides, writing that "the more I look them over the more enthusiastic I become about this method of studying geography."[46] For Huntington, the notion of zone life, so well illustrated in the technology of the stereograph, provided a tangible way of understanding different patterns of development. Ultimately, it would also prove the basis for his map of civilization.

In tropical zones, nature was too generous. It provided warmth, but discouraged labor. There was little need to develop complex shelters or modest, covering clothing. The colored races that contently remained in the tropics, nearest the site and climate of human origins, had never faced natural pressures to adapt. In the warmth and generous cornucopia of the tropics, they remained closest to the state of nature, to savagery. Near nakedness—for turn-of-the-twentieth-century observers one of the most obvious clues to savagery—was sufficient. Abundant food grew in the unchanging warmth. Industrious cultivation and tending of crops was unnecessary. As John R. Commons lamented, "Where nature lavishes food and winks at the neglect of clothing and shelter, there ignorance, superstition, physical prowess and sexual passion have an equal chance with intelligence, foresight, thought and self-control."[47]

The energetic, strenuous life so prized by the vocal architects of ideologies of muscular masculinity and by moral celebrants of labor was not meant for the tropics. The climate of the "torrid zones" created an environment of moral laxness that was linked to native laziness. Huntington, for example, lamented the indolence of the tropical native: "Mexico, with its happy-go-lucky peasants and banditti; Guatemala, with its unchanging, stolid Indians, who literally will not work so long as they have anything to eat." Emblematic of the laziness of the tropical native, conditions provided no stimulus to industry. "Progress," Huntington concluded, "is almost out of the question, since there is…nothing to promote energy or ambition."[48] The bodily movement of tropical natives, like the nature of their dancing, highlighted the moral laxness and indolence of the tropics. Some observers even suggested that domestic animals and beasts of burden were lazier and more docile in the tropics than in temperate zones.[49]

Industry in the tropics was seemingly impossible, but still necessary for the success of empire. As the American journalist and celebrant of British imperialism Alleyne Ireland put it, "What possible means are there of inducing the inhabitants of the tropics to undertake steady and continuous work, if local conditions are such that from the mere bounty of nature all the ambitions of the people can be gratified without any considerable amount of labor?"[50] Thus, Europeans first turned to African slavery. After the abolition of slavery in European empires, the British and, to a lesser extent, the Spanish looked to "coolie" labor—Chinese, Japanese, or Indian migrants. American planters in Hawaii also sought Japanese and, ironically, Filipino migrants. Filipinos might be incapable of labor in their own islands, but were still recruited for work in Hawaii's sugar plantations. Tropical development, for Ward, depended on Asian migrant labor—although his support for using migrant Asian labor, albeit tepid, was particularly surprising for one of the nation's leading immigration restrictionists.[51]

American observers, including those already in the islands, were divided about the suitability of Chinese migrant labor. Chinese migrants had long been essential workers in the Spanish colony. In fact, the numbers of Chinese were significant enough that Spanish rulers had accused them of fomenting an eighteenth-century revolt. That revolt, however, did not prevent the Spanish from later importing Chinese coolies (a word used broadly in the Philippine context to describe a range of Chinese migrant workers). In 1876 there were 30,797 Chinese in the archipelago. That number had risen to 99,152 a decade later.[52] Those estimates did not include the substantial population of mestizos—mixed-race Filipinos, including, most notably, Emilio Aguinaldo, the leader of the revolt against American rule. Spanish colonial officials had argued that the Chinese, unlike the natives, were capable of industrial, hard labor. Still, in the face of fears of Chinese immigration, the Spanish instituted an "industrial tax," a charge Chinese migrants had to pay to come and work in the islands.[53]

There was continuity between Spanish and American racial attitudes toward Chinese and Filipino workers. The pathologists Simon Flexner and L. F. Barker testified to the Philippine Commission that Americans should follow the warnings of the Spanish and English about the dangers of tropical climates: "I think a great many men would sicken, and if they tried it for two or three generations...they would 'peter out.'" Chinese labor, for many American observers, as for the Spanish before them, offered a possible solution. The Chinese, happy with "a few handfuls of rice and little clothing," alone seemed capable of heavy lifting and consistently hard labor.[54]

Yet Americans in the Philippines were not deaf to mainland campaigns for the restriction of Chinese immigration. Given debates about whether the Constitution followed the flag—that is, whether the annexation of the

Philippines made the islands a colony in the European model or an insepa-rable part of the United States—fears of Chinese migration resonated in the Philippines. The same language in the United States that cast Chinese immigration as a form of organized invasion traveled across the Pacific and arrived as American administrators sought to address the labor crisis of the islands. Such fears were magnified in the Philippines because of beliefs in a Chinese immunity to tropical climates. Under the Spanish, declared one American witness to the Commission, Chinese immigration represented a "sort of invasion." Another witness recalled fears that the islands might have become a "Chinese colony with the Spanish flag." The new American consul in Manila warned that Chinese coolies might come "swarming" to the new American colony and the merchant William Daland reported that while native workers were lazy, he still favored the exclusion of the Chinese who might otherwise arrive by the millions.[55]

Colonial administrators were forced to reconcile a mainland-bred con-cern about Chinese immigration with imperial notions of the lazy tropical native and of white occupiers as biologically incapable of labor in a tropi-cal climate. There was little consensus in testimonies to the Commission. One railway engineer supported the free entrance of coolies, but encour-aged laws to prohibit their movement from manual labor to trading and shopkeeping. A banker overcame his antipathy to the Chinese and declared they must be allowed to emigrate because natives simply would not work. Another banker suggested that Chinese migrants be restricted to urban labor. Ultimately, the Commission recommended that the Chinese should be restricted to urban and industrial labor, leaving field labor to natives. The only exception was the fledgling tea plantations where American ob-servers, echoing previous generations of British planters in India, declared that the Chinese alone were racially capable of tea production. The Com-mission also encouraged the use of the new technique of fingerprinting to supervise Chinese workers and to prevent them from turning from hard labor to shopkeeping.[56]

In the end, it was Congress and Gen. Elwell Stephen Otis, the islands' military governor, who slammed the door shut on the large-scale importa-tion of coolie labor to the Philippines. As Congress debated the labor cri-sis of the islands with its "native races…in a backward industrial state," it inevitably confronted the question of Chinese migration.[57] Congress and Otis remained wary of the prospect of massive Chinese migration. In 1899, Otis issued an order temporarily extending the Chinese Exclusion Act to the Philippines over the protest of the Chinese government and against the wishes of Adm. Dewey.[58] His actions would be validated by Congress three years later. Meanwhile, despite the push for coolie labor, in 1902, led by supporters of the anti-imperialist Senator George Hoar, Congress outlawed large-scale land sales in the Philippines and effectively closed the door on plantation coolie labor. But, even supporters of congressional action, such as the economist Theodore Boggs, recognized that Chinese labor was important "in order that the 'development of the islands' might proceed." His opposition to coolie labor was not an endorsement of native Filipinos as effective workers; rather, it was a recognition that tropical natives could not "compete … with the industrious, ambitious, and cheaply-living Chinese."[59] Conflict over Chinese labor did not disappear with congressional action. As late as 1907, an army surgeon declared that the extension of the exclusion was an error. He argued that the solution to the labor problems of the Philippines was "to follow the course of natural selection through the importation of the Chinamen."[60]

Industry and Civilization in the Temperate Zone

If the tropics discouraged the selection of the fit, the temperate zone demanded adaptation as the very price of survival. In temperate climes nature was stingy, giving up food only through arduous labor. Those unwilling to labor for food or to construct safe, sturdy shelter were bound to starve and freeze. Labor became not a fleeting necessity, as in the tropics, but dignified and industrious. The very preconditions of industry and the work ethic itself were born in the earliest struggle of northern races for survival. In the temperate zones, "all is activity, movement. The alterations of heat and cold, the changes of the seasons … incite man to a constant struggle.… A more economical Nature yields nothing except to the sweat of his brow." If the tropics made life too easy and polar zones rendered it too difficult, the temperate zones demanded constant "development" of clothing, food, and shelter. The price of life in the temperate zones was adaptation and "hard labour."[61] For Ward, temperate weather provided a "climatic discipline" that forced temperate races to favor sobriety, labor, and innovation. Cool seasons of fall and winter countered the "deadening" effects of hot summers, while the short, warm, growing seasons provided an inducement for arduous outdoor work. As Ward concluded, "Where work is a universal necessity, labour becomes dignified, well-paid, intelligent, independent."[62] Even hill stations, for Ward, could not truly reproduce the salubrious temperate zones. While making tropical summers bearable for "white men and women," the hill station could not re-create the valuable temperate winter: "the northern winter—disagreeable as it often its—has contributed much toward making our northern races what they are."[63]

Yet in the polar regions or on the highest mountains, where nature was stingiest, evolution was stunted. Humans could only hope to survive. Unlike in the temperate zones, where cold was a selective force that encouraged progressive evolution, in the polar regions the cold was omnipresent and brutal. Close examination of Eskimos in the far northern climes or Fuegians in the far south suggested that extreme cold was stultifying. Neither Eskimos nor Fuegians had achieved industrial development beyond the simple production of kitchen implements and weapons from animal bones. Because food gathering was a struggle that occupied most waking hours, existence was necessarily rude and nomadic.[64] Anthropologists surmised that regions of cold, whether in polar latitudes or in mountains, were the refuges of races that had failed in the racial struggle for survival. In their retreat to the poles or to forbidding mountains, they had fled the conquering migrations of more advanced races. Ripley, for example, described Europe's "Alpine race," which dwelt in the mountains of central Europe, as a people conquered by the migrations of Teutons. Blond Teutons occupied the fertile valleys, while dark-haired Alpines retreated to the hills and into isolation. As a rule, he concluded, "to the conquerors belong the plains, to the vanquished the hills."[65]

Of course, the celebrants of the temperate zones had to confront—and ultimately to explain away—races and peoples in temperate zones that they would refuse to recognize as civilized. In particular, Huntington admitted what he called the "anomaly of America" and the "anomaly of China." If America provided the ideal climate for industry for migrant white races, why did the Indians fail to forge a great civilization? Huntington turned to the ancient past of the Ice Age to argue that glaciations in Europe had a more profound selective effect in Europe than in America. Glaciation left European white races with a "better racial inheritance." At the same time, he blamed the effects of ancient migrations through the polar far north and Siberia for the inferior development of the American Indian. Because they passed through the polar zone, such migrations must have favored only the stolid, stubborn type—the very stereotype of the Indian—at the expense of the energetic adventurous type needed to develop new civilizations and to leave beyond a healthy racial inheritance.[66] Paradoxically, he also looked beyond climate to consider the importance of iron tools, horses, and cattle. The Indian, he suggested, had "developed his civilization as far as was possible without iron tools and beasts of burden" and then had stagnated. Without tools and animals to work the soil, they had been forced to remain in the hunting evolutionary stage. Although he wondered "what might have happened had iron tools and beasts of burden found their way to America," he argued that instead the Indian had "stagnated and died."[67]

The same easy argument, however, could not be made about temperate Asia. Still, American observers insisted that China had followed a different social evolutionary path, away from efficiency and industry and toward the world of contemplation and spirituality. Edward Ross's book *The Changing Chinese*, based on his voyages in China, analyzed the "race fiber of the Chinese." Ross sought to explain why the Chinese were capable of hard work, but had not realized industrial civilization. Like Huntington, Ross noted a paradox in Chinese racial evolution. The forces of natural selection, he argued, acted more harshly on the Chinese population than on their white counterparts, protected as they were by "scientific medicine and the knowledge of hygiene." Without medicine, natural selection carried off the weakest newborn children. Given this acute selection of the fittest, the Chinese "ought to possess greater vitality" than whites. They were, therefore, capable of hard work in harsh conditions.[68]

Ross, along with so many other observers, was not willing to consign the Chinese to the ranks of the lower races, despite their lack of energy and innovation. "Chinese conservatism, unlike the conservatism of the lower races," argued Ross, "is not merely an emotional attitude." Rather, it was the result of racial "atrophy." The Chinese race had, for some reason, simply stopped evolving since the glorious "civilization of the Middle Kingdom." Chinese evolution remained stalled at the same level as "ancient Babylonia or Egypt." Painting reproduced older styles and ignored the Western innovation of perspective. Their music did not try to achieve harmony. For Ross, this was ample evidence of a civilization that in worshiping past achievements no longer advanced. "As well expect the apple tree to blossom in October," Ross concluded, "as expect genius to bloom among a people convinced that the perfection of wisdom had been granted to the sages of antiquity."[69] Huntington, similarly, described a race adrift, characterized by parsimony, not thrift and economy; by wandering, not urban civilization; by immorality, daughter selling, and concubinage, not the nuclear family.[70]

Ancient migration, in addition, had advanced European races but had hindered the Chinese. As Huntington lamented, "The history of China consists of a repeated and dramatic cycle of invasion, migration, progress, decay, anarchy, and again invasion." From the ancient Mongols to the more recent Manchus, the Chinese had been devastated by new migratory invasions and new dynasties. As F. W. Williams, the noted American observer of China, argued, each migration swept southward from the frigid north and induced a period of progress, innovation, and energy. Like European migrations, invading races were driven by their "physical and mental vigor" to conquer new lands. Yet in China each upheaval also brought famine and, with that, decay and degradation.[71] Huntington and Williams agreed that the invading Mongols and Manchus were "relatively capable and vigorous people." Yet their invasions, in general, fostered anarchy, not evolution. Famine, caused by continual conflict, ravaged the Chinese and drove them from farms. They became a wandering people who survived by beggary, robbery, and the selling of female children. The forces of selection in famine created a race that persisted because of selfishness, not progress.[72] Nonetheless, as concern about Japanese military and industrial progress in the early decades of the twentieth century would demonstrate, European and American observers were acutely aware of the possibility of nonwhite industrial evolution.

Still, they believed that the climate of Europe and North America had ensured the continuing evolution of the European white races and specially privileged the Anglo race of the northern regions of the American continent. According to Huntington, the ideal temperature existed in a narrow range that matched the climates of England, the northwest Pacific coast, Patagonia, and the eastern United States and southern Canada. The ideal combination of temperature and storminess set England and eastern North America apart as the center of human activity, innovation, and civilization. Stormy weather, argued geographer Charles Kullmer, produced a rapid change in temperature that was mentally and physically stimulating. Monotonous weather with little temperature change led to indolence. Too many storms could be too stimulating and lead to nervousness and, even, to suicide. California, with its combination of warmth and frequent storms, was too "uniformly stimulating." Huntington noted that California had a comparatively high insanity rate and that its four largest cities—San Diego, Sacramento, San Francisco, and Los Angeles—had the nation's highest suicide rates. Even the American East Coast was too energizing. This led to remarkable innovation, but also to the characteristic American nervousness. England, alone, enjoyed the perfect combination of temperature and storms.[73]

Huntington's and Kullmer's analyses defined the ability to labor as central to civilization. Civilization was not the brilliance of painting and sculpture, the harmony of music, or the eloquence of prose. Rather, it was the energy and labor needed to realize these achievements. Art did not make civilization; rather, it was the energy to produce art that forged the glory of Europe and America. "Human energy," concluded Huntington, was the "essential" condition needed for civilization.[74] In locating civilization in labor, Huntington followed up his grand map of civilization with a closer analysis of the effects of climate on the racial ability to labor. By the 1910s, as entrenched American officials developed methods to confront the rigors of colonial life (and, later, constructed an apparatus of public health that would slowly alleviate fears of hot climates), debates about race, climate, and efficiency refocused domestically on questions of factory organization and scientific management.

Huntington again proved a trailblazer. He launched several intensive studies to examine under what conditions different groups of workers were most efficient. Beginning at the same time that he was preparing his map of civilization, these experiments traced the effects of climate on different sets of workers. Starting in 1911, Huntington examined the efficiency of workers at the Stanley Iron Works in nearby New Britain, Connecticut. With the help of factory managers, he prepared charts that detailed the productivity of different groups of workers in varying climatic conditions. Workers were broken down by craft, task, and racial background. One group, for example, was recorded as "Polish men who were born abroad. They are decidedly foreigners." Their task was listed as "polishing and buffing."[75] As a subsidiary study at the Stanley Works, Huntington linked higher piece rates to ideal levels of humidity.[76] By 1914, Huntington had looked to Georgia to complete a similar study at J. P. King and Co., a textile factory.[77]

Using these experiments, Huntington announced confidently that he could complete a global map of efficiency to complement his map of civilization. His map was based on five distinct climatological factors: mean monthly temperature, storminess, the range of temperature from the coldest to warmest months, the proportion of months in which the temperature stayed between 8.9°C and 18.3°C (which he discovered to be the ideal laboring temperature), and, finally, the extent to which mean temperature rose above 20°C and fell below −6°C. These factors determined the efficiency of workers, their ability to labor without undue stress for long periods, as well as levels of innovation. The final map correlated nicely with his map of civilization. The heart of the ideal industrial region lay firmly at the center of the civilized world. The map also confirmed that certain temperate zones were better suited for industry than others. Some were too stimulating and wore out workers both physically and mentally. Others were too dull and rendered productivity stable but low. Huntington, therefore, confirmed that, for example, "there are enormous differences between the civilization of Germany and of Turkestan." When he printed the two maps one above the other in his *Civilization and Climate,* Huntington highlighted the link between energy and civilization. The only difference between the two

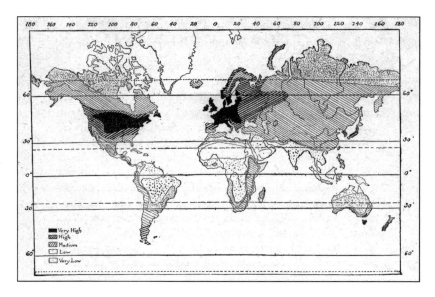

Figure 2.5. "The Map of Energy": Ellsworth Huntington traced this map based on his studies of climate and factory work. As he pointed out, this map mirrored his map of civilization almost exactly. Civilization was achieved by the ability to labor. Ellsworth Huntington, *Civilization and Climate* (1915).

maps was an abnormally high level of civilization in places such as India. This, for Huntington, was the understandable effect of European imperialism. However, even this colonization could not raise these tropical, lethargic regions to the levels of civilized Europe. In his deterministic view, efforts to civilize the savage necessarily foundered in hot climates and the civilized likely would degenerate in the process.[78]

Socialist commentators engaged equally with the effects of climate on the ability to labor. However, socialists were less interested than Huntington in efficiency. Rather, they argued that the temperate climate permitted the perfect balance between work and leisure. Climate, it seemed, marked the temperate zone as the ideal locale for socialism. The socialist Rand School even produced a syllabus for a children's class on climate and industry. Complete with directions to write the phrase "struggle for existence" on the blackboard, instructors aimed to show how temperate climates permitted the balance of eight hours of work and eight hours of leisure. Eskimos' struggle for existence was too extreme. By contrast, in tropical zones "the struggle for existence…is so easy that work is almost left out." Children were encouraged to deduce that tropical inhabitants were lazy and avoided clothing. "In the temperate regions, nature has done her part," the class

Figure 2.6. "Lazy Savages": A typical vision of indolent, half-naked savages set amid the vegetation of the torrid zones. Nature was too abundant and provided little incentive for labor. J. Howard Moore, *Savage Survivals* (1916).

concluded. J. Howard Moore, a socialist activist and biologist, provided the characteristic image of the indolent savage. According to Moore, primitives were imbued with an instinct for indolence. They shunned labor for laziness, "chocolate creams" for wild fruits, and Pullman cars for their own legs. "Many primitive peoples cannot be induced to do any kind of sustained labour," he noted, "unless they are driven either by hunger or the sex impulse." His image of "lazy savages" displayed half-dressed primitives relaxing on a torrid savanna. If workers in the temperate zones worked too hard, it was the fault of the capitalist classes. The working classes lived as Eskimos in a temperate zone: "the struggle for existence uses all their strength." With the continuing progress of civilization, Moore hoped that labor would become seen as natural, even pleasurable. Laziness would give way to efficiency, and "some day we human beings will be as naturally industrious as the bee."[79]

This was the shared dream of scientific managers—though for different ends. As historian Warwick Anderson has suggested, fears of life in warm climates had begun to dissipate in the Philippines by the 1920s. This imperial racial science, however, had an enduring influence on domestic shop floors. Huntington's studies of energy helped spawn a new industry that sought to make workers more effective and easier to manage through the control of factory temperature and humidity. Like Frederick Winslow

Taylor, who offered to remake factory management, this new industry offered advanced production through climate control. By the 1910s, the National Research Council, in association with the Weather Bureau, the Public Health Services, and life insurance companies, set up the Committee on the Atmosphere and Man. Operating at the domestic zenith of climatological thought, the Committee explored "problems relating to weather, climate, ventilation, migration, acclimatization, and the relation of disease, racial traits, biological functions, and mental activity to geographic environment." Between 1921 and 1926, the Committee conducted experiments in thirty-nine factories concerning the "industrial effects of the atmosphere." The experiments sought to determine "the atmospheric conditions under which the operatives show the greatest efficiency."

The Committee offered far-reaching conclusions on the effects of climate. In a report on *The Atmosphere and Man,* for example, the Committee argued that climate conditions "exert a strong selective effect whereby certain types are weeded out in unfavorable climates." Climate, therefore, was central to "migration...and...the racial changes which accompanied or caused the decline of ancient civilizations." More narrowly, the report offered hope both for the improvement of industrial efficiency and of colonial governance. Studying the effects of climate on racial inheritance "might even enable us to select certain types of Europeans whose biological constitution is such that they can colonize tropical countries without danger of deterioration either for themselves or their descendents." Similarly, the "study of atmosphere" promised to help determine the perfect conditions for labor and to help business owners "artificially to create indoor conditions which will largely neutralize the harm done by unfavorable atmospheric conditions out of doors."[80]

Even as the Committee offered such recommendations, a new industry of climate control was emerging. It promised to improve industry through the control of temperature and humidity. Huntington was inundated with brochures and letters offering new inventions to provide "warm, moistened, fresh air...to further efficiency from our staff," as one testimonial promised.[81] Among others, the Steamo and Vapo Air Moisteners, Excelso Quality Water Heaters, Twin City System Washed Air Heat, Zephyr Washed Air Systems, Universal Humidifying Company, Tinken Complete Air Conditioning Units, and the Edulator System all sent Huntington advertisements and testimonials. Company salesmen or directors reported having read *Civilization and Climate* and eagerly sought Huntington's expert endorsement.[82] By the 1920s, Huntington had converted his home into a living laboratory offering different companies the chance to install their products in his home for him to test. In 1927, for example, he offered the Universal Humidifying Company the opportunity to install a humidifier in his home when he was dissatisfied with another company's product.[83]

Despite his unhappiness with this one unnamed product, Huntington's enthusiasm for climate control was boundless. While he often echoed the praise of individual factory owners who reported improvements in worker efficiency, he considered these individual accomplishments part of the general trend of evolutionary progress. Huntington was especially effusive when he praised the humidifier of Rev. J. D. MacLachlan, a clergymen turned inventor who ran his small business out of his church basement: "It is quite possible that the improvements in heating houses by means of humidity and also by means of obtaining the right degree of variability may do much to move the center of civilization still further north."[84] He lauded the members of the American Society of Heating and Ventilating Engineers as "the logical successor[s] of the inventors of fires, clothing, houses and furnaces." Through their innovation, civilization, defined through human energy and regulated by climate, could spread. Climate control, he mused, might finally allow civilization to penetrate the tropical zones: "Can [the ventilating engineer] in time overcome the evils of too great heat and thus enable civilization to move south once more?"[85]

The Coldward March of Industry

Huntington was charitable with his praise. Yet his tributes were grounded in the belief that the tropics were detrimental to hard labor and that industrial evolution was accompanied by a steady march northward—coldward. The history of civilization that stretched back to prehistory could be characterized by a steady march northward to low-lying temperate zones. As the evolution of industry demanded new, more advanced, and efficient kind of people, whether managers of workers, it steadily moved northward, away from the tropics and away from tropical races. Over the course of human evolution, civilization inexorably moved northward to the areas where energy could be highest. As Semple admitted, the earliest civilizations occurred in tropical or subtropical areas where ample flowing water and annual floods allowed for irrigation and fertile soil. Civilization emerged through a stable food supply and an "escape from the...vicissitudes of an uncertain climate."[86] Warmth was necessary for the first civilization, but each human innovation, a step in the process toward industrialization, allowed civilization to move northward toward the ideal temperate zones. "Advancing civilization," wrote S. C. GilFillan, "has given us warmer clothes, tighter houses...stoves and furnaces." Society could thrive "northerly," closer to the places where human energy could be highest and progress faster. Advancing civilization northward allowed for "efficient and important work within doors." Moreover, the growing complexity of civilization demanded "more responsibility and stability, characteristics of cool climates."

Temperate, colder societies quickly overcame their older southern competitors. As GilFillan pointed out, all the ancient civilizations showed a steady march coldward as they advanced. The heart of Greek civilization moved from Crete to Constantinople, Italian civilization from Sicily to Rome to Milan, and German civilization from along the Rhine to the cold shores of the Baltic.[87] Similarly, the fall of ancient civilizations could be linked directly to their move southward. Egypt's fall into "barbarism" coincided with the shift of its capital southward along the Nile. Greece's decline corresponded with the move of its civilization from Athens to Rhodes, Sicily, and Alexandria. Retrograde, southern movement suggested a "slump in culture," disorder, and degeneration.[88]

The evidence of ancient civilizations highlighted links between energy, civilization, and race. Through the slow process of evolution, powered or hindered by climate, the white races especially had developed into temperate, industrial races. Nonwhite tropical races remained indolent and resistant to hard labor. Nonwhite temperate races were the victims of their own past and the peculiarities of their environment, which hindered their development into proper industrial races. Industry, more than historians have noted, was cast as a racial accomplishment. This concept allowed scholars such as Huntington to think broadly about the division of the globe into distinct racial zones, divided not by nation but by levels of civilization and efficiency, as well as about the specific and localized effects of climate on industrial workers.

In this context, national borders were reduced merely to artifacts, historical reminders of past migrations and past racial boundaries. They indicated little about the present or about the particular character of races. "Modern political boundaries will...avail us but little," wrote Ripley in his study of racial migrations. "They are entirely a superficial product; for, as we insist, nationality bears no constant or necessary relation whatever to race."[89] Scholars like Ripley and Huntington divided the world into naturally ordained zones of efficiency and civilization that closely mirrored perceived racial boundaries. Nation was secondary to race. In fact, Huntington, in sending out the cards for his map of civilization, explicitly encouraged his participants to think past national or state boundaries to consider the racial character of the region listed on each card. Moreover, the regions on his cards often crossed national boundaries.

Climate, race, and migration transcended the nation and permitted scholars to rationalize the racial kinship they felt with their European counterparts. As Huntington's maps of civilization and energy demonstrated, the heart of world civilization crossed national boundaries, but linked together peoples of the same race. His maps confirmed the industrial superiority of the white races. Equally, it helped scholars remap the world into zones of efficiency, again delineated by race and climatic zones, not by nation.[90] They also encouraged government census takers and policymakers to privilege racial categories at the expense of those of nation origin. By 1918, Huntington allied with the leading nativist Madison Grant to pressure the United States Census to record the racial background of new migrants. As he complained in a letter to the census director, Samuel Rogers, national background could reveal little about the character of immigrants, let alone about their ability to survive the rigors of temperate North American industry.[91] In fact, just as they launched their campaign, government officials had already begun to consider migrants as representatives of a race, rather than as products of a nation. Rogers replied quickly—within days. Agreeing with Huntington about the need to secure statistics about "the racial origin of the population of foreign birth or parentage," he pointed out the progress already made. The 1910 census collected data on mother tongue, a hint at racial background. He hoped for even more data in the 1920 census.

Rogers agreed that character was embedded in racial background and, therefore, downplayed the significance of national origin in understanding migration. After all, what could national background truly reveal when nations encompassed more than one racial group and, similarly, racial groups were arbitrarily divided by political borders?[92] The census should enumerate the foreign born "not only by country of origin, but also by language or race of origin, including the Jews as a race," urged the Home Missions Council, a religious reform organization.[93] In fact, as early as 1899 the Bureau of Immigration had begun to classify new arrivals based on race, not on country of origin. In 1901, the federal Industrial Commission of Immigration announced a marked shift in focus from national origins to racial origins. As one member of the commission declared, "The purposes of the new classification of immigrants according to race rather than nationality is to afford a fairer basis for judging the industrial character and effects of immigration." Proudly, the commission announced that its transformative efforts met little resistance, except from Jewish groups, largely led by established, wealthy Jews of German background concerned about being associated racially with their poor Eastern European co-religionists.[94] The Dillingham Commission in 1911 further codified this view by privileging racial background over national origin in its *Dictionary of Races or Peoples.*[95]

The *Dictionary* reminded its readers that national boundaries encompassed and, indeed, obscured centuries of racial migrations and race mixing. As its authors turned their attention to the Philippines, they reiterated the complex racial hierarchies, from "Malays" to "Negritos," so poignantly exhibited at the St. Louis World's Fair. In an implicit critique of Filipino nationalism, the Philippines became merely a "geographical" rather than a racial or "ethnographical" label. As the potential national cohesiveness of the islands dissolved in the face of racial theories, the *Dictionary* also noted the bitter irony of imperial law as it collided with the realities of immigration. As the *Dictionary* mourned, "But few words can be given to them in this dictionary, for they are not considered legally as immigrants upon coming to the United States." The turn to racial analysis reflected a desire to employ theories of race and science to know more about the character of immigrants arriving in the United States. Race, not nation, would be the primary category of analysis as observers came to confront the new reality of American immigration. The ideal climate had, indeed, created the energy for progress and innovation. But, the immigrants arriving in the United States were coming from the periphery of civilized Europe and from nonindustrial Asia. Could they withstand the rigor of American industrial work? The logic of climate helped observers conclude that some races, in particular African Americans, would have no place in American industrial life. However, industry seemed to have an insatiable demand for cheap labor in the form of European and Asian immigrants. Filipinos remained legally outside of the framework of racialized immigration law established in 1924. They were neither immigrant nor citizen and paradoxically would become the migrant workers of choice in the plantations and factory farms of California and Hawaii. Ironically, as the domestic met the imperial and the industrial met the tropical, Filipino workers, disdained by Americans in their homeland, ultimately seemed to lie awkwardly outside of the process of disaggregating race and nation.[96] As Huntington completed his map of civilization, he added his voice to a chorus wondering aloud whether the process of modern immigration and the demands of industry had reversed the selective nature of migration. Was migration too easy in the modern world? Were the less civilized inundating, perhaps colonizing, the civilized? Was immigration leading to the survival of the unfit?

3

THE OTHER COLONIES

Immigration, Race Conquest, and the Survival of the Unfit

"All men are created equal." So wrote Thomas Jefferson,
and so agreed with him the delegates from the American colonies.
But we must not press them too closely nor insist on the literal interpretation
of their words. They were not publishing a scientific treatise on human
nature nor describing the physical, intellectual, and moral qualities
of different races and different individuals.

John R. Commons, 1907

Northeastern Pennsylvania, the hard, or anthracite, coal region, produced the fuel that warmed the nation's hearths and fueled its factory furnaces. At the dawn of the new century, it was also a war zone.

Strikers, aligned with the United Mine Workers, squared off against soldiers and strikebreakers. Mine heads were wreathed in barbed wire. Railroad bridges collapsed in dynamite explosions. By October 1902 the entire Pennsylvania National Guard, the equivalent of one army division, had been ordered to the coalfields. The local sheriff requested that the governor declare martial law. As winter approached, even President Theodore Roosevelt joined a chorus calling for negotiation and arbitration. Shortages gripped the nation and newspapers reported the frantic search for substitute fuels. Some burnt asbestos bricks soaked in oil. In Philadelphia, sales of gas ranges skyrocketed. In New York, the poor purchased coconut shells from candy manufacturers. In Chicago, streets paved with wood were torn up and factories ran their furnaces with sawdust. As the autumn chill descended, the national relief was palpable when the strike ended with the appointment of a commission of arbitration.[1]

For the journalist Frank Julian Warne, who covered the 1902 strike (and an earlier strike in 1900) for the Philadelphia *Public Ledger,* the details of the final settlement were less important than the cause of the strike. In fact, the commission's recommendations essentially split the difference between the union's and the companies' demands. The "real cause of the miner's strike" was the arrival of new immigrant races in the anthracite region. The decline in wages and the actions of the mining companies were merely subsidiary to what Warne termed—without deliberate hyperbole—the "struggle for industrial supremacy." This was a battle for supremacy between races of workers, not between labor and capital. Strikers may have battled companies, but the real struggle pitted the "English-Speaking" race against the Slavic race.

According to Warne, until the 1890s miners in the anthracite region were part of the English-Speaking race and enjoyed—indeed, demanded—a high standard of living. Warne's image of the miner's life was shrouded in Victorian sentimentality. He evoked "the little red school-house on the hill," frame houses with porches, and well-dressed wives. The "decreasing tide of Teutonic Races" signaled the arrival of the less civilized Slav. If Warne's description of the English-Speaking races recalled a civilized ideal, his image of Slavic races echoed descriptions of the colonial native. The Slav would "wear almost anything that would clothe his nakedness." The mining capitalist was inevitably going to hire the "cheap man" with few needs and willing to work for low wages. More civilized workers struggled to contain the "Slav invasion" and the union and the strike were the last lines of defense against wage cuts associated with the arrival of new and cheap workers.[2] Yet regardless of whatever the commission may have recommended, theirs was a losing battle. Still, Warne worried that Slav racial supremacy would, in turn, prove transitory: "it will be only a question of time when the coming supremacy of the Slav will be attacked by still cheaper labor."[3] Immigration had created a race to the bottom, not only in wages but also in racial standards.

Warne joined a chorus of voices warning of the passing of a great race. For observers of American industry, the "invasion" of new immigrants marked the decline of a distinct American race. This race was shaped—like other superior races—by the forces of selection inherent in migration. The American race, formed in the early days of the colonies, was even more energetic and industrial than its European Anglo cousins. As migration became easier around the turn of the century through new innovations such as the railroad and steamship, the forces of selection that made migration a progressive process were removed. For critics like Warne, immigration, without its biological and natural constraints, became a movement of the weak, of those forced from their place of origin. The superior American race that had emerged out of the first European peopling of America was now pitted against inferior migrants in a new, subverted form of colonization. Inferior races were replacing their betters. The work of Huntington and other geographers of race and climate presented an optimistic vision that firmly grounded civilization in North America and Western Europe. In contrast, the close inspection of industrial workplaces led by scholars such as Warne raised significant questions about whether industrialization could forever foster progressive development. If industrialization marked boundaries between civilized and savage, it also drew the inferior to the civilized temperate world. Factories needed labor, a demand met by immigrants from the periphery of civilized Europe and from retrograde, stagnant Asia. Once an arduous task that favored the fittest, immigration had become a comparatively easy journey of the unfit. With modern immigrants carried on railroads and steamships, less civilized races confronted the more civilized in a struggle for jobs.

For these same worried critics, the lessons of climate also seemed to confirm that African Americans, still bearing all the negative traits of their tropical ancestors, had little place in industrial civilization. They seemed confident that African Americans were destined for extinction. As a result, they effectively evicted black Americans from discussions about race and industry; as descendents of tropical races, former slaves had little place in industrial civilization. Historians have often noted the intersections between the racialized debates about African Americans and about immigrants.[4] In fact, for many observers, the focus on industry as a principal terrain of race conflict worked to segregate racialized debates about African Americans from contemporaneous debates about European or Asian migrants.

The age of immigration was equally the age of colonization. Though historians tend to discuss immigration and colonization separately, for many turn-of-the-twentieth-century observers they were not distinct processes. They argued that both were the result of pressures that drove one race to confront another. One race would succeed, triumph, and flourish; the other would either perish completely or be reduced to a subordinate status. These observers came to understand both immigration and colonization as products of human evolution and as essentially identical forms of race struggle. Colonization, like migration, was the successful transplanting of one race to another part of the world accompanied by the amalgamation, extinction, or total subordination of another race. Migration was an even more complete form of race conquest than imperialism, which was largely the political or military control of foreign and often tropical territory. The "armies" of immigration were ultimately more successful than any imperial force.[5]

Colonization and the Immigrant

When observers of America's growing cities talked about immigrant "colonies," they truly meant places and sites of conquest, not simply close-knit and growing communities. William Z. Ripley declared that immigrant colonization was converting American cities into outposts of foreign races and nations. Because of its foreign colonies, he warned, Chicago had become the second-largest Bohemian city in the world, the third-largest Swedish, the fourth-largest Polish, and the fifth-largest German city. New York, meanwhile, had a larger Italian population than Rome itself. For observers such as Ripley, colonization suggested grouping—literally the forming of colonies like bacteria on a Petri dish. It also meant racial conquest through migration.[6] As settlement house workers and other social reformers turned their attention to the growing numbers of immigrants and their concentration in cities, they cast themselves as explorers of foreign or alien "colonies." Such investigators described immigrant neighborhoods as conquered areas and as beachheads of an immigrant invasion.[7]

The waves of new immigrants washing up on American shores reminded critics of the violence of ancient migrations. They bemoaned the effects of contemporary immigration by tracing past migrations. Henry Pratt Fairchild, a strident critic of immigration policy, began his 1913 study of contemporary immigration with a description of prehistoric migration. "The study of immigration is part of the study of the dispersion of the human race over the surface of the earth," he reminded his readers. Moreover, he was equally quick to connect immigration, invasion, conquest, and colonization: "All have this in common, they are reasoned movements arising after man had progressed far enough in the scale of civilization to have a fixed abiding place."[8]

Opponents of modern immigration often linked images of liquid waves of immigration to the waves of advancing armies. For critics such as James Davenport Whelpley, the appropriate metaphor for immigration was liquid: waves, eruptions, and floods that overflowed the dikes of restrictive legislation. "Two million people emigrated from Europe last year, and one-half of them came to the United States. This vast stream of human beings is unnatural," he warned. Equally dramatically, Ripley cast migration to the United States as "eruptive." It was "a lava-flow of population, suddenly cast forth from Europe and spread indiscriminately over a new continent."[9] Such liquid imagery conveyed a sense of immigration as inundation and inevitable conquest.

Images of floods merged neatly with warnings of immigrant armies. Modern conquest did not occur necessarily at the barrel of a gun or the point of a sword. It was accomplished by armies of individuals who together composed an invasion of millions. As one worried critic warned, "Was Attila's army one-half, even one-tenth, as large when it overran Gaul and Italy?...Or, did William the Conqueror lead any such army in the Norman invasion of England in the eleventh century?"[10] As Whelpley similarly declared, "Out of the remote and little known regions of northern, eastern, and southern Europe, for ever marches as a vast and endless army. Nondescript and ever-changing in personnel, without leaders or organization, this great force, moving at the rate of nearly 1,500,000 each year, is invading the civilized world."[11]

The titles of the many articles and books warning of the foreign invasion were deliberately alarmist, but they also reflected the application of theories of colonization to the question of immigration. Warne, for example, would use his study of the Slav invasion in the anthracite fields as the basis for his larger study of migration, *The Immigrant Invasion*.[12] The Boston lawyer and immigration critic Daniel Chauncey Brewer, likewise, described immigration to New England as *The Conquest of New England by the Immigrant*. His polemical text began by promising to describe "the invasion, the occupation, and the conquest of New England." The language of invasion and of militarism ran throughout the text. The earliest arrivals to New England, "their battle flags now bestowed," had, in their turn, conquered the Indians and seized the land. Armed with a vigorous spirit, the "Yankee" expanded into the west and took to the seas in merchant clipper ships. Yet the fifty years following the Civil War, according to Brewer, witnessed an occupation and conquest by new immigrants so complete that it could not have been equaled by "invading armies...in full military occupation of the penetrated country." "The industrial occupation of New England," he lamented, "was hardly begun by aliens...before it was accomplished."[13]

Urban observers and critics of immigration described an exotic colonial world emerging in the heart of American cities. In the mill towns of Bedford and Fall River, Massachusetts, Brewer insisted there were more Italians than in Perugia or Ravenna. In his beloved Boston, he decried the transformation of the old city into a small part of Italy. Where once "artisans and mechanics...vibrated to the appeals of Sam Adams," the streets around Copp's Hill and Old North Square had become "characteristically Italian." Once the "Yankee" tourist would have to travel to the Mediterranean to see the spectacle of Italian crowds, but "now the whole pageant glitters and be-comes vocal in bypaths that the austere Puritan trod."[14] Yet behind foreign spectacle and alien pageant lay the looming danger deliberately implied in the word "colony"—these were the beachheads of a foreign invasion. Brewer insisted on describing the transformation of Boston "for any one who thinks of the foreign invasion of New England as a passing phase."[15]

The fear of colonization would remain constant through the early de-cades of the twentieth century. It provides the cultural background to the successful efforts to restrict immigration in 1924. With the dawn of the new century, the South End House settlement compared the settlement house to an imperial outpost. Its location in the heart of Boston's slums allowed its residents to complete studies of races in surrounding "colonies." In similar language, the Immigration Restriction League, one of the earliest and most vocal organizations calling for the restriction of new immigration, railed against the growth of foreign colonies in urban centers. More than two decades later, James John Davis, the secretary of labor, continued to warn about the impending perils of the "immigrant invasion." Charles Daven-port, meanwhile, translated fears of immigration as conquest into eugenics. His landmark *Heredity in Relation to Eugenics* (1911) began its discussion of immigration with a survey of ancient migrations and ended with warnings about conquest and invasion. As late as 1939, when the eugenicist Harry Laughlin completed his survey of American immigration, he called it *Im-migration and Conquest*.[16]

The Steamship Problem and the Unmaking of the American Race

Before the arrival of the railroad and steamship, it was only the most fit who demonstrated the bravery, foresight, and energy to move to new and un-known lands, according to critics of recent migration. Migrations before the industrial age necessarily included only those willing to confront the perils of the journey.[17] Such migration proved selective, not only because it pitted stronger migrating races against weaker sedentary natives but also because movement itself was rife with perils that would carry off the weakest of the race. As the leading restrictionist, Prescott Hall, concluded, "The United States, from its geographical position, has an opportunity to perpetuate its unique advantage in having been founded and developed by a picked class of immigrants."[18]

Even opponents of modern immigration saluted past American migra-tion as a prime example of natural selection. The prototypical migrant to

North America was, according to the mine workers' leader, John Mitchell, writing in the pages of the Immigration Restriction League's journal, "a sturdy, adventurous pioneer, who was willing to undertake and withstand the struggles and the hardships incident to the development of a new and ofttimes dangerous country."[19] The rigor of migration, the impetus for moving, and the trials of the North American climate all aided in breeding a new American race. The first step in selecting the "modern Americans" began with the emergence of the Puritans. Making up only a small percentage of the English population, the Puritans "possessed certain highly valuable qualities." They were a "strong-minded, thoughtful, self-controlled, and self-sacrificing people." They were also willing to accept the perils of travel.[20]

The second stage in the natural selection of the American was the voyage out of England and overseas. The simplicity of technology made ocean voyages long and dangerous. This led to the further selection of the fit from this group of migrants. Fearful men, according to Huntington, resisted the voyage and timid women prevented their husbands from attempting the ocean trip. Thus, "it was the motor-minded, determined, conscientious, religious, adventurous, and physically strong who embarked in the *Mayflower*." The arduous nature of the journey and primitive conditions upon arrival led to further selection. "Here again it was the weak who perished, especially those who were weak in spirit as well as body and therefore gave up in encouragement."[21]

The frontier played a further role in forging a new American race. The frontier lured the most adventuresome and, of these, only the fittest survived. It was in this context that the groundbreaking work of Frederick Jackson Turner on the meaning of the frontier for American racial development merged with concerns about the effects and quality of recent immigration. For Turner, the frontier had been crucial to the selection of a new American type. It forced Americans to abandon the protective civilization of the Old World and even adopt the savage strength of the native Indian. "In short, at the frontier," Turner declared, "the environment is at first too strong for the man. He must accept the conditions which it furnishes, or perish."[22] Yet as the frontier finally closed, recent immigrants gravitated to cities, especially those along the East Coast tidewater. Life, though harsh in the slums, lacked the selective quality of the rural frontier.[23]

Observers tentatively suggested that the process of migration to North America and then west with the frontier had forged a new "American race." For Richmond Mayo-Smith and Henry Pratt Fairchild, this race was composed of Teutonic elements—Scotch-Irish and English—and was expansive, energetic, and conquering: "They were by nature typical pioneers....They were one race sufficiently unified, endowed with the spirit of liberty."[24] Lazy and weak elements had been weeded out in the process of making a distinct American race. The result was an English-speaking, Teutonic, and Protestant American race that was fit, creative, energetic, and industrially minded.[25] Some even worried that the rigors involved in the peopling of the continent had created an American race that might be too energetic and frenzied. As Turner would write, "Energy, incessant activity, became the lot of this new American."[26]

Modern migration, however, was different. No longer did migrants have to overcome long and perilous journeys. Technology and transport had countered the selective element of migration.[27] New migration, moreover, came from the "Slavic and Mediterranean branches of the Caucasian race." This "flood" of migration had rapidly and thoroughly transformed the racial makeup of the nation. Fairchild mourned that "the real dilution of the original American stock is a matter of scarcely half a generation."[28] Lower races were arriving because the forces that forged the American race had dissipated. The motivation for migration had changed; it was no longer encouraging the movement of the fittest. As critics of immigration noted, if the earliest migrants to North America came for reasons that would encourage only the most fit and courageous to migrate, modern migrants were driven solely by "economic lot." Francis Amasa Walker was especially strident in his belief that modern migration was catalyzed by failure: "They are beaten races, representing the worst failures in the struggle for existence."[29]

By the 1880s, at the moment when patterns of migration to the United States shifted to eastern and southern Europe, rail networks had penetrated even the most remote areas of Europe and laws to prevent emigration were limited and easily evaded. Of particular importance, steamship lines crowded the Atlantic shipping lanes. In 1887, as Mayo-Smith pointed out, steamers made 259 trips from Liverpool; 265 from Bremen, Hamburg, and Le Havre; and 106 from Antwerp and Rotterdam—among many other European ports. As steamships grew in size and competed fiercely for passengers, companies sought to fill the lower decks—steerage—by offering low fares. In 1885, in the face of competition among steamship lines, steerage fares fell from twelve to ten to, even, seven dollars. The largest ships offered well over a thousand steerage berths. Their voyages mocked the trials of earlier migrants. For Mayo-Smith, the disease, insecurity, expense, and dangers that older immigrants faced had diminished. Modern migrants simply purchased a ticket from a steamship agent in their village and traveled by rail to a port to await a steamer. The steamship trip took only eight to ten days. The voyage itself, from food to medical care to sleeping quarters, was regulated by law.[30]

Overland and overseas journeys were no longer forms of biological selection. As Walker warned, "Fifty, even thirty, years ago, there was a rightful presumption regarding the average immigrant that he was among the most enterprising, thrifty, alert, adventurous, and courageous, of the community from which he came." Yet the arrival of cheap railroad and steamship fares in the steerage decks meant that immigrants came from the "least thrifty and prosperous members of any European community."[31] Though conditions in the steerage bowels of steamships were uncomfortable, they lacked the qualities that helped make earlier migration a selective process.

Steerage remained a frequent target of criticism from both migrants and reformers from the 1880s to the 1920s. In their memoirs, migrants frequently recalled the tight quarters and miserable food of the voyage over. The Jewish immigrant Rose Cohen's description of steerage travel was typical for its vivid descriptions of illness and cramped surroundings. For Cohen, the trip began pleasantly with a smooth ocean. Soon, comfort gave way to three days of seasickness. All around her, she "heard voices screaming, entreating, praying."[32] Reformers also railed against conditions in the bowels of steamships. By the early decades of the new century, the cramped conditions of steerage became the subject of exposés in popular magazines.[33] Like their counterparts studying factory labor and tramping who assumed the dress and manners of their subjects for their research, steerage investigators joined migrants in steerage. They took on the role of migrants and generalized from their own trials of seasickness in the cramped conditions of steerage. In describing their experiences disguised as immigrants in steerage, they rendered accessible to native-born English-speaking readers the experiences of travel underneath the steamships' main decks.[34]

The changing character of immigration shaped the concerns of the Dillingham Commission, the most far-reaching study of the effects of immigration to the United States. Starting in 1907, Vermont senator William Dillingham organized a critical review of the effects of immigration. His committee, the United States Immigration Commission (known as the Dillingham Commission), printed forty-two volumes, which recorded almost two years of hearings and investigations. The Commission drew on the testimony of a wide range of social scientific experts and on the work of field investigators.[35] The Commission joined critics of migration in arguing that the technology of movement had made immigration too easy. For its report on steerage conditions, even the Dillingham Commission relied on investigators who secretly dressed up as migrants. They presented lurid details of sexual depravity and cramped and malodorous conditions.

The Commission sent agents, "in the guise of immigrants," in steerage in the twelve largest Atlantic steamship companies. Its investigators uncovered the "disgusting and demoralizing" conditions of steerage. One investigator arrived in an unnamed port, disguised as a single Bohemian woman. After a cursory medical examination, she descended into the bowels of the ship, into the crowded berths of steerage. The floors were never cleaned during the voyage, even though neither cuspidors nor cans for seasickness were provided. Men and women shared two washrooms. Food was poorly cooked and haphazardly served. "If the steerage passengers act like cattle at meals," she concluded, "it is undoubtedly because they are treated as such." Even worse, female passengers faced the constant threat of harassment. One evening, as the investigator retreated to her bunk with sickness, a nearby Polish migrant struggled to escape the grasp of a ship's steward. Only the timely arrival of other passengers saved her "virtue." The investigator's testimony was scathing: "During these twelve days in the steerage I lived in a disorder and in surroundings that offended every sense." She was surrounded by "sickening odors...the vile language of the men, the screams of the women defending themselves, [and] the crying of children....Every impression was offensive."[36]

Such undercover reports of moral turpitude and revolting conditions encouraged new laws regulating steerage and steamship companies. As one congressional report mourned, "Anyone who has ever witnessed the plight of those who have to travel in the steerage of the large steamships bringing people to America is bound to be shocked at the brutal and even murderous conditions under which many of them have to travel."[37] Congress passed the first laws regulating steerage as early as 1873 to recognize the transition from sail to steam in transatlantic passenger ships. By 1882, Congress passed new laws governing steerage. In 1907 and again in 1908, Congress introduced laws requiring improved conditions and more airspace for each passenger.[38] Still, steamship companies protested the new laws because they curtailed the number of immigrants they could carry. Appearing before congressional committees in 1908, the companies succeeded in reducing the scope of regulation.[39]

The critique of steerage by a host of reformers and government officials represented much more than simply well-meaning concern about the exploitation of poor migrants. Congressional critics of steerage praised their reforms partly because they would inevitably raise the price of emigration.[40] Higher prices, after all, might encourage a better class of migrants. Even in its worst incarnations, steerage helped the least civilized and the least fit to overcome the difficulties of movement. Immigrants might be sickened on their oceanic journey, but mortality was now the exception. Steerage might leave migrants with "body weakened and unfit for the hardships that are involved in the beginning of life in a new land," but hasty and fraudulent onboard medical examinations allowed all but the most obviously diseased and unfit to enter.[41] As the leading social work journal *Charities and Commons* would declare, "Out of the East, since the days when our own Aryan ancestors began to make history, have come immense hordes.... To-day the world is face to face with the same situation. The steamship, the railroad, the telegraph and the newspaper have made it a peaceful invasion." Warne, similarly, compared the arrival of immigrants in steerage to ancient invasions in warships: "Is not this immigration as much of an invasion as was the coming of the Saxons to Britain or of the Huns to Italy? Is it necessary that the invader should come in warships instead of in the steerage hold of steam vessels before the migration can be called an invasion?"[42]

Steamship companies emerged as a particular target for those worried about the racial dangers of recent immigration. Laws passed in 1882 and reinforced in 1907 outlawed steamship companies from promoting immigration and decreed that companies pay the return costs of rejected migrants.[43] Yet the laws were laced with loopholes as steamship companies could still print advertisements and circulars giving prices and sailing times. The Dillingham Commission decried the colorful posters printed by steamship lines and the activities of agents who might mingle with the population on market days to sell ship tickets. While the steamship companies did not "openly" flout immigration law, their agents certainly did: "It does not appear that steamship companies as a rule openly or directly violate the United States law, but through local agents and subagents of such companies it is violated persistently and continuously."[44] "More emigrants, more dollars" for the steamship companies, mourned Robert DeCourcey Ward. When the pool of potential migrants had dried up in an area, agents simply moved on to yet untapped—and less civilized—areas.[45] Steamship companies and their agents, in the search for profit, seemed responsible for the ever declining racial status of immigrants. Italy, Austria-Hungary, and Russia, at the edges of civilized Europe, were caught in the "drag-net" of the steamship companies and "drained of their human dregs through channels made easy by those seeking cargo for their ships."[46] Indeed, mourned Edward Ross, steamship companies had once looked only to southern and eastern Europe for possible migrants. By the 1910s, Ross accused them of inciting migration from Asia. During his explorations of rural China, he noted the omnipresence of labor agents hired by the steamship companies who sold ship tickets "as newspapers and apples are cried on our streets." They offered loans and mortgages and provided instructions on how to give answers to evade officials upon arrival in America.[47]

For the Commission and other immigration critics, such agents and the companies they represented transformed the biological conditions of migration. As Ward noted, "The majority of our aliens are no longer the 'pick'. Weaklings are no longer afraid to undertake the journey." Ward placed the blame squarely on the steamship companies and their agents. As evidence of the effects of steerage travel, Ripley pointed to his analysis of the stature of migrants. In the past, the law of "natural emigrants" equated physical vigor with an adventurous spirit. In the age when migration was perilous, migrants were taller than those they left behind. By contrast, even a brief visit to Ellis Island revealed immigrants who were limited in stature and broken in health.[48] In the era of the steamship, as the Immigration Restriction League declared, modern migration was countering the forces of natural selection.[49] The League demanded new laws and new regulations on steerage as needed protection for the American race.[50]

At the same time, medical inspections on American soil did little to mitigate the effects of steamship companies and to thwart their insatiable demand for steerage passengers. When they examined the immigrants in steerage, immigration officers insisted that they were fighting a losing battle. The League frequently reproduced the words of beleaguered immigration officials. The League quoted William Williams, the commissioner of immigration for New York: "The time has come … to put aside false sentimentality in dealing with the question of immigration and to give more consideration to its racial and economic aspects.… Our first duty is to our own country."[51] The League railed that while the size of steamships was increasing, the number of immigrants "debarred" from entering the country remained paltry. In 1892, it noted, only 0.5 percent of migrants were turned back. By 1907 that percentage had grown to only 1.1 percent, or 14,059 out of 1,285,349 arrivals.[52] Moreover, immigration officials could only turn away migrants for reasons of disease, insanity, or anarchist sympathies. The League reprinted Williams's letter of resignation. Williams complained about too few resources and too few deportations of immigrants from "backward races."[53]

The Immigrant Invasion and Race Substitution

Given the lack of success in regulating steamship companies, observers worried about an accelerating competition between the established virile

American race and the arriving immigrant races. Because recent immigration was the movement of wage earners prompted by economic concerns, this competition played out in the context of industrial America. The Dillingham Commission concluded that "recent immigration is responsible for many social and political problems. Its chief significance, however, is industrial, and of the industrial phases of the subject none is of greater importance than the effect of recent immigration from southern and eastern Europe upon native Americans." In the industrial struggle for survival the fit were not necessarily the most desirable or the most racially advanced. Rather, for Commons, immigrants had a natural advantage in the "competition of wage-earner against wage-earner." Driven by desperation, often unencumbered by family, and drawn from less civilized races, immigrants proved willing to work longer hours for lower wages than their native-born competitors.[54] In study after study, critics revealed that immigrants were replacing the native born. The older model of conflict through migration that favored racial superiors had been reversed in what the economist W. Jett Lauck called "racial substitution in our mines and manufacturing establishments."[55] New immigration was leading to a race war that pitted "Latin, Slave, Semitic, and Mongolian races" against "Teutonic Races."[56]

Warne's study of supremacy in the coalfields depended on the idea that racial character was intertwined with fitness for labor. As the work of Huntington amply demonstrated, the character of races—their efficiency, energy, and innovation—could be used to map civilization. Observers such as Commons similarly sought to gauge different races' ability to withstand the rigors of industry. "However hard one may work," he concluded, "he can only exercise the gifts with which nature has endowed him....These gifts are contributed by race."[57] Commons led the way in assigning to different races certain general industrial characteristics. Syrians featured an "intrinsically servile character." Armenians were shrewd merchants and moneylenders. French-Canadians reflected medieval France, untouched by revolution and content with a lower standard of living than the Irish or the Italians. Italians, meanwhile, were illiterate and subservient, but thrifty. Finns were sober and industrious, though primitive. Jews were highly individualistic, preferring the confusion of the sweatshop to the order of the factory.[58]

Commons's belief that the ability to work was a significant racial trait was echoed by numerous other observers of immigration, including government committees. The bulk of the Dillingham Commission's work traced the industrial effects of recent immigration; a full nineteen of the forty-two volumes were devoted to "immigrants in industries." These volumes were prefaced by the *Dictionary of Races or Peoples,* which linked racial traits and industrial character. Blending a wealth of ethnographic and anthropological work on racial difference, the *Dictionary* meshed descriptions of linguistic background, bodily color and size, and head shape with commentaries on industrial background and ability. Italians were divided into two races, partly based on industrial character. North Italians were "cool, deliberate, patient, practical," whereas South Italians were "excitable, impulsive, highly imaginative, impracticable." The Slav was closer to the Asiatic Tatar, the *Dictionary* insisted. The Slav displayed "carelessness as to the business virtues of punctuality and often honesty, periods of besotted drunkenness." Slavs were, however, placid and disposed to mining and manufacturing.[59]

As observers evaluated the ability of different races to labor, they also described their different standards of living. Superior races might be able to work more efficiently, but they also required higher standards of living. According to one congressional report, for example, Jewish immigrants in New York seemed content to live and work in filthy and cramped surroundings. Among rural coal miners, a congressional investigator found that Welsh and Irish miners insisted on "clean dwellings," unlike recent Slavic arrivals.[60] The different abilities of the races to labor and their contrasting life demands created an unequal and potentially devastating race competition. The Welsh miners studied by the Dillingham Commission were leaving for jobs in the West "because they could not stand the competition of the newcomers, with their low standard of living." When races had distinctly different standards of living, the industrial battlefield was bound to be littered with a defeated, but superior, race.[61]

Commons (and his students) traced "the displacement of earlier nationalities by newer types of immigrants with lower standards of living" in meatpacking, garment manufacture, and mining. In each case, the race characteristics of newer immigrants such as Jews, Italians, and Slavs allowed them to wrest employment from the hands of older migrant racial stock.[62] As Warne realized in the anthracite coalfields, although the battles of labor may have pitted workers against bosses, the real industrial war was being fought between old and new immigrant racial stock. Comparing the clothing, company store accounts, living quarters, and even methods of gathering firewood of different races in the coalfields, Warne concluded that race substitution by Slavs was inevitable. Moreover, the conflict that exploded so dramatically in Pennsylvania was also occurring in industries ranging from steel to cigar making to textile weaving.[63]

The Commission worried that as a result of the immigration of racial inferiors there were fewer of the old kind of migrants. The fit had few inducements to migrate. Recent migration had caused numerous social problems, concluded the Commission, but nowhere were its effects more noticeable than in the world of industry. Fewer old stock and more new migrants meant that "the racial composition of the industrial population of the country has within recent years undergone a complete change." A few of the old stock might become managers directing the haphazard work of their inferiors, but social advance was restricted only to the highest grades of the old stock. Others were displaced. Some took to the roads and rails as tramps, the detritus of race substitution. The earliest of nineteenth-century migrations of energetic new arrivals had spurred the industrial advance. Despite the growth of factories, the Commission warned, the pace of industrial evolution had slowed because of the character of recent migrants. In the age of steamship immigration, there were few new industries and little technological advance.[64]

Race conquest was geographic, as parts of cities were submerged in the immigrant flood. Colonization was also industrial. For the Commission, "the racial substitutions in, and the present racial composition of, the operating forces of the…coal mines of the country may be considered typical." Indeed, textile work, woolens, clothing, shoemaking, and glass had all, in turn, fallen into the hands of the unfit.[65] In one New England shoe factory, Irish and native-born Americans had given way to recent migrants. The factory, by the time the Commission investigators arrived, was 85 percent Greek and Jewish. A small Western Pennsylvania glass-making town faced a similar "racial change" as skilled English, German, and Belgian glassmakers were replaced by Poles, Russians, and Slovaks. In Paterson, New Jersey, innovative and skilled French and English workers developed the silk industry but ceded their place to Italians. In Wilmington, Delaware, Germans, Scandinavians, and Irish introduced new techniques and constructed new industries, only to lose their jobs to Italians and Poles. In each of these case studies cited by the Commission, race substitution was the victory of inferior recent migrants over their biological superiors and it signaled the end of a period of vigorous development, innovation, and growth. Race substitution was more than simply a transformation in work or the association of an industry with different immigrant communities; it was conquest and it was the slowing down of the pace of industrial evolution.[66]

To comprehend the shop-floor race struggle, economists transformed their understanding of the wage. The wage, in an age of immigrant colonization, was not determined solely by the availability of workers set against the demands of employers. As the Commission reported from their investigations of the American economy, greater racial efficiency and civilization did not translate into higher wages. Instead, lower standards of living determined access to employment.[67] Because employers were ready and able to hire workers with lower and lower standards of living, barely civilized races seemed to have the upper hand in the industrial struggle for survival.[68]

Tropical Labor in the Temperate Zone

Yet the industrial struggle for survival did not necessarily include all races. Inferior European immigrant races were clearly replacing their white betters and Asian migrants, held at bay by restrictive legislation, seemed to loom on the horizon, but what of the ancestors of tropical races already dwelling in the United States? Nonwhite races descended from forced migrations from the tropics still lived in the ideal temperate zone of North America.[69] As observers of American industry turned from describing race substitution among white races, they recognized the unique position of the descendents of African American slaves. African Americans were largely excluded from industrial jobs. Observers insisted that their absence from the industrial economy was the result of an ingrained inability of African Americans to withstand the demands of industry. It was not the effects of widespread racism that kept African Americans in the agricultural South or that monopolized factory jobs for the white races.

If whiteness was cast as the ability to industrialize and to countenance industry, then blackness was about indolence, laziness, and inefficiency—the heritage of a tropical past. As Commons insisted, in the context of industry, the white and black races were firmly and inexorably divided by their different heritages of climate. The African climate "render[s] this region hostile to continuous exertion. The torrid heat and the excessive humidity weaken the will and exterminate those who are too strenuous....Thus nature conspires to produce a race indolent, improvident, and contented." The legacy of slavery had merely reinforced the effects of climate. The "Negro" was simply not an industrious race and was destined to fall away in the face of mechanization and immigration. "The Negro...works three days and loafs three," wrote Commons. The "Italian, or Jewish immigrant works six days and saves the wages of three." Unlike the Negro, the European races—however debased and backward—could envision and work for the future. Saving and hard work, Commons concluded, was a mark of an "industrious race."[70]

Race may have been a hierarchy but, as the complex tables of the *Dictionary* highlighted, it had its hard ceilings. The legacy of nature itself ordained black races as incapable of industrial labor, marking them for extinction at worst and for a limited future based on agriculture at best. For the University of Wisconsin's Paul S. Reinsch, the Oriental seemed "ready to adopt our industrial methods." The Negro, by contrast, was biologically removed from white industrial civilization. The University of Chicago's W. I. Thomas, similarly, divided the world into "active and passive races." While the Japanese and other Asian races existed somewhere between the active and the passive, the Negro was passive. When it came to issues of efficiency, energy, and industry, for Thomas the white and black races were as different as "night" and "day."[71] African Americans reflected the indolence of their tropical ancestors, Commons argued: "Those races which have developed under a tropical sun are found to be indolent and fickle. From the standpoint of survival of the fittest, such vices are virtues.... Therefore, if such races are to adopt that industrious life which is a second nature to races of the temperate zones, it is only through some form of compulsion." The Negro could only labor effectively as a slave. As the South Carolina historian and social scientist Joseph Tillinghast declared, the African continent featured "conditions adverse to the growth of industrial efficiency; indeed, few regions are more hostile to such a development." Warm weather, abundant food, and fertile soil conspired against "motives to labor."[72]

The legacy of the tropics hindered the labor of manumitted slaves and their descendents. In his widely cited "Race Traits and Tendencies of the American Negro," the statistician Frederick Hoffman described a decline in the efficiency of former slaves. Trained in the violence and discipline of slavery, the newly manumitted slave was a "desirable" field laborer and held a monopoly on Southern paid labor. However, a new generation of African American workers, untouched by slavery, had reverted to their tropical heritage. "They became more lazy, thriftless and unreliable, until they will soon attain a condition of total depravity and utter worthlessness," Hoffman insisted. Drawing on the reports of different white farm owners, Hoffman reported that he did not find "any evidence that colored farm labor is improving in quality."[73] Similarly, given their tropical past, African Americans could hardly withstand the rigors of industrial work. Tillinghast, for example, insisted that they could not countenance the discipline and long hours of factory labor. Even generations living in a temperate (or subtropical) zone had not erased the tropical heredity of the former slaves: "The progenitors of our present negro population were indolent and wasteful, lacking mechanical ability, foresight and will." As evidence, he pointed to the experience of a Fayetteville, North Carolina, silk mill that hired black workers. It was set up by manufacturers from Paterson, New

Jersey. Its manager concluded that "no one can make success of a mill by applying white methods to colored people. With the latter there is but one rule to follow, that of the strictest discipline." In the case of this mill, that discipline meant whipping. Without such discipline, black workers seemed "indifferent" to labor. For Tillinghast, black industrial labor seemed possible only under conditions that recalled the past of slavery and even then could only involve the simplest of mechanical work. He concluded that "an overwhelming majority of the race in its new struggle for existence under the exacting conditions of American industry, is seriously handicapped by inherited characteristics."[74]

Of course, Africans had arrived in North America at the same moment as those whose progeny were the pride of the new American race. For Charles Woodruff, black arrivals in the Americas differed from the contemporary European white migrations, insofar as they could not be considered true colonizers. Rather, they were part of a forced migration in the hands of racial superiors who were looking for a solution to the acute problems of agricultural work. Woodruff agreed that on the plantation, under the direction of the brutal overseer, the African American could work acceptably. In the factory, by contrast, the African American could have little place.

If there was agreement among elite and white observers of African American labor that slavery had been the only way of making the tropical African work and that Reconstruction had ignored the immutable legacy of biology, there remained some disagreement about the fate of the African American in freedom. This dispute was grounded in divergent readings of (flawed and haphazardly collected) data from the eleventh and twelfth censuses (1890 and 1900). The nation's social science journals and popular magazines echoed with debates over the population trajectory of African Americans. Only a minority insisted that the African American population was increasing at the same rate as the white.[75] The loudest voices concluded that the African American was largely restricted to the rural South and inevitably moving toward extinction. Deaths would outnumber births and the race would ultimately disappear.[76]

Relying on census data, observers also noted that as European and Asian immigrants slowly made their way into the states of the former Confederacy, they seized whatever jobs had once been allotted to African Americans. The Mississippi planter and Southern historian Albert Stone, most notably, argued that African Americans were losing their place to recent European immigrants as industrial and agricultural workers. According to Stone, African Americans' retreat from industrial labor was due not to racism but to their low level of efficiency. Where the Italian would plow to the very edges of the field, the Negro seemed content to let such areas lie fallow. Similarly, lower levels of efficiency than Italians or Jews, not racism, meant that African Americans struggled to find industrial work.[77] The place of the African American in industry, concluded Commons, could be little more than that of a strikebreaker.[78]

Like the Indian before him, the African American seemed destined for extinction in the industrially evolving American nation. The words that William Graham Sumner used in 1887 to explain the decline of the Indian in the face of the evolutionary progress of farming seem eerily similar to those used from the 1880s to the early decades of the twentieth century to explain the apparent decline in the African American birthrate. "It appears clear that the Indians have never wanted to be civilized," decided Sumner. "They want to be left to sink down into squalor and neglect, and to die out, as they undoubtedly would, if the supporting hand of the whites were withdrawn."[79] In a similar tone, Tulane's William Benjamin Smith argued that, despite the help of Northern philanthropists, the African American was being left behind in the progress of industrial civilization: "But the heredity differences between white and black remained: this was his only essential handicap.... He is gradually receding before the more efficient and reliable white laborer.... Even in the South the tendency is to relegate him to menial service and the coarser occupations." Instead, "poor white" workers monopolized work in textile mills and immigrants seized railroad and mining jobs. "Meantime the streets and dives of our cities have been crowded with idle Negroes. Traveling through the South one sees gangs of idle Negroes hanging around the railway stations in every little town."[80]

In the aftermath of the twelfth census, which suggested a declining African American population, the American Economic Association assembled a blue-ribbon committee to examine the "economic position of the American Negro." For most of the committee, the declining population of African Americans was ample evidence that they had no role in the industrial future of the nation. At least until African American labor became essential with the temporary cessation of European immigration during World War I and its more permanent restriction after 1924, the discourse around race and industry was profoundly segregated. It was a scholarship firmly and consciously about white workers, white immigrants, and the perceived threat of Asian migrants. African Americans were defined out of this discourse and designated—with the American Indian—for extinction.[81]

Typically, scholars such as Commons, in their examinations of races and industry, included little detail on African American workers—beyond a statement of the black race's lack of fitness for industrial society. The

Industrial Commission and the Dillingham Commission also associated industrialization only with European and Asian migrants. As tropical races awkwardly caught in a temperate civilization rapidly entering its industrial stage, the descendents of former slaves were imperiled. Imperialism might provide the answer, however. As the United States debated the rise of industry alongside the future of the Philippines, some proposed "colonizing" African Americans in the islands. As a tropical race with several centuries of contact with temperate betters, they would thrive in the islands' lax standards and torrid climate. By 1900, Nathaniel Southgate Shaler, the dean of Harvard's Lawrence Scientific School, was promoting the use of African American soldiers in the Philippines. According to Shaler, they were better able than their white counterparts to countenance the harsh tropical climate. He hoped as well that black soldiers and their families would settle in the islands. His enthusiasm for black settlement in the Philippines was soon echoed by Alabama senator John Tyler Morgan, a fire-breathing advocate of Jim Crow. In 1901, Morgan proposed the formation of black colonies in the Philippines. He reasoned that African Americans were suited for the climate and could compete more successfully with indolent natives. Interestingly, where critics of immigration to the United States criticized steamship companies, Morgan cast them in the opposite role: as means to rid the nation of the unfit. He argued for the incorporation of steamship companies to transport black migrants from the U.S. mainland to the Philippine colonies. His plan, however unfeasible, received some traction, garnering the support of the secretary of war, Elihu Root. By 1903, the War Department proclaimed that the islands could become the home to the nations' black millions.[82]

The Black Worker and the Industrial Future

African American workers found few nonblack defenders of their place in industrial civilization. Socialists, for example, generally repeated condemnations of the black workers as "palpably ignorant and thriftless and immoral," as William Noyes declared—even as they condemned Southern racism, Jim Crow segregation, and lynching. Even one otherwise sympathetic writer in the *International Socialist Review* noted the declining population of African Americans. "The negro," he argued, "has degenerated both physically and morally." Typically, the party's standard-bearer, Eugene Debs, agreed that socialism could offer the African American nothing more than class solidarity. If he argued that black class interests merged with those of white workers, other socialists were not willing to afford African Americans equal evolutionary status. African American and white workers were at "widely different stages of evolution."[83]

Even as African Americans were being scientifically expelled from the nation's industrial future, African Americans who constructed themselves as race leaders vocally contested the conclusions of debates about biology, race, and industry. African American scholars and political activists were fully aware that to be left out of debates about labor and industry also meant exclusion from visions of the nation's future and, as the Philippines colonization experiment suggested, potentially from the metropole itself. Although W. E. B. Du Bois and Booker T. Washington sparred publicly about black responses to Jim Crow, they both responded to an evolutionary logic that evicted African Americans from industrialization. As historian Eric Anderson has argued, leading African American intellectuals remained wary of evolutionary thought—not surprisingly given its obvious role in the justification of racial hierarchy. Some, like Du Bois, regularly saluted evolution as the leading intellectual current of the day. He was even included in the American Economic Association's committee that analyzed the twelfth census. Yet Du Bois was typical of African American leaders in defining evolution narrowly as primarily a science of the nonhuman world. He warned of the overapplication of evolutionary thinking. Moreover, it played little role in his more theoretical writings on the origins of race and of human social development. Alone among leading sociologists of the day, he never talked of "social evolution." Though the intellectual foe of Du Bois, Booker T. Washington had an equally fraught relationship to evolutionary theory. Neither Washington nor the curriculum at his Tuskegee Institute displayed much interest in Darwinism specifically or evolutionary science in general. Yet his efforts to train a generation of skilled African American workers and artisans should be understood not simply as a tacit acceptance of the Southern racial order but as a deliberate response to a broader conversation about race that cast African Americans as a tropical race unfit for and incapable of energetic and demanding labor. Washington's celebration of a black work ethic directly responded to notions of black indolence that were grounded in the scientific language of race and climate as much as in the stock racist characters of popular vaudeville culture.[84]

African American leaders, such as Du Bois or Howard University's Kelly Miller, offered limited rebuttals of arguments about black extinction. In his lengthy response to Hoffman, Miller, for example, constructed his critique solely around a debunking of the census statistics.[85] As African American community leaders negotiated a fraught relationship to evolutionary

Figure 3.1. "Contractors and Builders": The artist Meta Warrick produced this diorama that represented African Americans as industrial actors, not as the indolent descendents of tropical savages. Giles B. Jackson and D. Webster Davis, *The Industrial History of the Negro Race of the United States* (1908).

science, they struggled to find ways of confronting the notion that the same people who sacrificed to build the agricultural foundation of the nation were unequipped to help build its industrial future. As the world's fairs in Chicago in 1893 and in St. Louis in 1904 proudly displayed the industrial progress of the nation in glorious white exhibition halls, African American companion or alternative exhibitions on a more modest scale sought to demonstrate black aptitude for arduous and industrial labor. Most notably, at the 1907 Jamestown Exhibition, the Negro Building featured an exhibit organized by the city's elite African Americans, who were intellectually allied with Washington. Featuring dioramas by the black artist Meta Warrick and memorialized in a book, *The Industrial History of the Negro Race of the United States,* the exhibition directly countered representations in other world's fairs of the lazy, contented "darkey." Dioramas such as "Contractors and Builders" cast the African American as a vital industrial actor.[86]

The Unparalleled Invasion

The same observers who agreed that African Americans could not survive the demands of industrial labor also recognized traits in Chinese, Japanese, and, less frequently, South Asian migrants that could aid them in the struggle for industrial supremacy. Despite the guarded optimism of scholars such as Huntington about the climatic forces that would ensure European racial superiority, the evidence drawn from industrial explorations suggested that racial substitution was not limited to white races. The evidence of the Dillingham Commission from its study of cannery, beet sugar, railroad, and lumber workers raised the alarming possibility that, despite restrictive legislation, Asian migrants could emerge as the ultimate victors in the industrial struggle for survival. Study after study cast these migrants as docile, hard-working, and provident, if unoriginal and physically weak. Even more than white migrants, Asians were deemed more likely to "colonize." In addition, the "yellow" or "Mongolian" races, unlike the "black," seemed capable of industrial labor. They could not as easily be evicted from debates about race and industry. The Commission considered Chinese, Japanese, and East Indian laborers together and described their competition with new and old European white stock. Because of their peculiar "Asiatic" race qualities, Asian migrants who were willing to work harder for less than even the most debased of white migrants were invading industries throughout the West. Although their standard of living seemed to confirm their racial inferiority, it also made them the most desirable of workers for beet sugar, logging and lumber mill, and cannery work, as well as industrial work such as mining or shoemaking.

As with European migrants, the Commission's investigators traced a distinct racial hierarchy among Asian migrants. Japanese workers generally sat at the top of this hierarchy, deemed the most "ambitious" and "versatile." In railway work, they were even considered more desirable workers than some Italians. The Chinese were equally stolid and low-living, but lacked the ambition of the Japanese. Still, in dirty and dangerous cannery work, the Commission reported that the Chinese were preferred: "Not only are the Japanese less experienced in the industry than the Chinese, but they are considered less reliable in contractual relations and do not have the highly developed instinct of workmanship." The Commission regarded East Indians as the least sought after of Asian migrants. In the Northwest lumber yards, where East Indians had worked since 1906, employers were unhappy with their labor: "In a few instances, they have been regarded as worth the wage paid them, but in most instances the employers have regarded them as dear labor at the price, because physically weak as compared to 'white men', slow to understand instructions, and requiring close supervision." Similarly, in railroad work "they have been found to be physically weak, unintelligent, and slow to acquire a knowledge of the work to be done."[87]

Such studies of Asian migrants connected an older nativism with the new science of industrial evolution. Anti-Asian fervor had a long history in the United States, dating to the anti-Chinese mobilizations in California in the 1870s.[88] As historian Alexander Saxton's work amply demonstrates, anti-Chinese mobilizations, grounded in the white working-class politics of California, anticipated later, more national concerns about race substitution. Nativist activists in California as early as the 1870s decried the Chinese for their low standard of living. They could undereat and undercut their white laboring competitors.[89] The terms of anti-Chinese nativism, voiced first in the rough world of California working-class and electoral politics, would have seemed familiar and comprehensible to a later generation of well-heeled immigration restriction activists. Looking back on decades of conflict, Jesse Frederick Steiner, in his influential study of Japanese modernization and migration, described the "racial struggle between East and West" on the streets and in the ballot boxes of California in the 1870s through the 1890s in terms of the economic struggle for survival: "As soon as American laborers realized that the presence of the Japanese was making their struggle for existence harder, their latent prejudice against them as Orientals broke out into open hostility."[90] Superior white races, it seemed, could not withstand competition from inferiors. As the physician Woods Hutchinson noted from his observations on the West Coast, "The Chinaman is the most industrious worker that walks the earth in human form. He doesn't seem to know how to get tired.... There is something positively uncanny about his affection for work." Even as he declared the "Mongolian" an inferior race, he admitted that "no class of white men will work with the unremitting persistency of the Chinese."[91]

By the 1880s, anti-Chinese nativism had produced a series of laws limiting, though by no means ending, Chinese immigration. The restriction of Japanese immigration proved harder to achieve and justify.[92] The fact that Japanese restriction was enacted by President Theodore Roosevelt only through the language of a "gentleman's agreement," as opposed to the power of legislative authority, highlighted the way the rapid advance of Japan on the world stage challenged domestic American conceptions of racial hierarchy.[93] The surprising victory of the Japanese over the Russians in 1904, coupled with the remarkable strides of the Japanese economy, represented a significant rupture in race theories that cast the Asian races as stagnant and incapable of reaching the industrial stage of civilization, even if they could outwork and under-live white workers.

Japan's victory confirmed its status not simply as a military force but as a newly industrialized nation. As the Clark University anthropologist Alexander Francis Chamberlain noted, early nineteenth-century Americans and Europeans who dismissed the Japanese as physically and mentally weak received a "rude shock" after the Japanese triumph. The Japanese could not be deemed biologically incapable of industrial labor, either at home in Japan or abroad as migrants.[94] In the aftermath of the war, Congressman William Redfield presented a reluctantly glowing report on "the evolution of Japanese industry." He recognized the prospects for Japanese development, but, perhaps as a comfort to his white American readers, insisted that Japanese industry remained behind its white and Western competitors. Japanese industry, he argued, remained in the handicraft stage—however "evolved and perfected"—while its heavy industry lagged. He also grudgingly praised the Japanese worker: "I suppose there is no more thrifty, able, capable worker than the average Japanese. He is accustomed to living to his satisfaction on the most limited scale."[95]

In American periodicals, Japanese authors boasted of their nation's industrial accomplishments. Alongside photographs of Japanese warships, the diplomat Jihei Hashiguchi praised the development of Japanese shipbuilding, textile production, railroad construction, steel production, and electrification. In a gesture to his American readers, Hashiguchi cited the Japanese debt to European and American innovation, but closed his article by proclaiming that the Japanese success on the battlefield was "the natural outcome of racial characteristics."[96] As T. Iyenaga, a Japanese lecturer at the University of Chicago, astutely noticed, the Japanese victory over the Russians undermined established notions of white racial superiority. This was a crisis that white scholars sought to solve through close analysis of language, anthropology, and the history of ancient migrations. "The logical consequence" of the Japanese victory, argued Iyenaga, "will be that the inquisitive West will search for light on the causes of the virility of the intellectual and moral fibre of the Japanese." He suggested that white scholars would explain the success of the yellow Japanese by finding "a few drops of white blood in their veins, owing to the mixture of races in prehistoric time."[97]

His suggestion was prophetic. Western observers sought to bolster older and now threatened racial theories by insisting through their study of ancient migrations that the Japanese were truly an Aryan, or at least a Caucasian, race.[98] In color, they might appear similar to the Chinese, but racially they had more in common with their European cousins. As Chamberlain noted, the consensus of anthropologists was that the "Mongolian" Japanese were not the first inhabitants of the islands. Rather, the Ainu peoples were the original Japanese. Anthropologists wondered whether the Ainu represented an "Asiatic branch of the Caucasic division."[99] William Elliot Griffis, a self-confessed admirer of Japan's "triumphs" and a former professor in the Imperial University of Japan, was the loudest American exponent of the theory of Japanese Aryan roots. For Griffis, the possibility of a Japanese Aryan past helped him reconcile his admiration for Japan with an ongoing belief in the racial inferiority of the Mongolian. At the beginning of his synthetic history of modern Japan, *The Japanese Nation in Evolution*, Griffis insisted that the Japanese are a "composite, and not a pure 'Mongolian' race. Their inheritance of blood and temperament partakes of the potencies of both Europe and Asia." Like other scholars, Griffis traced ties between the allegedly Aryan Ainu and modern Japanese. He found his evidence in language. He argued, for example, that "most names of places held most sacred in Japanese history are Aryan." The "basic stock" of the Japanese, he concluded, must be Aryan. The Dillingham Commission in its *Dictionary of Races or Peoples* equally declared that the Japanese race had a crucial "undoubted white strain." The Japanese might at first glance appear Mongolian, but closer inspection revealed less yellow skin and rounder eyes. The Ainu had left their mark.[100]

Sidney Gulick, perhaps the leading American expert on Japan and its "evolution," was equally insistent on proving the "Occidental" orientation of the Japanese, even if he paid less attention to the Aryan origins of the Ainu. He preferred to look past clothing and manners that might mark the Japanese as Oriental and instead pointed to "the military, the commercial, and the industrial virtues" of the race. For evidence, Gulick pointed to the hospitable treatment offered to Russian prisoners during the war. In lauding this "beautiful altruism," Gulick concluded that the Japan "has definitely passed out of exclusively oriental life and is to-day, in important respects occidental."[101]

For some, the apparent Aryan or Occidental roots of the Japanese presented ample evidence that Japanese immigration to the United States posed little threat. Perhaps they could, in fact, be more easily assimilated than African Americans and even the lowest European white immigrants. Mary Coolidge, a sociologist formerly at Stanford, sought to argue through a close analysis of wages in California that the arrival of Asian immigrants, especially the Japanese, did not result in a decline in conditions or wages, as was the case in the East with the arrival of European races.[102] While recognizing racial difference, Coolidge was part of a contingent of observers that stressed the strength of the Asian immigrant character, going so far as to counter the notion of Asian penury. They do not, she insisted, live on rice and the odd rat. Instead, they demonstrated an admirable sense of responsibility. Of course, her limited praise of Asian migrants depended only on their contrast with lowly recent European migrants.

More typically, even those who argued for the Aryan influence on Japanese civilization remained hostile to Japanese migrants. Many observers insisted only that the elites, and not poor migrants, showed the influence of the Occident. Most critics of Asian immigration tended to the shrill and the alarmist. They cast Asian immigration not as an added benefit but as an invasion and as race substitution. "Asia has found us out, too, and the flood from the Orient has started," warned one editorialist. "The reservoir that is tapped is limitless. Literally hundreds of millions of brown men, yellow men, and bronze men would now like to come to America....And, then— the deluge."[103] European immigration may have been a flood, but Asian immigration would be a deluge. Such alarmist critics sought to understand the process through which the Japanese had realized such a rapid industrial evolution and wondered whether a slowly awakening China might also experience its own form of development—largely through racial solidarity offered by Japan. Steiner, in his otherwise temperate and measured account of Japanese industrial evolution, became shrill when he warned of the need to keep Japan and China separate. A united Asia would only increase the economic, racial, and political effects of a migration that was already pressing at America's gates. "We seem to lose sight of the fact that the few thousand Orientals in our country," he wrote, "are but the vanguard of many millions" seeking to escape the harsh conditions of Asian life.[104]

Yet it was this "struggle for existence" that rendered Japanese and Chinese more solid, stolid, and resilient, though physically weaker, than their white counterparts. Edward Ross was typical in insisting that the brutal and temperate Asian environment and near total lack of knowledge about disease and hygiene had actually benefited the race as a whole. Those that did not perish horribly in almost constant famines or epidemics were quick to heal from even severe wounds and were much more resistant to disease than the white races.[105] Japan, therefore, through "occidental" influences, had entered the industrial stage of development. Yet it continued to draw on the natural advantages of the "oriental."[106]

Although many observers of Asian migration joined the Dillingham Commission in supporting existing restrictions on Chinese and Japanese labor and in calling for the exclusion of East Indians, they cautioned that such laws were flimsy at best. The Chinese invasion was held back by the weakest of laws, the Japanese only by the good graces of the Japanese government, and the East Indian by British imperial policy. Ross warned that the teeming millions of Asians might soon come to fight against American restrictions. "May they not make the rearing of such dikes a *casus belli*?"

he wondered. "Even now the Japanese show themselves restive in the presence of anything which savors of exclusion, and it is not hard to foresee a time when the peoples of India and China and Siam and Egypt may challenge the barriers which keep them out of all the more desirable markets for their labor."[107]

As Ross's warning suggests, it was in the case of the Japanese that the ties between concerns about migration and colonization were most fully realized. Not only did Steiner entitle his influential book *The Japanese Invasion* but also the *Annals of the American Academy of Political Science* published a special issue on the Asian "invasion." In concluding that the Asian immigration was in fact an Oriental invasion, John S. Chambers, the California state controller, returned to the imagery of the flood. In his contribution to the special issue, Chambers described the individual Japanese migrant as a gopher eating away at the levee of restrictive legislation. First a trickle of water, then a deluge, and the flood swallows the land. In this way, the Japanese would colonize California. Chambers warned of a process of "peaceful penetration, of conquest by colonization," or, as a professor at the University of California described it, a "bloodless struggle." California senator James Phelan, similarly, hinted that the Japanese government itself had targeted California for colonization. Japanese migrants seemed poised to "overwhelm the civilization which has been established by pioneers and patriots."[108] Shailer Mathews and Sidney Gulick concluded that fears whipped up by the Japanese victory in 1904 and by the possibility of increased migration were grounded in concerns that immigration was actually part of a deliberate plan of territorial expansion.[109]

Socialists stood apart from debates about race substitution caused by European migrants. They were notably silent and added European migrants to the ranks of the "American" worker. Yet they joined in the chorus warning of the "yellow peril." Socialists despaired of the "heathen Chinee" who "learned to live cheaply over the sea." The capitalist system, argued one socialist poet, favored Asian workers with their miserable standard of living: "And he is the fittest beyond dispute / The present competitive system to suit / Whose life comes nearest to that of a brute." The Chinese and Japanese were for another critic "an innumerable horde of . . . dread competitors" who would undermine the condition of the "American laborer" whether they arrived as migrants or labored at home with "English and American machinery." Because socialist observers recognized both the Asian potential for industrialization and the low standard of living, they were willing to jettison international solidarity. The socialist activist Cameron King, for example, argued for exclusion, despite fraternal solidarity. "What monstrous twisting of 'Workingmen of the World, Unite!," he declared, "gives us the slogan 'Workingmen of the World, Compete!" Instead, he insisted that "our feelings of brotherhood" toward the Asian worker had to wait. Class consciousness and solidarity dissolved in the face of "economic invasion."[110]

In fact, it was the socialist Jack London who provided the most dramatic depiction of Asian colonization as part of his *Strength of the Strong* collection. In his short story "An Unparalleled Invasion," London described how the Chinese slowly began to threaten the white world, not through their armies—Chinese military technology could hardly compete with that of Europe or the United States—but through migration, fueled by Chinese fecundity. Initially, the Chinese posed little threat to the white races. The development of the Japanese, however, helped adapt their co-racialists to the discipline of industry and, as the Chinese moved around the world, they created "colonies."

If colonization led by superior white races was accomplished at the point of the bayonet, the colonization by Asian and inferior white races succeeded through breeding and migration. China was "spilling over...with all the certainty and terrifying slow momentum of a glacier." London's story reads as a wild nativist fantasy, yet for his contemporary readers he articulated widespread fears of migration as race conquest. This was an invasion that, for London, could only be confronted militarily. Still, individually, the white nations in London's nightmare vision of the future unsuccessfully tried to defeat the Chinese through military invasions. The massive population of the Chinese nation simply swallowed the invading armies. White races were only saved when white scientists collected germs of the worst diseases in glass tubes, which were dropped from airships on the Chinese. Those that survived cholera fell victim to the plague; those that survived the plague were killed by yellow fever.[111] In the end, London left his readers wondering about his title the "unparalleled invasion." Did it mean this new invention of germ warfare or the Chinese migratory colonization?

Migration, for the turn-of-the century observer, was not the benign movement of people. Nor was colonization simply a process of expansion undertaken by Western industrialized nations. Rather, migration was a form of race conflict and colonization was a form of race conquest. These were biological processes, central to the larger process of industrial evolution. The evolution of industry, however, complicated the criteria for social selection. The civilized could be the weak, unable to compete at work and unwilling to breed. In the age of the steamship, the inferior now were the mobile, the fecund, and the winners in the struggle for work and survival. For some, like London, annihilation seemed the only answer to conquest.

London highlighted the links between fecundity and race substitution. Conquest, he reminded, had to be accomplished domestically, that is, not through political control, but through the force of sheer numbers. In the battle of races acted out in an industrial world, the fecundity of inferior races gave them an advantage. Race substitution thus merged with questions of gender, family, and race suicide.

4

CAVE GIRLS AND WORKING WOMEN

The White Man's World of Race Suicide

Woman...exists in the white man's world of practical and scientific
activity, but is excluded from full participation in it. Certain organic
conditions and historical incidents have, in fact, inclosed [*sic*] her
in habits which she neither can nor will fracture."

W. I. Thomas, 1907

When Nadara de la Valois, the daughter of French nobility, married Waldo
Emerson Smith-Jones, a scion of one of Boston's most aristocratic families,
the ceremony took place in Honolulu onboard his parent's yacht, the *Priscilla*. It was the most civilized of ceremonies for two former savages.

Their marriage was the final scene in Edgar Rice Burroughs's saga of
prehistoric life, shipwreck, rape, and love in the South Pacific. *The Cave Girl*
echoes themes from *Tarzan*, Burroughs's most famous novel. The romance
of Nadara and Waldo Emerson is a narrative about overcoming the feebleness that Burroughs associated with too much civilization. As his name
suggests, Waldo was raised in bookish Boston. His was a world of ancient
Greek and a wasting cough. His doting mother had taught him "to look
with contempt upon all that savored of muscular superiority—such things
were gross, brutal, primitive." His cough had led him on a curative ocean
voyage, but he was washed overboard and ashore on a tropical island. Now,
stranded in the jungle, this lanky Boston aristocrat needed to discover his
internal primitive strength when his life was endangered.

Happily, Nadara, a beautiful cave girl, rescued him from a clash with
primitive, brutal troglodytes. Given its tropical locale, the island's population was caught in the Stone Age. Waldo did not realize that in washing up
on this island he was also stepping back in evolutionary time to the moment when marriage was realized through rape and violence. Nadara had

fled her savage village to escape Flatfoot and Korth, two muscular brutes. They both intended to seize her as their bride. Because she wrongly imagined that he was brave, Nadara led Waldo back to her village to confront her suitors/would-be rapists. At the last minute, he fled back into the jungle. In the next six months, Waldo fortified his muscles and overcame both a panther and murderous savages. When he finally returned to her village, "he might have been a reincarnation of some primeval hunter from whose savage loins had sprung the warriors and the strong men of a world." In the ensuing battles with Korth and Flatfoot, Waldo, in his fury, lost "every particle of the civilization and culture and refinement that had required countless ages in the building, stripping him naked, age on age, down to the primordial beast that had begot his first human progenitor." Yet in his own courtship of the beautiful Nadara, Waldo made the mistake of returning to civilized behavior. She rebuffed his gentle words of love. Finally, he grabbed her. "He did not ask if she loved him, for he was...the cave man."

As her "brown" arms encircled Waldo's neck, Burroughs confronted the sexual reality of imperialism: miscegenation. Burroughs diffused the racial peril of their love by revealing—unknown to Nadara—that she was actually the daughter of a shipwrecked French countess. Her mother had perished in childbirth and Nadara had been raised by savages. The Stone Age and imperial contexts permitted erotic and manly fantasies of domination of women and lower races. In fact, Burroughs's narrative was typical of stories about the Stone Age and savages, what can be termed primitive fiction. Other novels and stories—including others by Burroughs—also offered lurid details of primitive marriage and mating. They reveled in episodes of abduction, violence, and rape, reinforced by lurid drawings. As in popular magazines, such as *National Geographic,* the primitive context provided a moral shelter for erotic imagery. Even the magazine cover that first introduced *The Cave Girl* was deliberately titillating. The viewer follows Nadara's gaze through a darkened cave toward Waldo. Waldo, dressed in the rags of his civilized clothing, swings his club at some hidden enemy. Nadara's racial status is ambiguous. Her darkened skin suggests a nonwhite savage, but she also stood in the shadows. The suggestion of savagery nonetheless permitted the display of her breasts, which are only slightly covered by her fingers and flowing hair. While a seminude image of an aristocratic white woman would have been unthinkable, the suggestion of her savagery enabled her eroticism.[1]

As much as Burroughs indulged fantasies of sexual and racial power, he still sought to distinguish between enfeebling over-civilization and tropical degeneration. Waldo regained his vigor through the competition for a mate.

Figure 4.1. The popular pulp magazine *All-Story* (1913) offered this image to illustrate Edgar Rice Burroughs's *The Cave Girl*. The shadow of the cave obscures the racial identity of the cave girl, while highlighting the hero's whiteness.

He recapitulated the evolutionary process that so many anthropologists believed rendered males the stronger sex. Yet if he had given in to his base instincts and raped Nadara like other troglodyte marriages, Waldo would have become an example of white degeneration in the tropics. Instead, he perplexed Nadara's tribe by refusing to consummate their marriage without the formalities of civilization. He was willing, nonetheless, to accept acclaim as their new king. His ability to win the most desirable mate marked him as the select; his desire for formal marriage made him civilized. Nadara and Waldo remained chaste. Only when they returned to civilization and she learned of her white, aristocratic heritage did they marry. Their marriage signaled their return from savagery.

Marriage by capture was the typical plot of the popular "romance of prehistoric times." Such stories intersected with dramas of innovation. Waldo's over-civilization is exemplified by his choice of college. He had enrolled in Harvard, not in an agricultural or mechanical school. He had rejected what the University of Chicago sociologist W. I. Thomas called the "white man's world of practical and scientific activity." His new virility was marked by his innovation. Like so many other fictional male heroes, Waldo invented the bow and arrow. Even as he cursed his lack of a practical education, Waldo spearheaded the industrial evolution of the primitive jungle. He constructed tools and taught rudimentary methods of farming.

Primitive-fiction novels gave license for erotic and racial fantasy. They also translated complex ideas about the place of marriage and the sexual division of labor in industrial evolution into popular entertainment.[2] Social scientists, including a group of feminist and socialist economists, linked the sexual division of labor and the primitive emergence of industry. Novelists turned its origins into dramas of rape, innovation, warfare, and romance.

The nature of marriage represented a marker of racial status and of the differing industrial roles of men and women. Studies of the emergence of modern marriage and the sexual division of labor made their way out of lengthy social scientific tomes and into everyday culture partly through fiction. This occurred at the same moment that the family became the focus of domestic social anxiety. As contemporary observers sought to understand marriage by capture and the emergence of gender roles through studies of primitives abroad, they simultaneously worried about the breakdown of the immigrant and working-class family at home. American critics examined "slum streets swarming with small bits of humanity." In immigrant colonies in industrial cities, they worried about the breakdown of a family division of labor, the rise of women's and children's wage labor, and the disproportionate size of immigrant families. The line separating fiction from academia was blurred, as social scientists and novelists both reached back to primitive mankind. Their long perspective turned contemporary discomfort with immigrants into apocalyptic fears of race suicide.

In the eyes of social reformers, socialists, and urban investigators, conquering immigrant races showed all the telltale signs of racial decline. A language of racial degeneration, born in Europe in the context of urban misery, had emigrated to the United States, where it was reformulated by a range of worried observers to describe a new American condition. Immigrants were degenerating and reproducing at an alarming rate. In vivid images of swarming, sweltering slums, critics employed animalistic imagery to express anxiety over the reproduction of immigrant races, no longer restrained by the requirements of civilized child rearing. The American race was threatened by increasing competition at the points of production (the factory) and of reproduction (the family).

The problems of industrial substitution, whether racial or gendered, were further heightened by the related issue of immigrant fecundity. As the immigrant family disintegrated in the heat of the factory and as women joined the industrial workforce in their own kind of "invasion," immigrants appeared to outbreed their racial betters. Were racial superiors, such as the effete Waldo Emerson, shrinking from race conflict and heading, therefore, toward race suicide and extinction? Nadara's South Pacific was a long way from Waldo's Boston, but the new couple settled in a stately mansion much admired by tourists. It is easy to imagine the young, aristocratic couple thinking about their own escape from primitivism and about their civilized marriage as they gazed at Boston's fecund immigrants. Had they truly left savagery behind in the tropical jungle?

Investigating the Immigrant Family

As the feminist lecturer and economist Charlotte Perkins Gilman insisted, the evolution of the family was intertwined with the process of industrial evolution itself. The primitive family was the first location of industrial production. Though assaulted by the pressing demands of urban and factory life, the family remained the key modern economic unit. In the age of industry, wages were received and pooled to ensure the survival, not of individual workers, but of their families, including their offspring. The working-class immigrant family was the primary focus, especially for urban investigators and socialist activists. Their studies paid particular attention to family life in tenements. Especially because of dramatic exposés by Jacob Riis, Lawrence Veiller, Robert DeForest, and B. O. Flower, among many others, tenements had become associated with the extreme deprivation of immigrant life.[3]

The tenement problem was displayed in an 1899 exhibition in New York. The exhibition illustrated the conditions in many large American cities in hundreds of photographs, maps, and charts. E. R. L. Gould of John Hopkins University joined the enormous crowds viewing the exhibition and resorted to notions of evolutionary survival to explain the evil effects of the tenement on domestic life. Tenement living caused the decay of the family, according to Gould: "Every member of the family, from earliest childhood, becomes an easy prey to the forces which drag down." In the end, he sounded the familiar refrain that the destruction of home and family led to "social degeneration."[4]

Investigators often relied on similar scripts to describe their visits to the murky tenements. The class line, they insisted, was a physical boundary that needed crossing. As they first "penetrated" the tenements, they related their repulsion. In particular, they focused on smells. They related being overwhelmed by the reek of close-cramped bodies, strange cooking odors, and festering garbage—even as their omnipresent working-class guides seemed impervious to the stench.[5] Through the dark, dank corridors of tenements, investigators entered apartments nearly bare of furniture, but filled with ragged, ill, and crying children and babies. "I have talked with them," noted B. O. Flower of the children of Boston's tenements. "I have seen the agony born of a fear that rests heavy on their souls stamped in their wrinkled faces and peering forth from great pathetic eyes."[6] In such families desperate to earn meager wages, children were left to play or—more likely—to cry alone, often among goods being produced at home. One investigator, Elizabeth Watson, described visiting a tenement of Italian migrants on a bitter and snowy March day. She discovered "a little shivering group of children…whimpering and huddling in the second floor hallway." The children confessed to having been locked out of the apartment while their mother finished garments. When confronted, the mother had little explanation: "Shrugging her shoulders, she asked: 'What, I must do? I maka the coats, my man he no gotta job. He walk this day for work. I lock a children in, they burn up. I lock a children out, they cry. What I must do?'"[7]

Even as William T. Elsing, a missionary working in the heart of New York's tenement district, criticized other observers for painting "too dark a picture," his own narrative of a visit to a tenement still traded in pathetic imagery.[8] He guided his reader/visitor past dark, dungeonlike and filthy hallways, into a tenement flat, redolent of all kinds of odors. Still, he presented counterexamples of tenement families living a Victorian ideal. His image of the "The Bright Side of Life in a Tenement-house" pictured a drawing room. Evocative of middle-class material culture, it was complete with comfortable furniture, lace, and hung pictures. The mother read to a baby, while a maid held the jacket of a plump younger boy. The gendered order of this family was maintained by the father's absence; he was, no doubt, at work.

Yet Elsing returned to notions of family breakdown in the companion image, "The Dark Side—under the Same Roof." The mother helplessly held a baby while a thin child cried in the folds of her dress. The accoutrements of labor are strewn around the dirty, dark room. Notably, the father is present as a slovenly fellow, staring with a mixture of curiosity and helplessness at his crying child. The disorder of the family is represented as much in the layout of the flat as in this private scene of pathos. Elsing related being asked to serve as the "referee" for just such a family on the verge of collapse. His was summoned at the moment that the wife sought to "break up housekeeping." Her husband worked only half-time in a cigar-box factory. Debts mounted and the wife sought to place the children in an orphan asylum.

Figure 4.2. "The Bright Side of Life in a Tenement-house": This idealized image of tenement life evokes middle-class material culture and suggests gendered order. *The Poor in Great Cities: Their Problems and What is Being Done to Solve Them* (1895).

She aimed to find work as a live-in servant. Only through the promises of Elsing to reduce their debt and his "verdict" that the family must stay to-gether for "the sake of the three little children" was the family saved.[9]

The narrative of the social investigation that focused on the breakdown of the proletarian family was predominant enough to cross into fictional representations of tenement life. Edward W. Townsend, one of the most

Figure 4.3. "The Dark Side—under the Same Roof": A classic image of the miserable tenement, it combines images of filth and of family disorder. *The Poor in Great Cities: Their Problems and What is Being Done to Solve Them* (1895).

prolific authors of fictional accounts of the working-class experience, narrated a tenement visit in his novel *A Daughter of the Tenements*. Townsend introduced the relationship between Carminella, a young Italian migrant, and Eleanor, her teacher. Both stand as symbolic characters. Hazelhurst embodied the well-meaning but sheltered reformer and Carminella the well-intentioned immigrant struggling against difficult odds. The story revolves around their visit to the filthiest of tenements in search of Lena, a young Jewish girl, absent from school for a few days. Carminella was Hazel-hurst's guide through the dangers and darkness. First, however, Hazelhurst physically had to cross the class divide. Townsend described her leaving behind her "liveried coachmen and footman" to cross into the immigrant quarter. Hazelhurst—through Townsend—was quickly overwhelmed. She resorted to images of insects to describe the "swarming inhabitants." With Carminella as her guide, she was led through dense crowds into a dark, dank, and fetid tenement: "Eleanor shivered as one who steps from sunlight into the silent, solemn shade of a vault." She was disgusted by a pile of snow that, despite the spring weather, remained cold under a pile of garbage. Then, she was so sickened by "the effluvia from the reeking wooden stairs, the odor of cooking cabbage" that she grabbed Carminella and implored: "It is silly, but I cannot go on now. How much further?" Her agony contrasted with the cool calm of Carminella, seemingly immune to the miasma. Finally, they entered a rear tenement (the worst and darkest kind).

And, there, Lena appeared, playing the characteristic role of the object of investigators' disgust as they confronted child labor. Lena sat pulling threads from a garment, even as she eyed her teacher. Her father, having abdicated his role as family provider, scowled to keep her at her task. Again, Hazel-hurst was nearly overcome: "the air of the hot, close room was lifeless: the odors from the perspiring, unclean bodies of the workers, the coarse fumes from the boiling cabbage." Just as she was about to flee, a "real" investigator, a health official, entered and diagnosed Lena's sister with typhus. Mean-while, all the children at work in the room called out "fourteen," thinking that the official was a truant officer come to check their age.[10]

The focus on the family by urban reformers and social investigators can be explained in relation to parallel efforts to explain the evolutionary origin of the family. Investigations of the crisis of the family in the industrial age frequently looked back to primitive tribes and, even further, to ape ancestors and the lower animals. Tribes offered evidence of the savage past of contemporary higher races. They suggested a slow transition from wife stealing and wife capture to polygamy to monogamy. In the *Popular Science Monthly*, A. B. Ellis, for example, reveled in a world of sexual primitivism. He looked to "those races of man which have made the least progress in civilization" for the roots of modern marriage and family. Ellis luridly described the conflict over the abduction of a woman from a primitive New Zealand tribe. Most extraordinary to Ellis, such sexual violence was accepted as inevitable; afterward, a feast was held.[11]

Vivid prose about sexual primitivism was wrapped in an uncomfortable garb of biology and anthropology. In this way, observers declared the family a product of evolution. As Edward Ross declared, "Science...has traced the origin of institutions and followed the upward path of man. It says...I will explain the origin of institutions like the household....I will show the rise of marriage-customs, ceremonies, social forms, and laws."[12] Drawing upon the influential work of Edward Westermarck, the British author of a three-volume history of the family, American observers traced the evolution of the family. Changing marital relations marked racial progress. The most primitive of humans exhibited few familial relations. "The witlessness and incapacity of savage parents is almost incredible," declared the sociologist Arthur J. Todd. "Parental discipline [is] a roaring farce or a tragedy of blood and iron." Rape or other (less violent) forms of temporary relations assured reproduction.[13]

F. Stuart Chapin, previously president of the American Sociological Association, placed the development of the family at the center of his theory of social evolution. He linked "wife capture" to the rise of tribal society. Wife capture and the strengthening of kinship and family bonds served economic functions as the focus of tribal life shifted from hunting and promiscuous life to herding, pastoralism, and family relations. Marriage by capture gradually evolved into marriage by purchase and then by courtship. The family was also solidified by the rise of patterns of ancestor worship.[14] Lester Frank Ward found contemporary vestiges of "marriage by rape." The contemporary "wedding tour" replicated the "marriage flight following wife-capture." He also recognized the industrial significance of the family. "Marriage," he declared, "is from the beginning an association dictated by economic needs."[15]

Ward believed that the emergence of romantic love, avidly described by primitive-fiction novelists, marked the ascendance of the stable, civilized family. For Gilman, the emergence of the family with a male head and the growing importance of monogamy ushered in "an entirely new phase." The family became the locus of political power within the tribe, clan, and, later, the monarchy. It was also the basis for personal and property rights. As the "vehicle of masculine power and pride," the family had become central to race progress, not only as a site of reproduction but also of production. The family was, for Gilman, the "birthplace of arts and crafts as well as of persons."[16] Industry and the manners of industrial life, Todd noted, were rooted in the family: "This is true not only of such things as agriculture, textiles, metal working, leather working and cookery, but also such less tangible but equally vital things as the arts of wise consumption and saving." A range of observers, from sociologists to socialist activists, agreed that production had moved outside of the home to the factory only in the recent past.[17] Ironically, as they worried that industrial production, especially by women and children, threatened the family, they also recognized the roots of production within the primitive family.[18]

Of Women and Insects

The origins of the family represented a major focus for sociological study. Outside of the academy, they were the grist for primitive-fiction novels. In fact, there was significant cross-fertilization between popular fiction and social science. The anthropologist Adolf Bandelier in his *The Delight-Makers* turned to fiction to explain his theories about pre-Columbian life in southwestern Pueblos. Stanley Waterloo used his *The Story of Ab* to argue against the scientific division between Paleolithic and Neolithic man. Ashton Hilliers built his narrative of *The Master-Girl* around an Oxford don's archeological discovery. He fictionalized the desire of all scientists for "one hour's genuine confab, *séance,* communication" with their ancient subjects.[19]

Primitive fiction turned anthropological debate about family structure, the sexual division of labor, and the process of invention and innovation into dramas of romance and adventure. Wife stealing, even the everyday division of labor, were central to popular plots. Waterloo's Ab became a great hunter and a tribal leader because of his invention of the bow. He abducted his wife.[20] Hilliers reversed this narrative of wife capture by imagining a situation in which a "wife-hunter" Pǔl Yūn had broken his ankle and was incapable of walking. He was discovered by Dêh-Yān, the master-girl. Though "well grounded in the domestic arts then practised by woman," she preferred hunting and fighting. The novel revolves around the subversion of marriage by capture and of "domestic relations." She alone decided that she would marry Pǔl Yūn. Pǔl Yūn assumed the "conservative" feminine role. He preformed women's tasks such as making needles. Dêh-Yān, by contrast, stands alone in primitive-fiction novels as a female inventor of the bow. Hilliers' master-girl emasculated Pǔl Yūn by defeating wild beasts while he looked on helplessly. She also scalped other would-be suitors/rapists who sought her capture. As he gazed at her trophies of bear teeth and male scalps, he faced a crisis. He even convinced himself that he was the one who had defeated their enemies. Dêh-Yān tried to restore the dynamic of savage marriage by begging him to beat and rape her. In the end, the two became more comrades then lovers. She bore no children and supplied him with other wives so he could pass on his heritage. At his death, the gendered order of the primitive family was maintained, when she immolated herself on his pyre. This was an act of *suttee* that would have been familiar to an imperial audience.[21]

Primitive fiction in its language and images ranged to the erotic, but its anthropological pretensions as the representation of the earliest forms of marriage eclipsed the pornographic. In P. B. McCord's *Wolf, The Memoirs of a Cave-Dweller,* the decline of Wolf from a tribal leader to forgotten old man coincided with the rape of his wife. Ironically, he had long ago stolen her from his own brother.[22] In describing the different rapes of Wolf's wife, McCord encouraged readers to engage in parallel erotic fantasies of civilization and savagery: the chivalric rescue of the white woman outraged and the primitive assertion of manly power. McCord even offered a drawing on the book's front page of Wolf's wife, a delicate, beautiful, and naked woman, as she was borne off by a "man-beast." Her eroticized rape served, at once, as a rejection of the over-civilized dynamics of modern courtship that seemed to be leading only to race suicide and as a familiar image of interracial rape.[23]

The different roles of men and women provided the erotica of human evolution for the novelist. For the social scientist, sexual difference was a marker of racial status.[24] Among "living inferior tribes," argued G. Delauney in the pages of *Popular Science Monthly,* differences between men and women were more muted. Women were almost as large as men and equal in bravery. "The women are equal or superior among certain African tribes—as in Dahomey, where they are the soldiers and have a higher official rank—among some of the hill tribes of India…and among the Morotokos of South America." Among Patagonians and the Bushmen, women and men had virtually the same stature. In the past, men and women of the higher races possessed more equal physical and social status. Teutonic and Gallic women had joined their menfolk as warriors and nearly equaled them in size. It was only with the Greeks and Romans that there appeared evidence of the "physical and intellectual predominance of the men." For Delauney, "the pre-eminence of the men of the higher races grows at a rate corresponding with the progress of evolution." For evidence, he pointed to measurements of brain size. The difference in the brain size of modern Parisian men and women doubled that of the difference of the brains of ancient Egyptian men and women.[25] In short, the measure of the race depended on levels of sexual inequality.[26]

Such a logic placed men at the forefront of evolution. For W. I. Thomas, women of the highest races should be compared physically and mentally to men of the lower races. Like the difference between higher and lower races, man was, for Thomas, "a more specialized animal than woman." As

Figure 4.4. "I saw his claws in her flesh as he bore her out into the moonlight." This eroticized image of wife stealing evoked ideas about racialized rape. The darker primitive "man-beast" drags off the naked white woman. P. B. McCord, *Wolf, The Memoirs of a Cave-Dweller* (1908).

evidence, he pointed to women's different healing powers. The lowest of animals, such as arthropods or mollusks, could easily restore a lost organ. The highest of mammals could not. Similarly, he reasoned, the lowest races, the lowest classes, and women showed a much greater tolerance of injury and surgery than civilized men.[27] He placed women alongside "the lower races" and in contrast to "the white man." The noted biologist Edward Drinker Cope proclaimed women closer to nonwhite savages than to their white male mates.[28] Hilliers, likewise, confessed to his readers (asking women to "thump me not") that women were "nearer to the savage."[29]

Cope and Delauney focused primarily—and positively—on the superiority of men in the higher race. By contrast, a small, but vocal, group of largely female, feminist, and, sometimes, socialist economists highlighted a history of women's superiority. The most vocal advocates included Eliza Burt Gamble and Charlotte Perkins Gilman. Neither Gamble nor Gilman had received formal training in economics, but both engaged in the field's debates about human evolution. As Gamble wrote, "After a careful reading of *The Descent of Man*…I first became impressed with the belief that the theory of evolution…furnishes much evidence going to show that the female among all the orders of life, man included, represents a higher stage of development than the male."[30] In their enthusiasm for evolutionary theory, Gamble and Gilman reached back to lower races and animals in order to understand the evolution of women's roles within the family. They located the origin of the mother instinct and of the sexual division of labor within the family far down the hierarchy of species, not with apes, but with insects. Among insects, Gamble noted, males existed solely for fertilization. They often did not even survive the sex act itself. "The male," concluded Gamble, "is the result of the cruder, less developed germ." Gilman's position—like her politics in general—was more strident. (Though she espoused socialism, she remained distant from organized socialist politics.) Pointing to bees, aphids, and spiders, Gilman noted that "in the two great activities of life, self-preservation and race-preservation, the female...is better equipped than the male for the first, and carries almost the whole burden of the second." According to Gilman, for much of evolution men struggled even for the basest of existences: "When the centuries of slavery and dishonor, of torture and death, of biting injustice and slow, suffocating repression, seem long to women, let them remember the geologic ages, the millions and millions of years when puny, pygmy, parasitic males struggled for existence, and were used...like a half-tried patent medicine."[31]

Among socialists, the *Socialist Woman* (and its successors) provided a forum to explore the place of women in socialist theories of evolution. Socialists also looked to living primitives to trace an evolutionary past of female superiority. Among Australian Aborigines, men only hunted, fished, and lounged; women claimed every other task. Socialist Party organizer Luella Krehbiel concluded, therefore, that "the female exercises the right of initiative throughout the animal kingdom, and she exercised this right in human history." Women in primitive society reproduced the race, while men's role was to compete for women's attention: "When nature first invented the male animal, his only occupation...for being was competition for the female."[32]

Even the most ardent defenders of male evolutionary superiority were forced to admit a biological past in which females reigned supreme. Although Delauney dismissed as "sentimental pretensions" the idea of women's intellectual equality with men, he still admitted the original superiority of females. Among butterflies, bees, wasps, and other "inferior species," the female was universally larger and heavier. He concluded that "the supremacy of the female is, then, the first term of the evolution which sexuality undergoes, while the supremacy of the male is the last term."[33] Delauney left unresolved the important question: When and why did females surrender primacy?

Gamble, Gilman, and socialist feminists found answers in the influential work of the self-trained biologist and sociologist Lester Frank Ward. Ward enjoyed a varied career that ranged from the 1880s until his death in 1913. For most of his career, Ward worked as a civil servant in positions that developed his interests in sociology, botany, paleontology, and biology. Ward was awarded a chair at Brown University only toward the end of his life.[34] Like many other contemporary sociologists, Ward focused especially on the origins of the sexual division of labor within the family.[35] He challenged the orthodoxy that assumed a primary evolutionary position for men, what he termed the "androcentric theory." Instead, he promoted his "gynaecocentric theory." The androcentric theory, according to Ward, seemed grounded in physical evidence. From birds to humans, the higher animals offered possible evidence of male superiority. Male birds could sing louder and were more brightly colored than females. Male deer featured antlers and horns. In most human races, men were larger than women and had heavier brains.[36]

However, waxing eloquent, Ward noted that science meant "supplanting the superficial and apparent by the fundamental and real." As scientists replaced the superficial theory of the sun revolving around the earth, so, too, would gynaecocentric replace androcentric theory.[37] Through his study of insects, Ward argued that all life began with the female, even at the stage in which reproduction was through parthenogenesis. The female, for Ward, was the fertile one, thus asexual reproduction had to be considered female reproduction. The development of the male—or the fertilizer—came later. He asserted that the female represented the "main trunk" of evolution: "The female sex, which existed from the beginning, continues unchanged, but the male sex, which did not exist at the beginning, makes its appearance at a certain stage...but never became universal."[38] The primacy of the female among insects actually reappeared among humans. Just as the female insect is many times larger than the male, the female human's ovum is three thousand times larger than the male sperm.

For Ward, male birds' adornments were evidence of the role of the female in selecting mates, not of male superiority. Because males were of little significance beyond fertilization, they became warriors. Their lives were of little value. Male physical superiority, whether in humans or birds, evolved through battles with other males for female attention. Male strength had not emerged in order to protect the female, as proponents of the androcentric theory argued. Ward concluded in eloquent prose that "the whole phenomenon of so-called male superiority bears a certain stamp of spuriousness and sham....It is pretentious, meretricious, quixotic; a sort of make-believe, play or sport of nature of an airy unsubstantial character. The male side of nature shot up and blossomed out in an unnatural, fantastic way.... I call it *male efflorescence*. It certainly is not male supremacy."[39]

Nonetheless, male humans at some point claimed supremacy. They birthed what Ward called "androcracy." Sexual selection had its unintended results when men used their physical strength to challenge women for power. As primitive men realized that they "had a part in the continuance of the race," they confronted the gynaecocracy. Women did not cede power easily, however. Ward cited evidence of primitive tribes—the Khasi in Assam and the Dyaks of Borneo, among others—to demonstrate the persistence of "amazonism," or some degree of female rule and supremacy. The rise of androcracy was recent. It emerged because of male physical strength, not as a biological imperative. In the end, he concluded, "the abuse of females by males is an exclusively human virtue," unknown elsewhere in the animal world.[40]

Ward was especially influential for Gilman and for socialist feminists. In glowing words of debt and admiration, Gilman dedicated her 1911 *The Man-Made World* to him. She praised his gynaecocentric theory as the most important contribution since the "Theory of Evolution."[41] Ward, in return, would write to Gilman that "no one is doing as much as you to propagate the truth about the sexes, as I have tried to set it forth."[42] Though she would reinterpret many of Ward's arguments, Gilman found in his theory a powerful explanation of why women had once been primary and why they had been relegated to inferior positions within the family. Echoing Ward, Gilman noted that it occurred to the savage male that it was "cheaper and easier to fight a little female" than to fight another male for her attentions, "so he instituted the custom of enslaving the female."[43] Women were reduced to a "subject class." Their enslavement even became the grist of primitive fiction. Jack London, in his *Before Adam*, for example, featured Red-Eye, an atavistic humanoid, who seized women at will. They invariably died at his hands.

For socialist feminists, the advent of private property marked the moment of female enslavement. "With the origin of private property, we can also trace the origin of women's age-long subjugation to man," noted one observer. Janet Pearl also described a primitive feminist utopia before private property: "The position of woman...was one of supremacy. Not only was woman regarded the equal of man, but his superior." Women dwelt in "joint tenement houses" and shared supplies, food, and child rearing. With private property, men removed women from cooperative living to the private home. She became merely "a breeder."[44]

Women were sheltered by their male enslavers and protected from the environment. Male protection stunted women's evolution: "She now met the influence of natural selection acting indirectly through the male." To guarantee her own survival she developed "the faculties required to secure and obtain a hold on him." In short, she developed a female version of "efflorescence." Women became more feminine, beautiful, and adorned in dress. In a poignant irony, Gamble noted that women—in an effort to attract mates—now adorned themselves with the feathers of male birds.[45] Similarly, the socialist Sara Kingsbury decried the "lady-like woman" as a "gross violation of natural law. She is a crime in nature." Alone in the animal kingdom, the civilized woman sought to charm the male. The ladylike woman had evolved into "bric-a-brac"—a mere ornament. Woman, superior in "prehistoric days," now spent her "energy upon acquiring qualities that would be pleasing to her new master."[46]

As women's well-being depended on the power of "sex-attraction," they were reduced to unproductive consumers and, especially among the ranks of the upper classes, divorced entirely from the world of work. Gilman remarked that "with the growth of civilization, we have gradually crystallized into law the visible necessity for feeding the helpless female....The personal profit of women bears but too close a relation to their power to win and hold the other sex."[47] For Gilman, the feminine female was a stunted member of the race and the race suffered. The race progress of recent centuries, she mourned, had been accomplished by men, not because of a biological inferiority of women, but because women's lives were dictated by the "sex-relation." Women played little part in the progressive world of "science, commerce, education, all that makes us human." "Man," she concluded, "is the human creature. Woman has been checked, starved...and the swelling forces of race-development have been driven back in each generation to work in her through sex-functions alone."[48] Ward in his *Dynamic Sociology* and Gilman in her *Women and Economics* and her fictional work tried to imagine a world beyond androcracy. Socialist feminists, in similar fashion, sought a co-operative commonwealth freed from the "dangers of exclusive masculinism."[49]

For Ward, women and men would share in what he termed "productive" work—that is, work, beyond child rearing and outside of the home. By contrast, Gilman revaluated mothering work. She remained ambivalent about whether women's paid industrial work represented an evolutionary advance or a threat to race progress. For Gilman, in the process of industrial evolution, men had seized the means of production and women workers were chattel. The *Socialist Woman*, meanwhile, recorded a divide among socialists about women's factory roles. Some observers celebrated industrial labor as economic independence; others decried industrial work as the capitalist invasion of the home. Kingsbury noted the contradiction: "As hard as is industrial and business life for women under the present insane system of industry, yet it is her first step towards freedom." When the journal's editor, Josephine Kaneko, imagined socialist equality, she described only the "perfect mother" who produced "the highest race of people."[50]

The Women's Invasion

Gilman, Ward, and Kaneko could all agree that industrial evolution—beginning with the creation of goods for the protection of children—was the result of female energy. "The natural origins of...industries," Gilman declared, "is in maternal energy."[51] Women were the first workers while men disdained productive labor in favor of warfare and hunting. It became axiomatic to feminists and many anthropologists and sociologists that industry had its origins in women's labors. Socialist feminist Elsa Unterman evoked primitive fiction when she declared that "in the distant shadowy past which is...pictured as a wilderness of dense forest and murky swamps alive with beasts, serpents, and birds...the female being sowed the seeds of...industry."[52] A distinguished range of male social scientists echoed her insistence that women were the "originators of industry." As Thorstein Veblen insisted, primitive life was largely predatory. Tribes competed for meager resources and for brides. Fighting—often including the seizure and rape of women—became men's work. Productive labor fell to women alone.[53] Thomas recognized in his examination of the "unadvanced stages of society" that women were the core of society. Men only hunted and fought in order to attract or abduct women. Women, engaged with the reproductive work of raising children, were left with the productive work of provision. "Women had been doing the general work and had developed the beginnings of many industries." As evidence, he connected the lower classes to primitive tribes. In both, "woman still retains a relation to industrial activities." In West Africa, young men in search of wives sought widows, who were well versed in the techniques of production. Similarly, the American lower classes preferred the "heavy, strong, patient, often dominant type" capable of hard labor. Even those, like Todd, who mourned women's engagement in wage labor, recognized that over the course of industrial evolution "the domestic arts became the industrial arts."[54]

This, then, was the paradox: women may have been the originators of industry, but observers worried about contemporary women's wage labor.

Gilman argued that while economic production had evolved, mothering work had not. Industry, she noted, was once domestic, but it had moved out of the home and was increasingly specialized in the hands of men. Bread was now made in bakeries and shirts in factories. "Vintner, brewer, baker, spinner, weaver, dyer, tallow-chandler, soapmaker, and all their congeners were socially evolved from the practicers of inchoate domestic industries." Women were restricted to mothering labor and to unproductive consumption. Family work retained its private and domestic "primitive labor status." While men were carried along in the main currents of evolution, women were relegated to its primitive and parasitic backwaters like the lower races. "When we say, *men, man, manly, manhood,*" she noted, "we have in the background of our minds a huge vague crowded picture of the world and all its activities…full of marching columns of men, of changing lines of men, of long processions of men; of men steering their ships into new seas…ploughing and sowing and reaping, toiling at the forge and furnace." But "women" evoked only sex.[55] According to May Wood Simons, a socialist journalist and member of the National Women's Committee of the Socialist Party, women gathered bark to make the first shelter. She sewed together skins of animals. She made the first pottery and planted the first grain. Yet as savagery gave way to civilization, she withdrew from industry. Secluded in the home, "child-bearing was…her chief occupation."[56]

Mothering work had become "woman-slavery." When Gilman imagined a female utopia in her 1915 novel *Herland,* she turned to metaphors of lower animals and to visions of industrial evolution without androcracy. Set in the highlands of South America, Herland was isolated from the androcentric world and, in fact, from all men through quirks of geography. The utopia was discovered by three male explorers, one of whom narrated the evolutionary history of Herland. The women of Herland reproduced through parthenogenesis, an obvious reference to the female origins of animal life. Herlanders rejected female efflorescence. Their clothing was durable, comfortable, and perfect for everyday labor. Female industrial evolution, meanwhile, had led to the socialization and specialization of mothering work—not to the factory. The different tasks of reproductive work, from bearing to rearing children, were in the hands of those most capable and most fit for such labor. Motherhood, moreover, was not an experience that isolated individuals, nor was it regarded by the inhabitants of Herland as something natural to women. Rather, it was a learned skill and an occupation. Motherhood and labor were joined in a public world of work—to the exasperation of the male narrator. In a telling statement, he expressed his rage about the collectivization of motherhood by comparing these utopian women to insects: "It was beyond me. To hear a lot of women talk about 'our children'! But I suppose that is the way ants and bees would talk—do talk maybe."[57]

Gilman's utopia critiqued the way women's adaptations, once necessary for the early history of the race, had been left behind in the rise of industrial society. Her utopia was a verdant and—in a gesture toward theories of industrial evolution—temperate land of groomed forests and gleaming communal homes. Yet it lacked workshops and smokestacks. The Herlanders labor as mothers, teachers, and foresters. The productive assembly line was as absent as fathers were in Herland. In the real world, Gilman hesitated to salute women wage laborers. She concentrated more on envisioning the socialization of domestic work, something she termed "cooperative housekeeping."[58]

Feminist economists who did, in fact, address questions of women's productive industrial roles were, at best, ambivalent about the effects of wage labor on women's social position. Nonsocialists promised to enshrine the home, above all through protective legislation. They sought laws that mitigated the effects of industry, especially by limiting the number of hours that women and children could labor. The University of Wisconsin economist Theresa McMahon, for example, traced women's roles in industrial evolution from the Stone Age to the present era of factories in order to advocate for protective legislation. Influenced by the work of both Ward and Westermarck, McMahon described the early industrial dominance of women, who turned to domestic production as part of the mothering instinct. Mothering, with its focus on domestic production for the good of offspring, was an adaptation that shielded women from the violence of primitive life. Although this helped catalyze industrial evolution, it also left women ill-prepared for the battle of the sexes that led to the rise of the androcentric family. In the "struggle for authority between the sexes," women lost power as "industry departed from the hearth." For McMahon, the growth of industrial production from prehistory to the age of the factory was inherently linked to the rise of the family, in which the male took the position of lord and master and later of breadwinner.[59]

For some, the increasing distance between home and work, reproduction and production, was a salutary trend in industrial evolution. By contrast, women's move into wage work hinted at evolutionary regression. Even feminist economists such as McMachon compared working women to racial inferiors. McMahon ended her study by comparing the lowly condition of working women to that of Australian Aborigines. She defined mothering work in an age of industry as being in conflict with wage work. Though not necessarily unsympathetic to the plight of women wage workers, she understood mothering work and wage work as irreconcilable. "Under the present industrial regime motherhood is not compatible" with "work outside the home," McMahon noted. Even the tireless advocate for protective legislation, Florence Kelley, lamented that the wage work of young women rendered them "unfit…for life in the home." She described the penetration of the home—especially the tenement apartment—by industrial production as the "degradation of the home by industry." The mother at work is "distracted from the care of the children through the invasion of the home." Turning to questions of reproductive health, the protective legislation activist Josephine Goldmark worried that industrial labor hampered the ability of women to bear children—let alone to raise them effectively.[60]

Socialist feminists, such as Simons and Kaneko, were dismissive of protective legislation. Simons, instead, urged working women to join with male comrades in the labor movement and in socialist organizing. She even militated against separate socialist organizations for women. Nonetheless, she worried about the effects of industrial work on the home. Capitalism rendered working women "unfit for wifehood or motherhood." The shocking condition of working women was leading to the "physical degeneration" of the next generation. The "future proletarian forces," warned Simons, "are in this way being gradually weakened and losing their power for intelligent revolt." Socialism would restore—not destroy—the home and protect the future of the race.[61]

Women, whose labor had catalyzed industrial evolution, had become foreigners in contemporary industrial civilization. According to Anna Spencer, a reformer whose work focused principally on child labor and women's suffrage, "primitive woman took the first steps on that dark path which led toward the higher industrial organization of later societies." Woman was driven to industry by the demands of family and child rearing. She was "pressed to her special tasks by the biologic push itself." By contrast, men had been pulled into industrial labor first through the violence of slavery and, later, through their claims to family lordship. By the dawn of the machine age, men had restricted women's labors to the home and claimed primacy in industrial work.[62] And, yet, noted McMahon, the "machine industry," with its repetitive and unspecialized nature, had returned women to the status of a "distinct economic factor in our industrial life."[63] When women reentered the industrial world, they did so at the lowest unskilled levels. Women who returned to the factory were not the most capable. Rather, they were forced to do so by desperation and, as a result, were unfit to handle the rigors of factory life. "Their homes spelled retrogression in the evolution of the race for they constituted the most unfit type in the industrial evolution," McMahon maintained. Observers such as McMahon or Spencer, sympathetic to the fate of the women of the tenements, could still condemn their wage labor by referring to the larger effects of their labor on the fate of the race.[64]

Such critics examined women's wage work alongside immigrant labor. In tracing similarities and connections between women's and immigrants' labor, they noted the impact of women's wage work on "industrial competition." Simons, for example, compared working women to "the lowest of the laborers—the 'downmost man.'"[65] Their arrival in the workplace led to a decline in skill levels and wages. The same metaphors used to warn of the effects of immigrant arrivals could be employed to illustrate the result of women's wage labor. The reformist journalists Rheta Dorr and William Hard lamented "the women's invasion." They borrowed words primarily used to describe the seizure of jobs by recent migrants. Woman was, they declared, "the white Chinaman of the industrial world. She wears a coiled-up queue, and wherever she goes she cheapens the worth of human labor."[66] Dorr and Hard described industrial decline and the substitution of male workers by women, of legitimate workers by the unfit. They pointed to "the spectacular triumph of women over men in the stogy factories of Pittsburg." As women took over an industry that had once been held by skilled male artisans, they ushered in cheap wages and unskilled machine work. "Woman hasn't risen to man's skill," they concluded. "Skill has been lowered to woman's level. Woman hasn't been masculinized. Work has been feminized."[67] Dorr and Hard resorted to the militarized language of armies and invasion to describe the industrial role of women. Women workers were part of an invading army, but they were not professional soldiers. Rather, the woman worker was "a guerrilla, a bushwhacker, entering the fight without training." To capture the notion of substitution and invasion, Dorr and Hard printed two photographs: first, of "the passing type," the skilled, but now unemployed, cigarmaker, and, second, the "bunch-breaker," the young female substitute who was aided by clumsy, if effective, machines.[68]

The industrial expert Helen Sumner described a similar pattern of industrial substitution in her report to the Senate as part of a nineteen-volume study on the wage work of women and children. She declared that women's wage work in the machine age "is a story, moreover, of underbidding, of strike breaking, of the lowering of standards for men breadwinners."[69] For Sumner, the age of industry was birthing "a hybrid class of women workers in whose lives there is a contradiction." No longer could working-class women look forward to labor linked to the "care of children"—like the original industrial work of primitive women. Instead, she worried about "the mothers of the nation." In warning of the threat of "wasted human lives" and the "loss of powers of body and mind required for efficient motherhood" and in appealing for effective protective legislation, Sumner lamented the dangers of racial degeneration. Her fears for race progress linked the perils of women's wage work, the need for protective legislation, and the dangers of child immigrant labor.[70]

Civilization and the Long Childhood

By the turn of the century, investigators wandered through the slums and tenements of urban working-class America and returned to bemoan the plight of immigrant children. In academic articles, government studies, and journalistic exposés, investigators worried about the character of child workers by focusing on their battered bodies, loose morals, and degraded manners.[71] They sympathized with working children. Yet they also lamented the biological effects of children's work on national racial development. Claire de Graffenried, a child labor investigator for the Bureau of Labor Statistics, decried both children's suffering and the biological effects of their work. Like so many other critics of child labor, she mourned the fate of "young children, who love play and hate work." Fearing punishment at home, they "creep in the black night or gray dawn from the wretched pallet…rushing breathlessly lest the shop be closed at seven and the tardy ones docked." She portrayed "girls and boys heavy-eyed, listless, dragging their tired limbs or asleep in the stupor of exhaustion." Opponents of child labor could have stressed the pathetic nobility of wasted youth. Instead, like de Graffenried, they lamented their degradation. She focused on the weakened bodies of child laborers as the outward physical manifestation of their inner degeneration. She described the "haggard, prematurely aged, debased looking faces of the waifs….Instead of the red cheeks and rounded limbs of healthy youth we behold pallor and glazed eyes, stunted bodies and narrow chests, stooping shoulders, early decay and death, or lifelong invalidism." Physical degradation hinted at moral decline. The atmosphere of the shops and factories weakened the child physically and morally. Confronted with "distressing depravity," they acquired "a prurient knowledge" of sex, crime, alcohol, and cursing. The "childish mind" was forever shaped by "duplicity and vice." The problem of child labor, for de Graffenried, was ultimately a problem of racial decline. "Families are disintegrated," she mourned, "physical deterioration of the race is going on, manhood and womanhood are degenerating…as an outcome of the abuse for mammon's sake of nature's laws of development."[72] Socialists similarly worried about children's moral and physical deterioration. Simons noted that "one has but to live in the slum of a modern city" to witness the degeneration of child workers. Children were "weak and pallid."[73]

In such investigations, child workers appeared as urban "savages," left behind by evolution. In this context, Stephen Crane's *Maggie, A Girl of the Streets* (1893) dramatized fears of immigrant children's street life. The book opened with Jimmie, "a very little boy," engaged in a street battle. The brawl recalled the primitive jungle. A crowd of children throwing rocks at Jimmie cried with "barbaric trebles." The yells of the childhood mob were "songs of triumphant savagery." Jimmie and his sister, Maggie, were the children of two drunk, useless, and violent parents and their father soon perished. Though still a child, Jimmie was already hardened to the savage life of crime, alcohol, and violence. He became a child teamster. Maggie, meanwhile, was seduced by Pete, a well-dressed bartender. Eventually abandoned by Pete, Maggie was rejected by her family and was left as a prostitute to wander the streets. More than a morality play wrapped in a naturalist veneer, *Maggie* stands as a commentary on child labor and the inevitable nefarious results of shortened childhoods.[74] Crane's story with its pathetic ending of Maggie on the streets consciously evoked themes of degeneration: moral depravity, child labor, atavistic tenement life, poverty, drink, and, linking them all, shortened childhoods.

His narrative of decline and degeneration echoed the warnings of leading opponents of child labor, such as Charles Loring Brace, the champion of efforts to remove children from city streets to purportedly healthy rural surroundings. Brace especially decried the condition of boys who hawked newspapers on the streets. For Brace, "newsies" were not enterprising youths doing their part to augment family income. Rather, they were street savages in need of quick salvation. Newsboys, he insisted, "make the selling of papers a cloak for leading depraved lives." They were, in reality, "vagrants and runaways, gamblers, cigarette fiends and hangers on in low dives."[75] Kaneko was no less critical. The "mind and soul" of the newsboy, she declared, was "black with the filth of the street."[76] Their condemnations resonated with studies of the relationship between childhood and civilization. The examination of the lowest animals, the most savage of tribes, and the most advanced of families linked the length of childhood dependency to levels of civilization. The labor of children was a sign of racial inferiority and of savagery, akin to the lowest of the tribes.

It was neither strange nor wholly original for feminist economists such as Gamble and Gilman to look to entomology for insights into questions of gender and the human family. Insects also provided a key for those seeking to understand the racial effects of child labor. As Alexander Sutherland in Great Britain and John Fiske in the United States both noted, the offspring of the lowest animals needed little parenting after their birth. Insects, fish, amphibians, and reptiles all produced offspring capable of fending for themselves at birth. Higher up the evolutionary ladder, however, progeny needed mothering. A chick can begin searching for food shortly after hatching, but the lion cub is helpless at birth. It is confined to the den and dependent on the protection of and food from the lioness. Similarly, among humans, higher races generally had fewer offspring and longer childhoods. The history of human evolution could be reduced to the story of lengthening childhood. Some critics wondered whether child workers did not need as long a childhood as their racial betters. Others noted that because of familial pressure to find employment, immigrant and working-class children still had abnormally shortened childhoods. Child workers were sheltered for as brief a period as the lowest savages. The reformer Margaret Byington celebrated the lengthening of childhood. Among apes, she argued, the child is virtually helpless. She described a "very simple family unit." The mother ape cares for the child and provides simple food, such as nuts, close to the home. The father provides protection and the main supply of food. The aborted childhood of the young laborer represented degeneration toward savagery.[77]

The longer the period of caring and motherhood, the farther a child could develop. As Felix Adler, the chairman of the National Child Labor Committee, insisted, "the long period of infancy and parental care is regarded by many as the origin of man's higher civilization, as a means of developing the unselfish instinct by which this civilization is graced." Byington, similarly, argued, "out of the prolongation of infancy in the shelter of family life, civilization has been made possible."[78] Again, the experience of primitives provided evidence, while the parallel between immigrants and savages suggested a racial threat. According to Sutherland, among the lowest savage races, children stayed in the home only until fourteen. In the most "civilised" races, the child remained dependent until twenty. Immigrant children, lower down on the evolutionary scale, experienced shorter childhoods similar to those of savages.[79]

The child psychologist G. Stanley Hall offered an explanation for the short childhoods of immigrant and savage children. A towering figure in American psychology and education, Hall spent much of his long career at Clark University. Hall and his many followers and students argued that children recapitulated the progress of the race. As developing fetuses, human children experienced nonhuman life, first as amphibians and fish and then as mammals. By birth, children were experiencing the life of savages. Their play and their games, Hall and his followers insisted, were similar to the rituals of savage races. Child development, in short, was individual evolution. Children passed from the state of savages to primitives and eventually to civilized adults. The long period of childhood, therefore, was crucial for the proper training of the child for civilization.[80] As a result, "premature toil," as Adler declared, "lowers the standard of civilization."[81]

Because immigrant races were lower on the evolutionary hierarchy, immigrant children reached adulthood comparatively quickly. They had less to recapitulate. However, according to opponents of child labor, immigrant progeny received little childhood at all. They were frozen in the state of savagery. The child worker—the little savage—craved drink, hated society, and turned to crime.[82] Edward Townsend's fictional reformer, Eleanor Hazelhurst, therefore, mourned the fate as child laborers because they were denied an appropriately long childhood: "This child...in another condition of society would be treated as only emerging from babyhood, as just maturing at best into a mental capacity to be safely vexed with the mysteries of A, B, and C." Instead, she was a "sweaters' slave." She never played but worked from dawn until late at night and fell asleep at her work. Her "bare, thin, dye-blackened arms" reflected her stolen childhood.[83] Echoing Hazelhurst, a settlement house pamphlet opposed child labor by warning of the racial dangers of abbreviated childhood: "Work for such long hours, is absolutely fatal to the mental, moral and physical development of the child.... The child of the tenement...can hardly escape slipping into visious [*sic*] habits." The result, for the child labor reformer William Willoughby, was degeneration: "Degraded and depraved...the whole field of labor is lowered."[84]

Race Suicide: The Life of an Idea

When critics such as Adler or Willoughby decried child labor or Dorr bemoaned women's work, they warned of decline and racial degeneration—not extinction. They worried that the breakdown of immigrant working-class families would lead not to their elimination but to their increase. Child labor, aborted childhood, and the "women's invasion" raised concerns not only about the *quality* of working-class, immigrant progeny but also about their *quantity*. Once more, the long view of evolution translated a contemporary problem into an apocalyptic threat, in this case, of what came to be termed "race suicide." Born from the fertile and pessimistic imagination of Edward Ross, the term "race suicide" gained currency. Ross argued that the highest races were contributing to their own extinction by eschewing childbirth. The lowest races were consolidating their conquests through fecundity. Race suicide linked a range of urban and industrial problems, from child labor to women's work to immigration, and tied them to the family behavior of elites. Child labor became more than the oppression of individual children by ruthless employers; the small families of elites became more than individual choices; and the size and working behavior of immigrant families became more than problems of stretched budgets. They were interlocking parts of the biological process of race suicide.

Policymakers and individual investigators turned to the animal world to explain why inferior immigrant races were producing large families while their superiors raised few children. Just as insects and the lower animals offered clues to the origins of industry, so too could they provide the basis for theories of race survival. Insects, fish, and other lower animals survived by producing massive numbers of offspring. Many of their offspring perished, but they produced enough children to guarantee the continuity of the species. The examination of the animal kingdom suggested that more advanced species produced fewer children. They had smaller numbers of offspring, but provided more care.[85]

A similar pattern emerged among human races. As Fiske argued, the most advanced races offered more care and longer childhoods, but to fewer children. Poor, immigrant races had more children, but they offered little parenting, as the evidence of gangs, prostitution, and begging confirmed. In debilitating slums, tenements, and factories, many of these children would perish young. Unlike among superior races, the death of immigrant children could not be mourned. One settlement worker, for example, contrasted the illness and death of an immigrant child and a native-born child. While the native-born family gathered around the sickbed of their son and without question paid the expensive bills for doctors and medicine, the immigrant child perished alone. The family continued to work and the doctor was never called.[86] Still, families continued to grow.

In the eyes of social observers, inferior immigrant races tended to bear many children, while the native born were moving toward smaller families. According to Ross, the higher races shrank from competition by birthrate with immigrants. Instead, they delayed marriage and restricted their family size. He blamed the invasion of the Irish and French Canadians for the shrinking size of existing New England families, the paragon of a superior race: "the higher race quietly and unmurmuringly eliminates itself."[87] Once again, the Dillingham Commission confirmed the alarming findings

of individual investigations. Starting with a study of Rhode Island and expanding to other regions of the country, the Commission confirmed the
inequality of births. Native-born white women below forty-five had a child
once every 5.3 years. Immigrant women bore a child every three years.
Among immigrants, the desirable English had a child only once every five
years and the lowly Polish every 2.3 years.

Some observers argued—with some satisfaction—that the greater perils
faced by the children of inferior races would overcome the problems of
large immigrant families. Their children would be carried off by disease
or by the effects of hard labor. Even the otherwise sympathetic reformer
Annie Marion MacLean declared that the high rate of infant morality "saves
the slums."[88] Ross, though, deflated such brutal optimism. Laissez-faire, the
serene confidence in the dynamics of evolution, could not be relied upon
for race preservation. In an age of industry, unlike past ages of warfare, the
forces of selection might lead to degeneration, but did not fully eliminate
the unfit. Moreover, medicine cured diseases that might otherwise have carried off more of the weak and undesirable. Dissimilar strategies of race survival tended to favor those with more children. As economist Frank Fetter
worried, "The ignorant, the improvident, the feeble-minded, are contributing far more than their quota to the next generation."[89]

The immigrant invasion was carried out in the sexual privacy of the
bedroom, as well as in the industrial public of the shop floor and mine
shaft where immigrants competed for jobs with the native born. "I can find
no word so apt as 'race suicide'," Ross declared. "There is no bloodshed, no
violence, no assault of the race that waxes upon the race that wanes." Ross
connected the decline in the birthrates among superior races to the growing
emancipation of middle- and upper-class women. Freed from the constraints
of the home, they no longer felt the need or desire to produce children.[90] Civilization, it seemed to Ross and other critics, caused a waning of the mothering instinct. They pointed with alarm to the birthrate of college-educated
women. Among graduates of Wesleyan University, in the 1901–10 period,
female graduates bore an average of only .69 children. Samuel J. Holmes,
a zoologist and pioneering eugenicist, conducted similar studies of the descendents of the Mayflower passengers, heralded as the naturally selected
parents of the American race. He found to his dismay that the average
number of children declined from 6.0 between 1810 and 1830 to just 1.5
between 1870 and 1880. He concluded that "we are probably losing the elements that belong to native American stock."[91] Civilization produced a type
of elite woman with little mothering instinct and who eschewed childbirth.
As the neurologist and criminologist Max Schlapp mourned, "Rich people

are the ones who have no children, and the poor have the greatest number of offspring. Any woman who does not desire offspring is abnormal. We have a large number…who do not want children."[92]

The declining birthrate of elites was linked to a reduction in the racial quality of immigrants. More advanced and less fecund immigrants were unwilling to undertake the migration journey only to confront competition from racial inferiors. Inferior immigrants produced larger families. Like Jack London in his "Unparalleled Invasion," Ross worried ultimately about the arrival of especially fecund Asian migrants. Ross believed that Asian migrants would outbreed their racial superiors: "The Asiatic will rear two children while his competitor feels able to rear but one. The Asiatic will increase his children to six under conditions that will not encourage the American to raise more than four."

Immigrants' fecundity was central to the colonization of the United States by inferior races. In the current age of industry, inventiveness or civilized status counted for little in determining racial success, argued Ross. The experience of immigrants, women, and child workers demonstrated that even the unfit could work at machines. "Each race must," he declared, "produce from its own loins; but in the industrial Armageddon to come it may be that the laurels will be won by a mediocre type of humanity, equipped with the science and appliances of the more brilliant and brain-fertile peoples." The body-fertile, not the brain-fertile, would win the race war. Ross combined fears of fecundity with concern about industrial substitution. He imagined the victory of the "Asiatic," not through force of guns, but through fertility and low standards: "The standards may remain distinct, the rates of increase unequal, and the silent replacement of Americans by Asiatics go on unopposed until the latter monopolize all industrial occupations." The native-born American would be "hopelessly beaten and displaced as a race."[93]

As industrial jobs fell to new immigrants and as birthrates declined among the elite, some even wondered whether the Anglo-Saxon race could survive in America. Concerns about race suicide spread from the pages of academic journals to the larger political dialogue.[94] A chorus of voices warned of the dangers of race suicide and blamed wealthy native-born Americans for their lack of race consciousness when they bore so few children: "[The] birth-rate is decreasing among all civilized people.…The 'lower' classes will the more rapidly become the upper classes."[95] Most notably, President Theodore Roosevelt castigated elite American women for their unwillingness to have more children. The end result, he warned, was race suicide.[96] Socialists met Roosevelt's warnings with predictable derision. Yet they criticized the president only for the persecution of birth control activists, especially

Margaret Sanger. "Only a walk down the Lower East Side in New York, or down Halsted Street in Chicago" served as "propaganda" for "fewer and better children." Like Roosevelt, they worried that the lower classes "breed like animals."[97]

The Perils of Degeneration

Race suicide proved that the decline of racial inferiors in American slums and tenements led to their increase, not to their extinction. In this context, race suicide was the alarming companion to racial degeneration. Racial degeneration was more a general fear than a specific malady and race suicide proved that decline did not lead to extinction.[98] The term "racial degeneration" was defined widely by different observers and often in contradictory ways. The historical significance of the idea of degeneration lies less in its specific definition than in its frequent application. As A. J. McKelway, the assistant secretary of the National Child Labor Committee, declared, "Certainly there could befall a people no greater catastrophe than race degeneracy. It is sufficient to say here that this catastrophe is not only threatening but already impending."[99] Concerns about immigration, industry, and urban life were linked together in oft-repeated fears of racial degeneration, the idea that race progress in an age of industry was being reversed.

Like the theory of evolution itself, notions of degeneration were first articulated in Europe and only later imported to the United States. On the western shores of the Atlantic, they became part of the vocabulary used to debate the effects of industrialization, urbanization, and immigration. As scholars have noted, notions of degeneration were employed widely in the European and American contexts. Degeneration broadly shaped the turn-of-the-twentieth-century cultural imagination. It guided gothic and naturalist fiction, which was popular on both sides of the Atlantic and often focused on the urban abyss. In the realm of social science, politics, literature, socialism, and social reform, degeneration loomed as the real possibility that, through human action or human indifference, progressive evolution could stutter, stop, and reverse. Races could advance, it seemed. Or, they might retreat.

Savagery lingered, not as a past that had irrevocably been left behind, but as a state that a race might return to in the worst of conditions. Such concern even influenced the way observers understood contemporary races designated as savage or barbarian. Byington, for example, wondered whether the primitive races she studied to form her argument about the "normal family" had simply never advanced beyond savagery or, ominously, had reverted to past social forms. It was in this context that social trends, from child labor to women's work to immigrant labor, were scrutinized for what they might reveal about the looming perils of degeneration.[100]

By the beginning of the twentieth century, conservative observers such as Dartmouth's D. Colin Wells and socialists such as Simons were ringing the alarm bells about degeneration. In a deeply pessimistic and wide-ranging article, Wells worried about the effects of women's education and emancipation, child labor, philanthropy, immigration, and the birthrate. In the end, his conclusions focused on the perils of degeneration. "Are we multiplying the unfit and increasing their proportion in the community?" he asked. Moreover, he worried that the conditions of modern industrial and urban life had fundamentally and forever altered the process of natural selection. "The human species and its foremost races developed under a vigorous weeding-out of the weak," but in modern society the weak survived and bred more rapidly than their betters. Wells, in the final analysis, mourned that degeneration seemed "to be the biological consequences of our vaunted industrial system."[101]

For evidence of the "effect of the factory system upon the degeneration of the race," as Willoughby put it, critics looked to the older industrial societies of Europe where fears of degeneration were first articulated. Willoughby, for example, argued that child labor was inevitably leading to degeneration by pointing out the experience of the British during the Boer War. As Britain marshaled men into the armed forces, its elites discovered that the industrial classes, destined to fill the ranks of the army, were degenerating. Few were even fit for service. For British observers, the experience of the Boer War recruitment was shocking evidence of working-class degeneration. For Americans like Willoughby, British evidence could be marshaled to highlight more hidden processes in the United States. As McKelway put it, "the nation awoke to the fact that its vigor was sapped." Britain's working classes showed up for battle a "hooligan, anemic, neurotic, emaciated [and] degenerate" race. The racial degeneration that claimed the headlines in Britain had arrived in America, according to McKelway. One needed to look no further than the American "factory type"—sallow, pallid, diseased, and ignorant of family life.[102] The lessons of European armies suggested to Simons the degeneration of the proletariat. Every year, she noted, the European powers lowered their standards for military service. How could socialism succeed, she wondered, with a "spiritless, indifferent proletariat?"[103]

Claims about degeneration, backed up by English statistical and American anecdotal evidence, set the stage for a generation of social reformers and socialists as well as a following generation of eugenicists. Even as reform, socialism, and later eugenics penetrated so many quotidian aspects of immigrant lives, they were aimed most directly at reversing the process of degeneration that the lives of immigrants had set in motion.[104] The concept of degeneration gave reform, in particular, a wide and comprehensive mandate to engage with issues as far flung as child labor, immigrant health, women's wage work, and proletarian physique, among many others. It also provided an unparalleled urgency. Thus, McKelway railed against child labor for "the welfare of the coming race." Similarly, the *Survey,* the leading American journal of social reform, began the second decade of the twentieth century with a call for a national battle against degeneracy, "to study the causes and the remedies for race suicide, infant mortality, orphanage, illegitimacy, child dependence, juvenile delinquency, illiteracy, and all the other forces that work for racial degeneration." Reform efforts, in short, were central to improving the racial stock of the nation. They worked "to discover and promote the influences that make for the improvement of the stock, for the strengthening of character, for the preservation of the body from disease and deterioration and of the soul from destruction." For the journal, reform meant confronting "the vital problems of birth, nurture, degeneracy, and racial progress."[105]

Yet even as reform wrapped itself in the battle flags of the campaign against degeneration, it could not address the contradiction that would hasten its eventual eclipse by the forces of eugenics. Reform, even as it focused on the future of the race, not on the plight of the individual, was accused by critics such as Wells of keeping the unfit alive to breed the next generation of degenerates. Because of the effects of charity, reform, and philanthropy, the "stocks of . . . degeneracy do not die out." Instead, they bred in ever-increasing numbers.

Similarly, socialists, such as the politician Raphael Buck, railed against a capitalist system that produced only "degeneration." He promised that socialism would create a "race of geniuses who shall be far more superior to us than we are to the savages of the Andaman Islands."[106] Yet he also recognized that abolishing the "old brutal struggle for existence" might preserve the "numerous less fit." Socialists struggled with the desire to preserve and the need to maintain the "struggle for existence." London, in particular, agonized over the potential that socialism might preserve the unfit. "All the social forces are driving man onto a time when the old selective law will be annulled," he warned. In an alarmist essay, 'Wanted: A New Law of Development," London recognized that while capitalism led to degradation and degeneration, "the race was purged of its weak and inefficient members."

Under socialism, the weak and their children would survive. As a result, "the average strength of each generation…begins to decline." With an obvious pun about the dangers of fecundity, London described the "pregnant problem" of socialism. How could socialism—like reform—prevent the procreation of the unfit? His concern was underlined by the essay's conclusion; London offered no solution. He merely suggested that the "answer rests with the common man. Dare he answer?" The University of Pennsylvania's Carl Kelsey suggested one answer. He summarized the political choices that confronted the emerging generations of reformers, socialists, and eugenicists: "society faces…the problem of the degenerate…the problem of the physically unfit." He advocated the development of a scientific philanthropy to apply to the problem of degeneracy. The only other solution was to prevent the reproduction of the unfit, perhaps by "knocking the defectives on the head."[107]

5

EXPLORING THE ABYSS AND SEGREGATING SAVAGERY

Abroad at Home in the Immigrant Colony

Ah, the hideous ugliness of the race to which we belong.

Walter Wyckoff, 1899

Furthermore, it is clear that the poverty which undermines the
workers is the great and constantly active cause of the fixed states of
degeneracy represented by the pauper, the vagrant...etc....When the
working people...once fall into the abyss, they so hate the life of their former
struggles and disappointments and sorrows that almost no one...can
induce them to take it up again. In the abyss they become merely
breeders of children, who persist in the degeneration into
which their fathers have fallen."

Robert Hunter, 1904

Beginning in 1907, some of the leading social reform experts in the nation collaborated to produce a complex portrait of life in an American industrial city. In six magisterial volumes, collectively called the Pittsburgh
Survey, they described and pictured grimy cellars, fearsome furnaces, exotic
immigrants, and downtrodden workers. The project was the brainchild of
Alice B. Montgomery, the chief probation officer of the Allegheny Juvenile
Court. Paul U. Kellogg, the editor of *Charities and the Commons,* nurtured
it into fruition. The work depended on academics, including John R. Commons; social reformers, such as Margaret Byington; numerous urban investigators; photographers; and artists. The volumes explored the steel mills
and neighborhoods of the region, its women workers, the workplaces of
Pittsburgh itself, and the perils of industrial accidents.[1]

The editors and authors of the Pittsburgh *Survey* promised a uniquely
detailed description of life in immigrant colonies at the heart of one of the

grittiest industrial metropolises. What they produced was unprecedented in its scope and, even as it proved a model for other efforts at social investigation, it was never fully replicated. Yet behind its claims to innovation lay a long genealogy. The *Survey* was the apotheosis of a turn-of-the-twentieth-century American urge to describe life in urban immigrant colonies. In the years preceding its publication, a range of authors, from social scientists to urban reformers, printed lurid descriptions of the immigrant colonies. Their publications were primarily destined for an English-speaking and native-born middle-class and elite audience. The scenery they described was bleak yet exotic and the people were alluring but primitive. Much of this literature depended on narratives of exploration. Narrators placed themselves centrally in their stories of life in the immigrant colony. They were heroes adventuring in the "dark places" of industrial America. The visitor to the immigrant colony was not merely an author but an explorer.

Slum investigation was an urban safari. Imperial adventure, as historian Matthew Frye Jacobson provocatively suggests, created a vocabulary to describe the foreignness of the slum. Narratives about journeys to the foreign quarters of cities often implicitly and sometimes explicitly echoed the popular genre of descriptions of travels to imperial hinterlands. Conversations about imperialism abroad and immigrant colonies at home often merged. As a result, the literature on urban America paralleled descriptions of foreign exploration. Studies like the Pittsburgh *Survey* were part of a broad genre of exploration writing. For the explorer visiting distant and darkest Africa, the published book might stand in for the actual collection of authentic specimens of people, fauna, or artifacts that characterized public exhibitions. Much the same was the case for urban explorations. Writers returned with samples and specimens, whether drawings or photographs, that illuminated the darkness of the slum. And each revealed with breathtaking clarity the dangers of degeneration that lurked in the dark places.

Imperial adventure helped shape the study of the American immigrant working class. American observers also borrowed the methods of English poverty experts. The authors of the Pittsburgh *Survey* admitted their debt to the groundbreaking work of Charles Booth and his complex analysis of London's poor. Booth suggested that the working class could be divided into a hierarchy, with wage earners at the top and paupers and degenerates at the bottom. For American observers influenced by Booth, explorations of the immigrant "colony" provided an opportunity for tourist visitors and social reformers alike to return with details on immigrant working-class specimens, that is, characteristic ethnic and racial types. These specimens were captured with cameras or prose and refined for exhibition or publication in drawings and etchings. The practice of specimen hunting, not unlike that filling the displays of natural history museums, helped adapt English theories of poverty to the American reality of an immigrant working class. Notions of characteristic types provided the evidence employed by elite observers and reformers, like those who compiled the Pittsburgh *Survey,* to articulate a hierarchy within the American working class. This hierarchy mirrored biological hierarchies of race and species.[2] In fact, in the study of immigrant colonies, racial and class hierarchies merged. At the bottom of these hierarchies lurked urban "savages."

In marrying imperial adventure to urban narratives of poverty, American observers found evidence that a European disorder—degeneration—had arrived on American shores. Degeneration was the disease of civilization. Already despised figures—the tramp, the prostitute, and the pauper— joined new outcasts—the feebleminded—as the gendered archetypes of the degenerate. The evidence of degeneration was crucial for an emerging generation of reformers. First, it brought new attention to the critical nature of urban reform. The Pittsburgh *Survey* was, above all, a demonstration of the depth of poverty and degeneration in the center of American urban industry. Second, it distinguished between those groups of European immigrants that could be saved through strategies of reform aimed at altering degenerative environments and those that could not be saved. For reform to work, these lowest sorts must be helped toward extinction by being segregated into "colonies" of their own.

Abroad at Home: Urban Exploration

The authors of the Pittsburgh *Survey* described their work as the dispassionate application of recent social-scientific knowledge to the problems spawned in immigrant colonies. The *Survey* was also the surrogate child of a much older American tradition of "class tourism." In the antebellum United States, the development of the first slums, such as the infamous Five Points in lower Manhattan, sparked the rise of the popular entertainment of "slumming." Elite visitors, men and women, traveled through the streets of the slum, often accompanied by policemen or tour guides, who could lead visitors safely into and out of the dankest, most sordid haunts of the neighborhood. It is impossible to know exactly how many guides and policemen were involved in leading slum tours or to gauge how many joined such tours. However, the sheer volume of mentions of slumming and its place in popular culture attests to its cultural significance.

The popularity of slumming highlights the growing segmentation of urban space in the nineteenth century. Prosperous neighborhoods were separated from poverty-stricken slums. As the wealthy traversed areas they would not otherwise enter, movement within the city gained new meanings as leisure. Slumming helped the elite articulate physical, cultural, and often racial differences that they believed separated them from the poor—the very classes they so openly observed. Equally, it provided an escape from the stifling class mores of Victorian America. The slums featured ample opportunities for sex, drink, and drama. Taverns, vaudeville theaters, brothels, and penny museums, like the midways of later world's fairs, offered elites, in the guise of participant/tourists, spaces and opportunities to engage in behavior unacceptable in other spaces of the city.[3]

Urban tourism remained popular through the turn of the new century. In fact, historian Catherine Cocks has noted the growing appeal of tourism in the United States in the decades after the Civil War. Cities became much more than the destination of immigrant workers; they become attractions for tourists. New cultural institutions, such as museums funded by elite donors instead of those located in seedy neighborhoods and privately owned by entrepreneurs such as P. T. Barnum, appealed to new visitors. Immigrants and their quarters also became popular destinations. The new phenomenon of guidebooks encouraged visitors to see the sights of the slums. As late as the 1920s, Chicago featured a tour of its slums with a guide who provided a helpful glossary of proletarian slang.[4]

As working-class neighborhoods drew curious tourists, so too did the factories where immigrants labored. The novelist and journalist Howard Vincent O'Brien turned the experience of class tourism into reformist literature. Written in the first person, his short story "The City" recalled a visit to a slaughterhouse and a sweatshop and, then, to the homes of immigrant workers: "Strange odors and stranger tongues assailed my ears. I could not read the signs on the shops and the passers-by were very foreign." A mysterious guide led the narrator on a tour in order to challenge his narrow vision of beauty. The narrator revealed his identity only at the story's conclusion: "The ancient Greeks called me Poetry." As they stood gazing at the city at night, the mysterious guide revealed the beauty of the industrial city, despite the abused bodies of workers and the mean streets of their neighborhoods: "As darkness fell, blotting out the harsh ugliness of the day, a sense of peace seemed to fall with it. The roar of industry was stilled and quiet brooded over the haunts of man." O'Brien married tourism to reform.[5]

Among the most popular destinations for urban tourists to Chicago were the meatpacking houses, perhaps the most degrading of industrial workplaces. As early as 1878, one guidebook encouraged visits to the stockyards: "To walk through one of the great packing houses when in full operation is something to be enjoyed, and most visitors to the city avail themselves of the opportunity."[6] Julian Street, in his travelogue of tourism in urban America, even devoted an entire chapter to the stockyards. He insisted on the visit in order to see its immense slaughterhouses and signs in surrounding shops in every language from English to Hungarian to Norwegian to Greek. Standing on the tourists' walkway, he was revolted but chillingly awed. He regarded both immigrant workers and the dying swine: "As I stood there, studying the temperament of pigs, I saw the butcher looking up at me [as] he wiped his long, thin blade. He was a rawboned Slav with a pale face, high cheek bones, and large brown eyes.... I have never seen such eyes.... They seemed to look alike on man and pig."[7]

For those who could not afford to travel to Chicago, W. Joseph Grand's 1901 illustrated history of the stockyards reproduced the pleasures of urban tourism. Grand described a tourist site with fascinating characters set against the never-ending slaughter of the killing floor. A vivid description of the cattle kill joined an anecdote of a worker killed at work ("the latest funny story"). In between, Grand sandwiched a portrait of a Joe Getler and his cat, Smut, who happily fetched rats. The midday "Can Rush" of exhausted workers released for a half-hour lunch became another attraction. To cement the idea of the stockyards as a site of tourism, Grand featured a photograph of well-clad visitors walking the muddy paths of the yards; the women hold up their skirts.[8]

In fact, when packinghouse owners in 1906 were defending themselves in Congress against charges of secretly selling tainted meat, they pointed to tours as evidence of the openness of the slaughterhouses. A million visitors trooped through the slaughterhouses with a guide each year. Every department, from the hog kill to the malodorous fertilizer rooms, was open to public view, they declared. Even the architecture of the slaughterhouses reflected their double life as tourist attractions. A raised walkway—complete with safety rails—gave visitors a panoramic view of the hog kill.[9] The packinghouse owners were using the evidence of tours to counter charges that first emerged from the socialist Upton Sinclair's muckraking novel, *The Jungle*. Ironically, Sinclair also employed the trope of the tour to provide his readers a first glimpse of the packinghouses. His newly arrived immigrant, Jurgis, joined a middle-class tour group as it wound its way through the packinghouse. Sinclair focused equally on the subject of the tour—the meatpacking disassembly line and its workers—and on the tourists themselves and their reactions to the horrors of the shop floor. In putting work,

Figure 5.1. "Visitors at the Packing-Houses": Chicago's stockyards were popular tourist sites. This image from a touristic history of the stockyards depicts elite visitors. The focus is especially on the women who lift their skirts to avoid the muddy and probably bloody walkway. W. Joseph Grand, *Illustrated History of the Union Stockyards: Sketch-Book of Familiar Faces and Places at the Yards* (1901).

immigrants, and tourists on display, Sinclair blurred the line between his own socialist sojourn in Packingtown and class tourism.[10]

The "home colony"—as Sinclair termed it—was exotic, foreign, and dangerous. Slum tourism, in its turn-of-the-century incarnation, meant penetrating a foreign world. Labeling immigrant neighborhoods "colonies" both highlighted fears of the immigrant invasion and suggested parallels between imperial safaris and urban tourism. Urban tourists were, as Street put it, "abroad at home."[11] The popularity of slum tourism would affect an emerging genre of literature on urban slums. This literature blurred boundaries between fiction and nonfiction. Short stories featuring immigrant characters often included complex details of life and work in ethnic colonies. Meanwhile, nonfictional exposés of immigrant workplaces or tenement homes included all the drama of novels. They evoked a sense of adventure and exploration by transforming the reformer/narrator into the slum explorer and the reader into a tourist.

In the early nineteenth century, slumming was a key way to understand an economically divided city. Later incarnations of slum tourism aimed at making sense of the urban slum and its strange races. By the late nineteenth and early twentieth centuries, slum explorations were drawing from and, no doubt, affecting, descriptions of imperial adventure. As Jacobson has noted, imperial adventure narratives were popular American literature. Readers devoured wealthy explorers' (allegedly) nonfiction accounts of confrontations with dangerous beasts and exotic savages. Popular fiction, such as Edgar Rice Burroughs's *Tarzan,* also capitalized on a public fascination with the exotic lure of imperial hinterlands. No less influential, though less publicly accessible, were government reports that explored new U.S. colonies and dependencies. Within a few short years of the Spanish-American War, the U.S. Geological Survey printed "gazetteers" that described the racial and geographic features of new colonies and dependencies.[12]

The industrial city, like the jungle, could be the site of virile adventure. Even gazetteers found their parallels in reports such as those of the Dillingham Commission. Each offered accounts, based on individual investigation, of races and peoples cast as exotic and inferior. Equally, even Burroughs transplanted his classic narratives of adventure, romance, and courageous manhood to the industrial city and exotic slum. In his *Efficiency Expert,* Burroughs substituted his stranded male white hero with Jimmy Torrance, a college graduate boxing champion seeking a job in Chicago. Like Tarzan lost in the jungle, Torrance was adrift in the industrial city. Unwilling to take help from his industrialist father, Torrance tumbled toward the bottom of the city's class hierarchy. He drifted between jobs and befriended a thief and a prostitute. Eventually, he found a job by pretending to be an efficiency expert—a scientific manager. Then, the adventure began, as a labor union leader and the corrupt and effete factory manager framed Torrance for murder. In the end, Torrance not only proved his virile manhood and his innocence but also married the owner's daughter.[13]

Public fascination with the foreign and the exotic transformed slum tourism, which became urban safari. The reformist journalist Hutchins Hapgood, for example, described himself as an "intellectual and esthetic adventurer." His adventures were "among people who are generally regarded as 'low.'" He recognized the thrill of urban adventures: "There is nothing more exhilarating than to turn yourself loose among the people. I know two or three men of education who … go out at night, alone, and wander for hours about the lowliest streets in the city." The urban adventure might lead to the Yiddish theater, to the sweatshop, to a saloon to confront the pickpocket or "bum," or to a working-class Italian restaurant.[14] Narratives of such adventures in immigrant colonies mimicked the style and exoticism of nonfictional accounts of imperial colonies. Sometimes narratives jumped seamlessly from urban slums to tropical locales. Hapgood interrupted his adventures in the New York slums with a jaunt to Waikiki. Other studies deliberately referenced the colonial safari. Just as "Stanley visited the heart of the Dark Continent and wrote 'In Darkest Africa'," the authors of *Chicago's Dark Places* claimed to "have gone into many of the dark 'dens' and 'black holes' of Chicago."[15] Descriptions of the "dark places" of American cities enjoyed a similar popularity to Henry Morton Stanley's narrative. Jacob Riis, for example, struggled as a printer and newspaper publisher until he began giving public magic lantern shows that illustrated his travels into New York's slums. He would convert these slide shows into popular books.[16]

Riis may be the most familiar of the urban explorers, but he was certainly not unique. Among many others, Walter Wyckoff, a young Princeton assistant professor, took to the roads and rails with an assortment of disguises in order to depict the lives of American workers. His journeys produced two dramatic books, replete with details of the brutality, misery, and violence of working-class life. He joined the ranks of the unemployed, drank in saloons, and sweated in factories. In the guise of a worker, Wyckoff adopted the character of the proletarian. In his Princeton office, he could retell his experience of savagery in the industrial jungle. For example, he described waiting in Chicago's cold among a gang of desperate unemployed men. He vented his anger in the most primitive fashion upon a porter guarding a factory gate: "I was strong and warm in the wild joy of the lust for blood. With one hand gripping his hairy throat I was pounding the porter's eyes with my right [fist]." Reflecting later, with his disguise removed, he recalled his surprise at the primal pleasure of the fight. Wyckoff's books were simultaneously political exposé, social science, and dramatic adventure. His account of the fight with the porter was illustrated with a picture that emphasized his personal lust for blood, the bleak cold, and the tattered clothing of his fellow unemployed.[17]

Often by relying on disguises, popular authors described life in urban colonies in fiction and nonfiction texts. Stephen Crane assumed the disguise of an unemployed transient to reveal the sordidness of municipal lodging houses. In claiming journalistic accuracy, Crane still relied on the same dramatic tone as his works of pure fiction set in working-class slums. Jack London's short story "South of the Slot" was first published in a socialist magazine. It is particularly illustrative as a fictional story about a nonfiction writer exploring San Francisco's slums. London's central character, Freddie Drummond, was a noted academic. Drummond was renowned for his popular descriptions of slum life. Like Wyckoff, he assumed the garb of a worker in order to explore the slums. Through his worker persona— "Big" Bill Totts—Drummond, the morally upright professor, reveled in

Figure 5.2. "It was a fierce battle": Jack London's short story "South of the Slot" fictionalized slum explorations. A socialist author, London still emphasized the primal fury of his working-class subjects. Jack London, "South of the Slot," *International Socialist Review* (1914).

the brazen sexuality and violence of the colonies "south of the slot," the cable car tracks that divided the city's wealthy and poorer neighborhoods. London offered a socialist critique of the hypocrisy of reformist urban adventurers. He described Drummond in a chauffeured auto next to his elite fiancée. She talked about settlement work, but criticized striking workers as "savages." Still, London offered titillating descriptions of slum exoticism— each emphasized by the original drawings. "A fierce battle" displayed muscle-bound proletarians caught in the primal fury of combat. Totts appeared in a saloon; he was "a trifle inclined to late hours."[18]

When London fictionalized the very narrative genre he himself was producing, he revealed the essential similarity between socialism and social reform. Both depended on touristic voyages, perhaps even more than book learning. The settlement worker Mary Simkhovitch, for example, denied the significance of books for reform. Nothing could match the time spent within immigrant working-class homes and workplaces: "Experience has led me to discard much book knowledge as untrue, more as irrelevant, and most as anaemic in the face of life itself."[19] Instead, reformers learned from guided visits to immigrant colonies. Like a slum tourist, B. O. Flower joined policemen when he visited Boston's slums. In the "ethnological" explorations that would produce his *Old World in the New,* Edward Ross relied on local nurses, social workers, clergy, and policemen as guides. Riis, similarly, followed the New York City police commissioner and future president Theodore Roosevelt on his jaunts through New York's immigrant colonies.

Descriptions of danger, used to equal effect in the foreign colonial and domestic reform travel literature, positioned the socialist narrator as a hero. Jack London studied the slums by traveling to the British capital. Like slum exploration in general, he blurred distinctions between home and abroad and highlighted his position as an outsider and tourist by studying London. Still, his commentaries were not specific to London, but offered an exploration of degeneracy at the heart of civilization. "I went down into the underworld of London with an attitude of mind which I may best liken to that of the explorer," London boasted. He ignored the concern of his friends. He disguised himself in the rags of a London slum dweller: "the vast and malodorous sea had welled up and over me, or I had slipped gently into it." His adventures led him from cheap boardinghouses to dirty workhouses.[20]

Similarly, reformist authors asserted their "manly" courage in describing their willingness to "penetrate" the barriers of race, class, and biology that separated the immigrant colony from the rest of the city. Flower boasted about his confrontations with "the under-world" where "we see the murderer, the thief, the burglar, the gambler, the courtesan, and the confidence man; the bully...the common drunkard." In employing the first person "we," his reader is transformed into the armchair tourist, less willing than Flower to face the "horrors of the abyss."[21] According to the settlement worker Robert A. Woods, the literature that revealed the "city wilderness" was the product of daring physical feats. The task of the urban explorer was "to draw one long breath and to make one and another daring, open-eyed dive beneath the social surface, to tell with vivid, breathless candor, of what he saw and felt and did before his panting return to the upper air and light."[22]

When they described urban exploration as perilous, authors recast the gendered meanings of political work in slums. Although reform remained open to middle-class women, narratives of danger defined the world of reform in terms of what Flower called "male heroism." Especially in the context of the stockyards, male authors/guides suggested that urban tourism displayed too much brutality for refined middle-class women. Both Grand and Sinclair described women fainting at the sight of the killing floor. One packinghouse official even admitted that female visitors should avoid the dank, moist rooms in the bowels of the slaughterhouses. To claim urban space as an arena for women's reform—as much as male courage—female authors, such as Annie Marion MacLean or Helen Campbell, also relied on a language of brazen heroism. They, in effect, had to adapt narratives of manly heroism to the cause of female reform. One female tenement inspector claimed that women's traits of sympathy and intuition were necessary for slum work. Women were especially equipped to confront "life at its barest and hardest, to grapple with cold physical facts." As she clambered over rooftops and explored cellars, she proudly reported Riis's surprise about intrepid female reformers. "How the other half lives" gained new meaning: "I never should have dreamed of a female being reduced to do such work." Campbell therefore cast her study of women's and children's labor as a "long quest" in the "foul city…a ghastly heap of fermenting brickwork."[23] Jessie Davis (a pseudonym) similarly recounted her "vacation in a woolen mill" as a courageous decision to learn from experience what her women's college education had omitted. Forgetting "most of my vocabulary," she ventured into a New Jersey mill town. Her description veered toward literary drama: "I landed in the…mill town on the fag end of a dreary winter's day. Its main street was filled with uninviting shops, its side streets crowded with inartistic hovels, and its outlying sections were dotted with forbidding factories." In her prose, the mill became a "medieval castle." Only a previous summer of canoeing and swimming endowed her with the muscles needed to withstand the demanding work of weaving.[24]

Slum exploration narratives transformed the reformer into an explorer. In the process, reform literature doubled as guidebooks. Chicago's *Dark Places,* for example, urged the "visitor to Chicago" to look beyond the city's lavish mansions to its immigrant slums. The visitor who marveled at the city's magnificent buildings and mansions should also gape at "tumble down, rickety, wretched frame houses—alleys full of reeking filth—the refuse of stables, ash-piles, decaying vegetable matter…[and] blear-eyed, bloated-bodied, semi-palsied, dejected, debased, degraded men and women."[25] Similarly, Riis's *How the Other Half Lives,* born from his magic lantern shows, guided readers, in the guise of tourists, through different immigrant slums. For such narratives, the stated audience was the "visitor," not the reader.

Urban exploration recalled the foreign colonial expedition in the practice of returning with specimens. Expeditions to Africa, Asia, and the Indian country of the West collected examples of primitive artistry that filled the display cases of natural history museums, alongside dinosaur and cavemen fossils. They returned as well with natives for display.[26] Slum life was equally highlighted in museums and exhibitions. The St. Louis World's Fair of 1904 featured popular displays of Filipino tribal life. The fair's Department of Social Science also offered exhibitions on the problem of tramps, the threat of poverty, and the work of Charles Booth and other English poverty experts. In 1908, just as the *Survey*'s authors were beginning their work, New York's Museum of Natural History hosted an exhibit on the problem of urban congestion. Complete with maps, charts, statistics, and photographs, the exhibit focused especially on Manhattan's slums, which were populated by Jewish and Italian migrants. On a less grand scale, the Chicago settlement house, Hull House, created the Industrial Museum as one of its earliest initiatives. The museum displayed the evolution of different industries as part of the march of civilization.[27]

Not unlike the expeditions that built up museum collections, urban explorers returned with descriptions and artifacts that documented immigrant life. When MacLean departed for a summer of work in a sweatshop, she went armed with an imagined camera. She declared herself a hunter, seeking to fix "a figurative Kodak" on sweatshop life. The slum explorer went hunting with the camera. Riis also made the photograph central; his text seemed to play only a supporting role. Sometimes, he captured the personality of his subjects by describing their reaction to his photographs. One street gang, for example, happily displayed their thieving ability for his camera. At the same time, his camera ambushed his working-class subjects and surprised them with their blinding light.[28] Helen Campbell also brought a photographer along on her explorations of New York's slums. She converted photo images, with all their promised objectivity, into etchings and drawings for her 1892 book *Darkness and Daylight*. Her book advertised "thrilling anecdotes and incidents" for the armchair tourist.

Campbell's act of converting photos into etchings fixed the viewer's gaze on specific elements of the slum environment. Etchings claimed to depend on the camera's eye, thereby retaining the political power of objectivity. The pathos they evoked must be real. At the same time, picture taking—like hunting—was a perilous activity that demanded bravery and stealth in order to capture working-class specimens in their true environment. As Campbell reminded her readers, cameras had to be set up covertly. After the picture, the photographer confronted the threats of prey roused by the magnesium flash. She claimed to present "faithful" images "in the vilest slums;…in cheap lodging houses and cellars; in back streets and alleys; in dens of infamy and crime, where the dangerous classes congregate."[29]

Graphic representations of slum explorations appeared in the same venues as reports of imperial adventures. Popular magazines such as *National*

Geographic or *World's Work* were guides for the imperial and industrial age. They provided photographic adventures for readers with little hope of traveling themselves.[30] They also focused inward to the exotic colonies within American borders. In *National Geographic,* imperial subjects were denied the dignity of the formal portrait. Rather, photographers presented them within their perceived natural habitat or in full-body images that stressed the exoticism of their dress. In this way, individuals lost the uniqueness of personality and became scientific specimens. Photographers used the same methods to depict working-class life. They focused on foreignness and exoticism through the presentation of specimens captured on camera. In May 1907, alongside articles on insects and far-off lands, the *National Geographic* pictured "some of our immigrants." It sought to display "different types" among the "great hordes" of immigrants. The pictures identified immigrants only by race. The article opened with a picture of three imposing Cossack immigrants clad in native dress and sporting swords. To capture the ominous numbers and fecundity of immigrants, photographs featured large families posed, unsmiling and in native dress, in a line from shortest to tallest. One typical image featured a "Hebrew" family of eight. A caption warned that 153,748 Hebrews "were admitted in 1906" and that even more would arrive in 1907.[31]

The touristic gaze inherent in such magazines was apparent as well in journals aimed at a dedicated reform readership. *Charities and Commons,* the social work journal that would launch and publicize the Pittsburgh *Survey* (and eventually even rename itself the *Survey*), depended on similar presentations of working-class types and specimens. In characteristic images of folk dancing, for example, the journal cast immigrant individuals at play as ethnic specimens. It celebrated the 1909 National Play Congress as a source of vitality for immigrant workers by picturing immigrants in "peasant garb." By returning immigrants to their perceived authentic roots as agrarian peasants, the magazine's photographers suggested that the Congress could reverse the dislocation of slum life. The Congress's elite organizers, meanwhile, appeared in formal portraits.[32] By contrast, when reformers sought to demonstrate "types" of dislocated immigrant workers, they eschewed photos of authentic peasants. Instead, the photographs of Lewis Hine or the drawings of Jacob Epstein in the journal and in the Pittsburgh *Survey* presented workers with physical features that revealed their racial origins. They were dressed in crumpled and dirty clothing that exposed their exploitation. Epstein's drawings were accompanied by text that invited the reader into "the busiest corners in Pittsburgh." There, "you can see groups

Figure 5.3. "Typical Russian Hebrew Family": *National Geographic,* better known for its representations of foreign peoples and animals, often presented immigrants as racial types. Typically, such images denied immigrants the respectability of formal portraits and focused instead on native dress and large families. "Some of Our Immigrants," *National Geographic* (1907).

of people who bear unmistakably the Slavic physiognomy." For readers unwilling or unable to tour the city's Slavic colonies, the written article presented Slavic "Pittsburgh types."[33]

Whether in *National Geographic* or *Charities and Commons,* such images recast individuals as representatives of their race and class. Hine's photographs of "immigrant types in the steel district" featured captions that only listed racial background. Absent any other text, Hine implicitly demanded that the viewer search physical features for evidence of racial character and garb for hints at biography.[34] Whether through etchings, photographs, or verbal descriptions, reformist urban explorations captured the working-class or immigrant subject. Reform narratives allowed readers and viewers to engage in the same close scientific examination that explorers had completed during their expeditions. Even as I. L. Nascher marveled at the numbers of "slumming guides" to New York's immigrant colonies, he also described himself as a "guide" and his reader as a "visitor." Inspired by the work of Riis in particular, Nascher depicted the populace of "povertyville" through a set of "types." He even included drawings that the visitor could use as a key for identification during their own visits. Hapgood, similarly, organized his study of the New York slums around anecdotes about different types, including the bum, the tramp, and the Bowery girl.[35]

Likewise, London's study of the people of the abyss featured anecdotes about the impoverished in different locations in the slums. He chronicled the lives of the homeless in the workhouses, tramps sleeping in parks, prostitutes in cheap brothels, and owners of squalid boardinghouses. The image that London presented focused on the ugliness of the slums and its denizens. Socialist commentators, such as London, frequently denigrated the savage ugliness of slum inhabitants. In fact, such commentaries were crucial to their condemnation of capitalism. They could, at once, be sympathetic to the plight of workers while still highlighting their degeneracy. London, for example, resorted to the image of the zoo to describe the poor in the abyss: "It was a menagerie of garmented bipeds that looked something like humans and more like beasts, and to complete the picture, brass-buttoned keepers kept order among them when they snarled too fiercely." The slums, in this context, became the "jungle." Struggles in the slum jungle were primal fights. London's description of drunken women fighting as the "barking and snarling of dogs" echoed Wyckoff's own primal reversion. London's accompanying photograph of two women caught in savage combat also recalled Wyckoff's drawing of his own bloody fight.[36]

Through the examination of the urban specimen—in person, for the explorer, and in text, for the reader and viewer—the biological evidence of class difference emerged. Ross used his explorations in industrial America to observe the bodies of immigrants. In industrial city after city, he was shocked by the physical decrepitude and ugliness of immigrant workers. Beauty, far from being a subjective measure, became a marker of racial inferiority. In River Falls, Wisconsin, he was "struck with the vast amount of asymmetry, facial asymmetry, facial disharmony and heavily moulded features." Elsewhere, he recorded "immigrants with asymetrical faces...blunt or low bridge noses, heavy unmoulded jaws, low foreheads and squat figure[s]." Only in pure-blooded American communities in New England and Old Virginia did he notice beautiful women. In immigrant communities, he noted "a great falling off in good looks." His critiques, recorded in fieldwork notebooks and later published in text, were much more than pejorative commentaries on immigrants. Rather, they were the ethnographic conclusions of a social explorer well versed in a science that read character and racial status from physical features. Beauty and ugliness were in the eye of the explorer and they revealed an inferior immigrant working class.[37] As

Wyckoff was moved to declare after his journeys in working-class America, "Ah, the hideous ugliness of the race to which we belong."

The Survey Movement

"What, then, is a social class?" demanded Franklin Giddings to the National Conference on Social Work. Too often, he mourned, "philanthropists, reformers, and legislators" relied on unscientific definitions of class in "theoretical interpretations of the phenomena of progress, social unrest, degeneration, pauperism, and crime." Instead, he promoted "our knowledge of social evolution." He urged social reformers and sociologists alike to employ "vast observation and statistical material" accumulated through urban exploration to trace a scientific system of classification. In this context, he offered new understandings of poverty and wealth, not as merely markers of status but as outward signs of biological class difference. The pauper, he insisted, was biologically distinct from the reformer who might be tempted to offer succor. Giddings advised his social worker audience before setting off on a program of unscientific charity to understand the biological distinctions of class that divided those who might "make positive contributions to…society" from the "unequally endowed."[38]

The movement that the *Survey* helped launch suggested that reformers heeded Giddings's call for scientific classification. Their explorations and their texts argued for gradations within the American working class, each illustrated and personified by specimens. Giddings's methodological inspiration was also the basis for the *Survey*. Both depended on the work of Charles Booth and other English observers who sought to apply the best of social science to the problem of urban poverty. Between 1886 and 1902, Booth and his investigators created a series of maps and studies that painted a detailed picture of urban poverty in London. Booth described a complex working-class hierarchy with "loafers" and day laborers at the bottom and skilled workers at the top. Alarmingly, Booth, along with other English experts, such as Benjamin Seebohm Rowntree, estimated that almost a third of the British urban population was poverty stricken.

Booth warned that the intensity of industrial labor was undermining workers' efficiency, pushing them lower down on the class hierarchy.[39] The *Survey*, like the work of Booth, depended on mapping and the accumulation of statistics in order to demonstrate the prevalence of poverty across the Pittsburgh region. The *Survey* authors admitted their debt to Booth. Yet their focus on immigrant Pittsburgh necessarily inserted the category of race, something that received little attention in the work of Booth, into the study of class. In adapting the methods of English investigators to the American immigrant experience, the *Survey* authors considered workers not simply for their position on a class hierarchy but also for their racial character.[40]

In its study of race and immigration, the *Survey* featured a touristic presentation of ethnic types.[41] Types were presented in a literary walking tour. They seemed to pass the narrator on a walk down a Pittsburgh street: "The life of the city has become intricate and rich in the picturesque....Here is 'Belle', who exchanges her winter in the workhouse for a summer in a jo-boat....Here is the gallery of miners pounding their grimy fists at a speech by [IWW leader Bill] Haywood...and here is a bunch of half-sobered Slavs." Drawings of a Slav on a bread line, an exhausted worker, a miner, a trainman, day laborers, and a "genuine American" graphically personified class and race types. Kellogg insisted that the touristic gaze was crucial to understanding the immigrant city: "You do not know the Pittsburgh District until you have heard the Italians twanging their mandolins round a construction campfire, and seen the mad whirling of a Slovak dance in a mill town lodge hall."[42] The *Survey* represented more than a social science study. It became an exhibition, true to the desires of its editors. Figuratively, it was a "structural exhibit" of the Pittsburgh industrial community. It was also, literally, an exhibit of maps, photographs, drawings, and plans displayed at reform and scientific centers, such as the Carnegie Institute and the National Municipal League, and social scientific meetings, including the American Economic Association and the American Sociological Society.

The class tourism of the *Survey* provided a new way of viewing the industrial city and its immigrant habitat as an innate hierarchy. In the process, it completed a slow revision of an earlier religiously driven practice of urban exploration that described elite neighborhoods in terms of daylight and working-class slums in terms of darkness. For the urban adventurers of the 1870s and early 1880s, binaries of light and shadow, daylight and darkness, sunlight and gaslight spatially divided the city simply into rich and poor.[43] Notions of a class hierarchy and the adoption of social scientific methods of urban exploration by the 1890s gave new relevance to images of light and shadows, daylight and darkness. Campbell appropriated such images for her title and adapted them to new visions of class hierarchy. Darkness and daylight no longer suggested a simple spatial divide; they suggested a hierarchy in which pitch darkness represented the degeneration at the lowest levels. As images of darkness were used to describe the perils of the African colonial jungle, so, too, were they employed to portray the urban slum. Rand McNally's 1912 guide to New York, for example, used a "ramble at night" to introduce the immigrant and tenement quarter. It coyly held out the possibility of danger and debauch: "If you are in search of evil, in order to take part in it—don't look here for guidance. This book merely proposes to give some hints of how the dark, crowded, hard-working, and sometimes criminal portions of the city look at night." The guide then narrated a walking tour of the slums, past the saloons of Mulberry Bend and the opium dens of Chinatown. Darkness, in the language of the guidebook, became real (in the context of a nighttime tour) and metaphorical (in debauch, degeneration, and decay).[44]

Urban explorers insisted that they carried the light of reform to the darkened slums, all the while revealing its darkness to titillated viewers. As Riis suggested, the camera's flash replicated the act of reform by enlightening the darkness of the slums. Riis's photos should be understood not only for their grim subjects, such as the dirty criminal-infested alley or the dusty cellar apartment, but also for their contrast between light and darkness. In the case of Riis's photos, such an analysis is not simply a matter of aesthetics; it reveals the very politics of his slum exploration. In the photograph, he—literally—was shinning the light of civilization. The imagery of darkness and daylight allowed for the easy articulation of a hierarchy of class that blended religious dogma and scientific authority. The hierarchy that ranged from degenerate darkness to moral daylight could be imagined, as Campbell does, as mirroring the stages of human development. Those closest to the troglodyte dwelt in cavelike cellars and shrank before the real flashlight of the explorer and before the metaphorical light of the reformer. For Campbell, in fact, the flash of the camera was nothing less than the light of "respectability." The magnesium flash shed the light of civilization "literally 'in darkest New York', where the light of day never penetrates."[45]

For socialists, the imagery of darkness and daylight was critical. It allowed them to condemn industrial capitalism while depicting in romantic fiction the coming stage of socialism. In this context, the dream of socialism played the role of the reformers' camera flash. George Allan England's *Darkness and Dawn* presented a postapocalyptic world, destroyed by the creation of a massive chasm at the center of the United States. However, the novel is equally a thinly disguised commentary on industrial slums. In calling the chasm an "abyss"—the word used frequently by socialists and nonsocialists alike to describe working-class slums—England presented a parable about current degeneration. At the bottom of the abyss dwelt the Merucaans; their name, alone, confirmed their degeneration from true Americans. They lived in savage squalor. The dark and steamy abyss itself

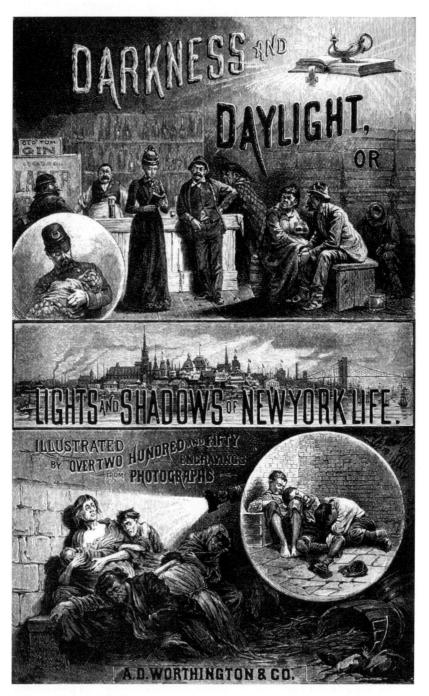

Figure 5.4. Helen Campbell relied on familiar images of darkness and daylight to describe slum degeneracy. At the bottom of this image, the most miserable of proletarians resemble primitive cavemen who shrink from the light of the slum explorer. Helen Campbell, *Darkness and Daylight: Lights and Shadows of New York Life* (1892).

evoked images of the slum. England's socialist parable about the fate of the working class is pejorative. The Merucaans were ugly, savage, and murderous. They had even lost the ability to tolerate light. The novel became a contemporary parable because of the use of the familiar term "abyss" and the imagery of darkness and daylight. The socialist hero Stern led the Merucaans through a recapitulation of industrial evolution. He carried them in his airplane—a classic symbol of modernity—out of the abyss and into dwellings closer to the light. Stern played a role analogous to the socialist activist or social reformer by bringing the degenerate into the light. Appropriately, the Merucaans first civilized habitations—the first steps away from savagery—were called "settlement caves," evoking the social settlement.[46]

The *Survey* highlighted the pathos of working-class life and continued this tradition of shedding light on the darkness of that life. Its visual imagery especially focused on industrial Pittsburgh at night. For its principal artist, Joseph Stella, Pittsburgh was a "black mysterious mass" and the "most stirring infernal regions sung by Dante."[47] His drawings shrouded his working-class figures in chiaroscuro. His *Rescue Workers,* notably, were depicted in an early morning fog, their wan faces only barely visible. This drawing was part of Stella's "Pittsburgh Types" series, which effectively married older tropes of darkness and daylight with newer ideas of class hierarchy.

American Degeneration

As Kellogg looked back on the Pittsburgh *Survey,* he compared it to the work of a surveyor, a journalist, and a geologist. Using all these methods of investigation, the authors created a gazetteer that graphically went below the "surface" of the industrial city. The work of the *Survey* persisted not only in the journal of the same name but also in parallel efforts in other industrial centers. The use of European methods in such surveys helped demonstrate to American observers that Continental disorders had arrived on the western shores of the Atlantic. Kellogg and the *Survey* echoed European fears of a degenerating working class. The work of Booth and Rowntree confirmed a growing problem of pauperism and degeneration. Nearly contemporaneous studies by Cesare Lombroso, Max Nordau, and Robert Reid Rentoul, among many others, helped identify degenerate types.[48] American theorists of degeneration, heavily indebted to their European counterparts, used the idea of degenerate types to provide new and more ominous meanings to already despised figures: the tramp, the prostitute, and the pauper. As the methods of the social survey updated moral and religious ideas of darkness and daylight, theories of degeneration gave new scientific meaning to the individual targets of older religious ire and moral reform. The personification of degeneration was focused on a few gendered types. As industrialization had evolved different roles for men and women, so too would it present different forms of degeneration.

The work of the noted Italian criminologist Cesare Lombroso on prostitutes was translated from Italian into English in 1895. Its English publication helped American authors, regardless of their fidelity to his theories, refocus debates about prostitution from the sense of moral crisis prevalent in the 1870s and 1880s to a biologically driven concern about degeneration by the dawn of the new century. The surgeon G. Frank Lydston, one of the most vocal American popularizers of Lombroso, sought to render moral outrage secondary to "modern scientific thought" and "evolutionary law as applied to biology." The task of the criminologist, as of the sociologist, was "to reduce the subject to a material, scientific, and...evolutionary basis." The degenerate, for Lydston, must be understood less in terms of moral failing than for its place at the bottom of the evolutionary ladder where hierarchies of race and class converged.[49] Vice commissions, formed in large American cities, often by elite civic leaders, sought to fulfill Lydston's dictate—even as they less rigorously followed Lombroso's theories. New York's Committee of Fifteen (and the Committee of Fourteen that succeeded it), as part of a report that used European theories and experience to understand New York's prostitution problem, noted a changing attitude toward the problem from one of moral outrage to one of scientific concern about degeneration. As the noted sociologist and the Committee's secretary Edwin Seligman noted, "In America to-day, we find not only special associations devoted to this matter, but also its more frequent appearance on the programmes of many of our great scientific associations." The committee's report located modern prostitution within the "industrial conditions" of the "social organism." However ancient the roots of prostitution, the committee focused on the "Social Evil" as "a creature of civilisation."[50]

Lombroso claimed that the prostitute—and the degenerate generally—was produced by a combination of physical heredity and environment. Therefore, he focused attention on the prostitute's body and surroundings. He advocated the close examination of the prostitute's anatomy: brain size, skull shape, and physical features. The same method of measurement applied to immigrants was also employed for degenerates. American investigators adopted the technique of examining the prostitute within the broader context of their slum explorations. Yet such a gaze presented problems. Explorations of the life of prostitutes might "gratify...prurient curiosity," as the authors of *Chicago's Dark Places* put it, but, unlike investigations of tramps and the unemployed, even disguise could not morally insulate elite observers.[51] Instead, New York's Committee of Fourteen paid working-class men to assume the disguise of men of their own class seeking out the city's vice.[52] Their findings replicated Lombroso's evolutionary hierarchy of the prostitute. Lombroso argued that some prostitutes were simply born into the profession; poverty acted merely as a "catalyst" that transformed moral deficiency into degeneration. Others were "occasional prostitutes," whose degeneracy could be blamed largely on their dangerous environment.[53] The Committee of Fifteen echoed Lombroso when it suggested that many prostitutes were "a type which varies little with time and place." In addition to these biologically ordained "real prostitutes," the committee described a hierarchy of other women seduced into the Social Evil. Some women were attracted to prostitution because the wages of prostitutes trumped those offered by wage labor. Others became prostitutes because their environment pushed them toward degeneracy. They were "contaminated by constant familiarity with vice in its lowest forms." At the top of hierarchy were "occasional prostitutes," driven temporarily toward the Social Evil because of destitution. Their position as "occasional," however, was fraught with peril. Degeneration loomed and "many of them drift gradually into professional prostitution."[54]

Unlike elite vice commissions, the socialist muckraking author Reginald Wright Kauffman was willing to live within the sordid world of prostitutes. He and his wife spent a year among New York's prostitutes to write his 1910 novel *House of Bondage*. Like many socialist critics of prostitution, he linked an attack on the "vice trust" to hierarchies of prostitutes. Typically, Kauffman included victims of white slavery in his pantheon of prostitute types. Although the novel focused on Mary, a victim of white slavery, it introduced a variety of prostitute types. Fritzie chose prostitution as an alternative to wage work. Wanda was an immigrant victim of seduction. Eveyln had trod the "descending steps" of degeneration. Celeste, alone, was "temperamentally predetermined" for prostitution. Most socialist observers minimized the numbers of women naturally inclined toward prostitution—even as they relied heavily on the statistics of reform vice commissions. Theresa Malkiel, for example, declared that "congenital sexual perverts...form only a negligible fraction of the entire number of prostitutes." Instead, Kauffman, Malkiel, and other socialist observers described the degradation of women as a symptom of capitalism. The "prostitute is a production of civilization," concluded the *Socialist Woman*. "And the capitalist brand is the worst humanity has ever known."[55]

Socialists cast prostitutes as capitalist victims by stressing their odiousness. Malkiel mourned our "unfortunate sisters," but her sympathy was restrained. Prostitutes were still "miserable creatures…left to rot in their own vice." The normally staid Malkiel turned to the lurid language of primitivism to describe prostitutes. They became "hordes, like beasts driven from their lairs." The aging prostitute Old Frances, likewise, was a "victim of the system" and also an example of "repulsive womanhood." In a socialist parable, she spent money for medicine on alcohol.[56]

Tramps emerged as the male equivalent of the despised prostitute. The focus on male tramps was less European in its origin. American fears of tramps coalesced in the aftermath of the great strikes of 1877, one of the first shocks of the industrial age. For elite observers the tramp came to represent, at once, the dangerous tendencies of the disaffected unemployed, the zeal of revolutionaries, and the murderous moral failings of the inebriate. Lee Harris's 1878 novel, *The Man Who Tramps: A Story of To-Day,* notably cast tramps as leaders of a conspiracy to foment tension between capital and labor. In dramatic, reactionary prose his novel echoed a chorus of voices accusing tramps of stirring up the culturally jarring strikes.[57]

Even as the memory of the strikes receded, the tramp continued to evoke alcoholism, uncontrolled sexuality, criminality, and indolence. Increasingly, the tramp came to be represented as a degenerate, a consummate threat to civilization.[58] By the late 1870s and 1880s many states had begun drafting vagrancy laws.[59] Such laws, while offering a series of punishments for vagrancy, also proposed definitions of the tramp. Tramps appeared in these laws less as conspiratorial revolutionaries than as degenerates living on misguided charity and unwilling to work. L. L. Barbour's 1881 definition of the tramp was typical: "He is an indigent, idle wanderer who has nothing to do and wants nothing to do—no trade, no business, no aim in life but to satisfy his daily hankerings at the expense of society." By 1886, the National Conference on Social Welfare had launched a survey to identify the causes of "trampery." Using reports from thirty states describing trampery as an "inherited mental condition" or—more colloquially—as "pure cussedness," the Conference characterized tramping as laziness, drink and vice, unemployment, and depravity.[60]

The tramp may have been reviled, but he was also a focus of fascinated study, as historian Frank Tobias Higbie has noted.[61] By assuming disguises, social investigators sought to penetrate the urban habitat of the tramp in order to understand the symptoms of his degeneracy. The moderate socialist and settlement worker Robert Hunter pointed to physical evidence gathered from the Chicago Municipal Lodging House to prove the degeneracy of the tramp. In a remarkable confession of voyeurism, he described watching vagrants taking spray baths. He noted their potbellies and pigeon breasts, curved spines, and degraded muscles. From these "physical signs of degeneracy," he read their character. Such tramps possessed a "childlike love of petty adventure," but little energy or efficiency. They were beyond redemption.[62] Similarly, Stuart Rice, an investigator for the New York Commissioner of Charities, studied the city's vagrants after donning the "outfit of hobo." As an investigator, he boasted that he often assumed different disguises in order to work in the most dangerous of environments. He had labored among immigrants mucking a railroad tunnel and slept in a rough bunkhouse. As a tramp, he experienced not simply the desperation of empty pockets but also the very process of degeneration itself: "Have you felt the insidious, downward pull of the undertow, the loosening of your moral grip, a deterioration of your character which you seemed powerless to prevent?"[63]

Such explorations helped observers outline a hierarchy of tramps and vagrants. Like hierarchies of prostitutes, the classification of tramps divided those who had sunk to the lowest levels of degeneracy from those afflicted by poverty. Stuart Rice and Alvan Sanborn both used their experiences in the guise of a tramp to study vagrancy "introspectively." Not only could they experience tramp life for themselves but they also could physically examine other tramps in the close quarters of the lodging house. Rice described the different categories of tramps, including the tramp temporarily "down on his luck" and the professional beggar. He even donned fake splints and bandages to join the ranks of vagrants unwilling to work. They preyed on the sympathy of the civilized.[64] The lecturer and self-proclaimed tramp expert Edmund Kelly built on the notion of tramp types to insist on the classification of tramps as the first step in their "elimination." In a complex hierarchy, he divided vagrants into five categories, including youths seized by "wanderlust," the born degenerate whose tramping was a symptom of his condition, the able-bodied, and the non–able-bodied.[65]

The gendered classification of degenerates—the tramp and the prostitute—produced a new kind of "poverty knowledge," as historian Alice O'Connor calls it. In particular, this hierarchical view helped distinguish poverty from pauperism.[66] The pauper, like the tramp or prostitute, was distinct from the poverty-stricken unfortunate.[67] Pauperism was recast as racial atavism, the final stage of degeneration. In the urban context, savagery by the 1890s came to mean the lowest levels of both class and race hierarchies.[68] The bodily decrepitude of paupers was the sign not only of their individual misery and fate but also of their biological unfitness and racial primitiveness. Even Edward Devine, the long-time editor of *Charities and Commons,* argued that "biologically, pauperism represents a primitive type, surviving in the struggle for existence only by parasitism."[69] Male and female degenerates led a life that was, as Devine argued, "suitable to an earlier and more primitive stage of existence, but out of place in the modern world."[70] The fall from poverty to pauperism was recidivism and economic free fall; it was also racial degeneration, as the semicivilized impoverished worker became an urban savage. John Commons worried therefore that the impoverished might join the parasitic ranks of "the criminal, the pauper, the vicious, the indolent, and the vagrant, who, like the industrial classes, seek the cities."[71]

Socialist commentators equally cast paupers and degenerates as primitives. Jack London compared those in the abyss to the primitive Inuits he had encountered in his travels in the Klondike. Residents of the slums, for London, were "the unfit and the unneeded! The miserable and despised and forgotten, dying in the social shambles. The progeny of prostitution—of the prostitution of men and women and children, of flesh and blood, and sparkle and spirit; in brief, the prostitution of labour." The evocation of savagery allowed socialists to undermine capitalist claims to civilization. Yet socialists did not abandon the comparison of civilization and savagery. "If this is the best that civilisation can do for the human," noted London, "then give us howling and naked savagery. Far better to be a people of the wilderness and desert, of the cave and the squatting-place, than to be a people of the machine and the Abyss." The life of the savage, dwelling in caves, was preferable to the squalor of the slums. The poet Ernest Crosby evoked "bleached and stifled and enervated" laborers and "the army of tramps" in a poem ironically entitled "Civilization."[72]

Socialist and reformer observers of prostitutes and tramps cast degenerates as primitives. Some were born degenerate whereas others tumbled down the slippery slope of degeneracy in the slum environment. The degenerate dwellers of the abyss, for London, lived "like swine, enfeebled by chronic innutrition, being sapped mentally, morally, and physically, what chance have they to crawl up and out of the Abyss into which they were born falling?"[73] When women tumbled into the abyss, they fell toward prostitution; degenerating men became tramps. Charlotte Perkins Gilman described women's turn to prostitution not as a moral fall but as racial degeneration toward primitivism. The prostitute "naturally deteriorates in racial development."[74] Likewise, a leading social worker, William H. Allen, characterized the tramp as a "swaggering, ill conditioned, irreclaimable, incorrigible, utterly depraved savage." For the noted reformer of tramps

John J. McCook, tramps had rejected the rigors and rules of industrial society and surrendered to primitivism. Tramps, like "aborigines," lived outdoors and relished the "savage life." Social worker A. O. Wright similarly declared that tramping was simply "a reversion toward the savage type" and Hunter argued that the tramps he watched in the shower had all the characteristics of the "savage." Like tropical savages, they lacked foresight with only "maudlin" dreams of the future.[75]

Degeneration gripped the lowest levels of the class hierarchy. However, it was not safely confined to the bottom rungs of the American abyss. It might spread because of the effects of the slum environment in which degenerates dwelt in close proximity to the poor and desperate. Degeneration also passed from parents to children. Hunter noted after years of exploring and living in the slums that "children, bred into the ways of pauperism, nearly always took up the vices of their parents." Girls learned promiscuity and boys learned to tramp. To explain degeneration in the environment of slums, Hunter appropriately turned to the image of the primordial jungle. He portrayed degeneracy in the form of predators preying on the unfit and unlucky. For those in the slum "the abyss of vice, crime, pauperism and vagrancy was beneath them, a tiny hope above them. Flitting before them was the leopard persistently trying to win them from their almost hopeless task by charms of sensuality, debauch, and idleness. The lion, predatory and brutal, threatened to devour them.... Some were won from their toil by sensual pleasures, some were torn from their footholds by economic disorders, others were too weak and hungry to keep up the fight." For Hunter, the urban and colonial jungles were thousands of miles apart. Racially and morally, they were frighteningly close.[76]

The Other Colonies: Segregation

Notions of hierarchy ordered views of what had seemed the anarchy of American racial and class diversity. They were crucial, as well, in justifying the seemingly contradictory ways reformers approached the problems of the poor. Some at the top of the hierarchy might be saved. The degenerate were to be eliminated. Historians have understood segregation simply as the dividing lines of blackness and whiteness, as something rooted in Southern Jim Crow culture. The move toward Jim Crow around the turn of the century occurred alongside another contemporary campaign for "segregation": the isolation of the degenerate and their eventual extinction through the regulation of reproduction. Segregation, most broadly, in turn-of-the-last-century America was obviously not about separate but equal, but nor was it purely about separation. Like observers of African Americans who imagined their eventual extinction, those who advocated the segregation of tramps, the feebleminded, and prostitutes aimed at their elimination, biological selection, and extinction. As Franklin Giddings would demand, "Give them the fat of the land if we must, build them separate cottages…but put a hedge and a ditch around their garden and prevent them mingling with untainted children and youth."[77]

The idea of segregation was applied to a range of degenerates, from tramps to prostitutes to the feebleminded to paupers. The segregation of delinquent women has received the most historical attention. Ruth Alexander, for example, has described the incarceration of adolescent women in two New York State reformatories.[78] The experience of these women, while providing a glimpse of the confrontation of young working-class women with changing sexual mores, also hints at the way segregation potentially replaced regulation as the means to combat female degradation and degeneracy. Thus, the Committee of Fifteen's report criticized simple regulation. It argued that regulation only perpetuated the existence of prostitution in tenement houses and poor neighborhoods. Instead, the committee recommended the formation of a "morals police" charged with the formal investigation—indeed, the "surveillance"—of prostitution. Along with a Chicago vice commission, the committee sought to arrest degeneration by separating the "semi-delinquent from delinquent girls" and by creating "an industrial home" that would segregate confirmed prostitutes.[79]

Concurrently, reformers sought to segregate tramps. They advocated the creation of model municipal lodging houses that would provide an alternative to city jails or to the private lodging houses castigated by Sanborn. Such model lodging houses, in addition to providing a sober environment, food, and a bath, would aid in the very process of classification. Inmates would be subjected to a "work-test"; those capable of work would be forced to labor. Advocates also described the municipal lodging house as a substitute for harmful almsgiving. Instead of loose change, a beggar might be presented with a ticket for a night at the municipal lodging house. Chicago's Municipal Lodging House created a ticket that the well-meaning could give to a passing tramp. Indeed, the text of the ticket, explaining the work of the house and its labor test, was aimed primarily at "citizens and housewives of Chicago," not at tramps. The ticket worked not only to segregate the tramp but also to replace, reform, and improve the practice of charity.[80]

Massive institutions, labeled "colonies" and removed from cities, represented the most ambitious plans for the segregation of tramps, as well as the feebleminded. "Colonies" for tramps and the feebleminded took their inspiration not from the nineteenth-century poorhouse (which they resembled in general form, but not in intent) but from European examples.[81] American observers admired European experiments in segregation, such as Mexplas, a Belgian colony for the "waste of humanity" that included the feebleminded and vagabonds. Rice argued that colonies might not only separate tramps from those they might contaminate and dissuade others from vagrancy but that they also improved the efficacy of philanthropy. With the colony, charitable aid could be directed toward those who might actually climb the class hierarchy. By the turn of the century, colonies became an essential part of the American campaign for social reform and settlement work. For Kelly and tramp expert Orlando Lewis, colonies were key to remaking reform along scientific lines. With the tramp removed, there was less temptation for what Kelly denigrated as "indiscriminate almsgiving and such…charities as shelters, soup kitchens, etc." Lewis, likewise, denigrated the mere imprisonment of the tramp. Once released from prison, the tramp remained at large and became "a teacher of parasitism." In the compulsory colony, vagrants could be put to work producing their own food, while being segregated from the rest of the population.[82]

At the same moment that Kelly and Lewis advocated the creation of tramp colonies, leading reformers also clamored for the construction of colonies to segregate the feebleminded, the label for a newly defined category of congenital mental degenerates. The segregation of the feebleminded, argued the Committee on Colonies for and Segregation of Defectives, would relieve future generations of the burden of their degeneracy. Yet their segregation, the committee said, was simply part of a movement for the segregation of all degenerates, including "prostitutes, tramps and…many habitual paupers." The colony housed classes of degenerates "who, if they mingled with the world at large, would be useless or mischievous."[83]

The term "colony" was affixed with yet another meaning, alongside its imperial and immigration contexts: the idea of removal. Calls for the construction of colonies along European models echoed in reform and even socialist circles.[84] It was, for example, central to efforts to remake New York's system of philanthropy. In 1911, with great support from reformers and social workers, the New York legislature approved a bill to fund the construction of a tramp colony. Reformers boasted that able-bodied tramps could be forced to work to grow their own food. They would meanwhile be removed from local jails and city streets.[85]

Segregation was particularly central to reform. Armed with the evidence gathered from slum exploration, reformers set out by the early 1900s to reform the proletarian environment. Naturally, they clamored for better housing and safe streets; they also sought the elimination of degeneracy. In 1901, the South End House, Boston's leading settlement house, proudly reported the eviction of a drunk tramp from its men's reading room. The settlement, it declared, was not "a resort of 'bums.'" The settlement's leaders even saluted the refurbishment of the reading room with new paint and open windows as part of a process of closing the city's working-class colonies to tramps. Something as simple as new paint might become part of a larger, if brutal, process of segregating and eliminating the degenerate: "Thus, death itself is the final factor in this process of social regeneration. The morally fit survive, and the morally unfit drop away." "Social degeneracy," the House declared, demanded isolation. The degenerate was a "carrier" whose presence could pass on their affliction to the desperate poor. As a result, the House urged the creation of lodging houses and colonies to put Boston on the "tramp's black list." Social reformers, the inheritors of the survey and slum exploration tradition, were dedicated to the positive amelioration of the social environment as well as to the segregation and extinction of degenerates. As the South End House reported, "From the beginning of its career the South End House strongly urged and earnestly striven for the gradual segregation from the community of its degenerate and degraded types."[86]

Social Darwinists such as William Graham Sumner had long been suspicious of reform as charity that led to the survival of the unfit. Reformers were similarly concerned about the biological effects of their work. To balance the preservation of those who might otherwise have been eliminated in the social struggle, reform depended on a program of segregation. "Charity...must not itself multiply the occasion for its exercise," editorialized the *Survey* journal in 1909. But, in its inherent contradictions, the balance between segregation and amelioration was impossible to achieve. As the new century wore on, definitions of degeneracy expanded and programs of elimination and segregation necessarily proliferated. As plans for segregation and elimination broadened in their viciousness to include sterilization and immigration restriction, social reformers would find themselves questioning their ability to combine uplift with extinction. As early as 1913, one settlement worker decried the expanding definition of degeneracy. "Care must be taken to guard the border line between the fit and the unfit," she warned. When should reformers "halt" the elimination of those "who carry the germs of degeneracy?"[87]

6

DREDGING THE ABYSS

Babies, Boys, and Civilization

If the Settlements did nothing else they would have a
scientific value as ingenious instruments for deep-sea dredging in
the ocean of humanity. Anyone who thinks that they can bring up nothing
but slime is pitifully mistaken. Many a rare and exquisite jewel of character;
many a transparent and lovely nature; generosities and heroisms that
might well put to shame the pall products of clearer waters—such things
are almost the commonplaces of discovery in the work of the
Settlements.... [I]t would be worth while to find again among
the often "forgotten half" nobilities of soul that increase
one's belief in and hope for the race of man.

R. W. Gilder, 1897

1913 was the year of the baby. In January, the Western Live Stock Exposition held a Baby Health Contest alongside its usual cattle and livestock
judging shows. Anna Steese Richardson, the national chairman of the Department of Hygiene, Congress of Mothers and Parent-Teachers Association,
set off to Denver to cover the show for the *Women's Home Companion*. Richardson had turned to journalism out of necessity when her husband died
and left her with three small children. She found a job first with a small Iowa
newspaper and later with the *Women's Home Companion*. She worked for
the magazine for three decades and supplemented her activism for suffrage
and for children as an author of articles on motherhood and marriage.[1]

As a journalist, she sensed "attractive 'copy.'" As an activist, she found
a campaign. Using the *Women's Home Companion* as a national soapbox,
Richardson publicized baby contests as a way to improve the future of the
race. Healthier infants meant healthier children, which, in turn, meant
healthier adults. Richardson brought national attention in 1913 to the baby
contests, but enthusiasm for the contests had already been growing, not only
on the state fair circuit but also in urban reform circles. Starting in 1912,

settlement houses and related reform organizations held numerous better baby contests, spreading the gospel of safe milk, scientific upbringing, and moral living to the urban immigrant working class. The following year saw even more contests.

Richardson's campaign for better babies spread from the pages of the *Women's Home Companion* to other women's magazines to the most respected newspapers in the nation. Baby contests graduated from state fairs to urban centers. It was, noted *Women's Home Companion* editor Gertrude Lane, "a year of almost unbelievable endeavor."[2] There were older models of the baby contest. In 1855, P. T. Barnum's American Museum had hosted a wildly popular baby show. In the first four days, 60,920 visitors watched babies and children up to five years old compete for prizes. Other museum owners copied Barnum's phenomenal success. Shows appeared in Boston, Philadelphia, Pittsburgh, and Albany.[3] But if Barnum's show was purely for entertainment—visitors could gawk at the "fattest baby" prizewinner—the baby shows that swept the nation in 1913 spread the gospel of health, hygiene, and science. They had more in common with baby contests sponsored by British imperial authorities in Africa. Such contests promoted the health policies of the empire among the backward races. Like those shows, the better baby contests in North America increasingly were aimed at the foreign and the inferior. Baby contest shows spread from country fairs, where native-born children were judged alongside cattle and pigs, to settlement houses in the industrial cities of the east. Native-born nurses and doctors evaluated immigrant children. In the end, Richardson estimated that one hundred thousand babies were judged in the contests of 1913.

As shows moved from rural fairs to urban settlement house clinics, the vocabulary and method remained similar. At the first baby shows, the form used to judge children was crudely adapted from the checklist to grade cattle. The same form made its way to cities. Where county fairs might judge the bloodlines of farmyard stock, the urban better baby contest considered the heredity of "immigrant stock."[4] Richardson tried to use the power of the shows and the authority of motherhood to intercede in the divisive debate over the science of evolution. "Science wrangles over the rival importance of heredity and environment," she admitted, "but we women *know* what effects" prenatal care and maternal care can have. Richardson was not alone in her faith that a stronger race could be bred through the intervention of experts and the application of science, as the success of her campaign attests.

She was one of a host of reformers and socialists at the end of the nineteenth century and the first two decades of the twentieth century who sought to apply the logic of philanthropy to questions of heredity. While some degenerates—the tramp and pauper most noticeably—were beyond hope and their extinction encouraged, the immigrant poor might be saved from degeneration. And, given their belief that the traits of civilization might be passed on from parents to children, these reformers and socialists believed that their work would not have to be repeated in each generation.

The baby contests combined hope for future generations with a concern about racial degeneration. In this way, they highlight a fundamental contradiction in progressive reform as well as in socialism that echoed in denunciations of charity. Did reform and socialism assuage the public conscience while preserving those who would otherwise have perished in the social struggle? Settlement house work and socialism were two of the most prominent examples of efforts to reconcile concerns about the survival of the unfit with the desire for reform and charity. There were obvious differences between socialists and more moderate reformers. Some socialists mocked the halfhearted efforts of reformers. On the other side, Robert A. Woods, the pioneering founder of Boston's South End House, one of the nation's earliest and most active settlement houses, was profoundly hostile to the socialist scourge. Still, even Woods noted that many settlement workers, from Robert Hunter to William English Walling to Maurice Parmelee, were also dedicated socialists.[5] Both socialists and settlement house workers articulated their ideologies in a broad language of evolutionary progress. Settlement work sought to spread the fruits of existing civilization to the forgotten half whereas socialists labored to realize the true civilization that would soon replace capitalism. Settlement house workers were graced with larger budgets and a more secure social position than their socialist counterparts. They also left behind copious details of their efforts. Socialists, by contrast, produced a wealth of theoretical literature that explored the place of socialism in the broader history of evolution.

Settlement house workers and socialists, no less than eugenicists waiting in the political wings, defined unfitness and degeneration. The biological mind-set of reformers and socialists helped them argue that the problems of degeneration were expressed in gendered terms, as a crisis in masculinity and femininity. Preserving manhood and womanhood meant arresting degeneration. Manhood and womanhood, in their perfection, were the physical mark of civilization. Reform, especially, was a physical process, literally the re-forming of the embattled immigrant and proletarian from the moment of birth to adulthood.

In an age of industry, empire, slums, and immigration, evolution needed help to remain on its progressive path.[6] Socialists and settlement house workers launched often intersecting efforts to defend the urge to philanthropy and social protection as an evolutionary development. Reformers saw themselves as guardians of evolution, weeding the possibly fit from the unfit, who needed to be isolated and eliminated. The ranks of the immigrant working class stood on a precipice with degeneracy looming in the abyss on the other side. To understand this degeneracy, reformers turned to a language of civilization that divided the world's population into savage and civilized.[7] Social reform sought to civilize the savage—whether in Manila, Havana, or New York—and to prevent the fall of the partly civilized into savagery. The processes of racial uplift or racial degeneration made sense only because of a kind of neo-Lamarckianism that posited that acquired characteristics were passed on to future generations. Such a notion of acquired characteristics suggested that the work of reform or the ravages of degeneration left their mark on heredity.[8]

The Evolution of Philanthropy

As progressives and socialists struggled to reconcile their fears about the survival of the unfit and their dedication to reform, they came to argue that their politics of philanthropy and mutual aid were evolutionary developments. If the ancients and the uncivilized left their infirm, weak, elderly, and unfit to perish, at the heart of civilization was an instinct for mutual preservation. The rise of the settlement house or the socialist movement could be justified not simply as the well-meaning gestures of a few but as the advance of the race itself.

As the history of industry had to be traced from the age of cavemen and the world of the lesser animals, so, too, could the evolution of what the journalist Alexander Sutherland called "the moral instinct." Human bonds that through social evolution took the forms of family, tribe, race, nation, and class were held together by the moral instinct, "the noblest feature as yet visible on this ancient earth of ours." This instinct first appeared with the lowly savage and was at its most evolved in the civilized.[9] Sutherland began his analysis with the lowest of animals, such as insects, who offered little care for the weak or for their progeny. They simply bred fast enough to perpetuate the species. Even the most savage of humans offered some protection for children, despite the persistence of infanticide among savages.[10]

Still, among the savage races the moral instinct was limited to families. The tribe could hardly be burdened with caring for the weak. Using the evidence gathered from imperial anthropology, European and American observers argued that primitive races simply left the sick, handicapped, or elderly to die in solitude.[11] Savages, like animals, could not afford the luxury of sympathy or conscience. The rise of sympathy, in fact, was central to the evolution of the social organism itself. As social units expanded from the family to the tribe and then to the race, forces worked to weed out the less social and less sympathetic. They could have no place in an increasingly complex industrial society that valued interdependence as much as personal strength. As Sutherland argued, "The whole course of history…features the emergence of the…social as against the unsocial races in the long struggle for existence."[12]

The progress from savagery to civilization could be measured by the growth of social sympathy. As the savage tribe left the aged and sick to die, those races entering the stage of civilization responded with kindness and charity. The result, of course, was the preservation of the unfit. Civilization, curiously, seemed to breed new depths of misery and poverty alongside its new comforts. The ironic mark of progress, poverty was simply evidence that those who in the past would have been "crushed out of existence" were helped to survive, though not to thrive. Almsgiving, the first form of social sympathy among the lower civilized races, "tends…[to] fill the land with crowds of those who, but for it, were doomed to an early disappearance. The blind, the dumb, the deformed, the idiotic, the imbecile, the incompetent, the incorrigibly lazy are preserved."[13]

Charity was also evolving from its earliest development by the lowest civilized races to its practice by the most civilized. As Europe and North America entered the "age of benevolence," the character of charity changed. Once a pure expression of sympathy, best represented by religious goodness, in an age of industry it had become the calculated application of science informed by knowledge of biology.[14] The celebrants of the new philanthropy of social work and settlement houses believed that they represented an evolutionary advance. The charity of their predecessors served only to assuage a larger sense of social concern, but it left behind a mass of unfit. Modern, scientific philanthropy, by contrast, sought to preserve the unfortunate and to confront the forces of degeneration. Through the careful application of science, one could seek to restrict the reproduction of the pauper and tramp and to reverse the degeneration of the poor immigrant. Of course, recalled Woods, "tramps, chronic loafers, beggars, cheats, ne'er-do-wells sought to make the gain of the optimism of young reformers." These "various camp-followers of poverty" almost converted the scientific reform movement into yet another form of charity that allowed the survival of the unfit. It was only through the effective and disciplined application of science that settlement house workers believed that they were able to overcome sentimentality and to reach "beyond forms of poverty which are obvious expression of physical and moral degeneracy."[15] Social reform confronted, in the words of Franklin Giddings, the "bastard gospel of the salvation of the unfit." It supported the notion that "men and women shall earn their bread by useful industry and not live as parasites upon others."[16]

The settlement house, then, was an evolutionary advance in constant struggle against older models of relief and wasteful charity. Settlement house workers proudly recited the triumphant story of the rise of their movement. Like so many other Americans dedicated to the value of social science, American settlement house workers were proud of their historic ties to England. "At each of the settlements in American cities were men and women who not only had visited London houses, but had learned at first hand other vital phases of English social readjustment," boasted Woods.[17] In particular, Americans lauded the achievements of Toynbee Hall, founded in 1884, in honor of the reformer Arnold Toynbee. It was from Toynbee Hall that Charles Booth launched his surveys of London's poor and paupers.[18]

By the turn of the century, American social reformers such as Woods and Jane Addams had made pilgrimages to Toynbee Hall. They described in great detail their almost religious conversion to the social scientific ideology of British settlement work. Just as the Jewish immigrants who settled around Toynbee Hall in London's Whitechapel were also arriving in the poor slums of America's largest cities, the settlement ideal was crossing the Atlantic to America. As the settlement appealed to the "best product" of English universities and society, so too could it excite the best products of America. As Woods insisted, the settlements "made an invincible appeal to the combined spirit of culture and of moral adventure which was coming to be quite as eager in American life as in the maturer civilization of the mother country."[19] The settlement house called on America's most cultured to explore the slums and to plant the seeds of civilization.

Setting biology in opposition to charity, settlement house workers cast reform as the progress of civilization within an industrial society. New York mayor Abram Hewitt told the city's University Settlement Society that the industrial growth of the nineteenth century "liberated the forces of nature" in a process of rapid development that produced both "higher civilization" and great poverty. The great insight of Arnold Toynbee and the first English settlement house workers was to recognize "the evolution of industry, and...the fruits of higher civilization." As industry developed, so, too, did political economy evolve from its early laws of fierce competition and survival of the fittest to enlightened altruism and mutual aid. Hewitt argued that "there was something else beside the elementary laws of political economy which would have to be applied to the social organization in order to satisfy the conscience of an enlightened man."[20]

Settlement house workers, well versed in the methods and ideas of biology, could reinforce the harmony of cities, which were cast as complex social organisms. Woods praised the early residents of settlement houses as educated, scientific academics and graduate students rather than well-meaning, but blundering, missionaries. "They represented the first generation of students," he wrote, "whose thinking was molded by the principle of natural science that the mass is to be identified and affected through the molecule or atom, and the living organism through the cell." They understood that the city was a social organism, not just a conglomeration of people. Therefore, they brought to bear the insights of biology to the challenges of social problems.[21] Just in case, the Andover Seminary, which supplied select graduates to the South End House, introduced a curriculum on "Social Structure," including classes on the "Family, the Social Evolution of Labor, Pauperism, and Crime." Boston's Lincoln House, similarly, boasted that the desire of the settlement was to improve the "social organism," to fill it "with a splendid enthusiastic spirit."[22]

The settlement sought to forge "sympathetic, militant co-operation" within the social organism of the city. It was emphatically not a sentimental charity that, as Hewitt put it, simply sought to "feed the hungry or cure the sick"—or, even worse, sought to preserve society's worst degenerates.[23] Settlement work was a constant rejection of earlier forms of charity. It applied science to the problems of poverty and pauperism. The University Settlement Society declared that the Society "is in no sense a 'charity.' It is a vigorous and intelligent effort to seek out and modify the causes which lie at the roots of pauperism and of the 'social evil.'"[24] Settlement house workers cast religious charity as the opposite of their campaign. Religious charity was an earlier, more primitive form of altruism that did little to advance civilization itself.[25]

The settlement was wary of providing free food or shelter. Settlement workers believed that such relief recalled earlier forms of altruism and philanthropy. They gazed beyond the borders of the modern city to the broader history of civilization and race development. The University Settlement Society, for example, turned to the case of the Indian: "The Indian, whose sole purpose is to get his daily bread, never rose in the scale of civilization." Similarly, it would not help the "wretched people" of the city—urban primitives—merely to receive relief.[26] Thus, when the University Settlement Society appealed to donors for money, it assured them that the

settlement was opposed to the "distribution of alms." "One more word—
perhaps the most important," the Society's chairman, Henry W. Taft, wrote.
"This work is not a 'pauperizing of youth', for it does not lavish on them
privileges which ought to be made really valuable by the effort it costs to attain them." In fact, not even membership in the Society's children's clubs or
classes was free. Like "students of Harvard or Yale," the settlement's patrons
paid tuition.[27]

Even in the harsh winter of 1893, as unemployment ravaged Boston, the
Andover House (soon to be renamed the South End House) was reluctant
to help provide relief. "The House," it reported, "under all ordinary circumstances, does not aim to give any considerable share of attention to charitable relief work." Yet, given the dire circumstances of the depression, the
House could not remain "aloof." Even then, as the House grudgingly turned
to relief, it proudly reported that it applied the finest standards of scientific
research and emotional detachment. It sought to quantify precisely how
many unemployed people lived in Boston and submitted its method to the
scrutiny of "two statistical authorities." The House worked with the Central
Labor Union Relief Committee and the Associated Charities to distribute
clothing and to find work placements for women. House workers paid special attention to unemployed men, who might be tramps. The House created a "suitable work test" to help separate the worthy from the degenerate.
The test could help the poor as well identify tramps. The House sought to
"dispose of all found to be tramps, so that they shall not return and prey
upon the community."[28] Yet even as the Andover House resorted to relief
activity, Woods still remembered the depression years as a "baptism of fire."
It was, ironically, in the midst of bitter winters of despair that settlement
house workers became convinced of the danger of relief if it encouraged degeneration and preserved the unfit. "Any lurking element of sentimentality,
of superficiality, of mere palliation," he declared, "was burned away." What
remained was the scientific settlement.

In a similar fashion, socialist intellectuals sought to cast socialism as part
of the evolutionary development of altruism. As historian Mark Pittenger
has demonstrated, American and European socialists, as far back as Karl
Marx himself, drew heavily from evolutionary theory. For many socialists,
the notion of evolution helped refashion socialism, not as a political goal,
but as a natural, inevitable development. The *International Socialist Review* highlighted the importance of evolutionary theory for socialists of all
stripes. Charles Darwin's *Origin of Species* joined Karl Marx's *Capital* in
the back-page advertisements for critical books for socialist readers. The
journal also devoted surprising numbers of pages to articles that explained

evolution to socialist readers. Starting in March 1905 and continuing for another six months, Ernest Untermann traced the "evolution of the theory of evolution" from the earliest Greek philosophers through Lamarck to Darwin and Marx. As Untermann declared, "Human history is not only economic history, but also natural history." Born in Germany in 1864, Untermann worked as a sailor before settling in Milwaukee around the turn of the century. One of the founders of the American Socialist Party, Untermann worked equally as a socialist activist and as an artist painting "lifelike" reconstructions of dinosaurs and prehistoric mankind. Untermann declared himself "a proletarian and a scientist."[29] Robert Rives La Monte, one of the most prolific American socialist writers and pamphleteers, similarly placed Marx in a hierarchy of evolutionary theorists that included Herbert Spencer, Charles Darwin, and Alfred Wallace.[30]

Untermann and Rives La Monte both argued that the interconnection between Marx's science of socialism and Darwin's theory of evolution confirmed that socialism was not mere politics; it was pure science. Untermann defined a proletarian biology based on the "vital connection between economic and biological facts." The struggle for survival within a harsh nature was accompanied by the struggle among classes. In the end, Untermann argued, mankind would unite "all his social and individual forces for the struggle against nature" in order to create abundance for all. Untermann boasted that even the bourgeois "avatar of evolution" Herbert Spencer "admitted but grudgingly that the evolution of society tended inevitably toward socialism."[31] Untermann insisted that evolution must occur in economics as in the "conscious mind." The shaping of class consciousness through socialist organizing was actually guiding biological evolution toward real socialism, rather than toward "industrial monopoly." Untermann declared that "no amount of development in industrial monopoly will free a nation, if the proletariat is not educated to such an extent as to understand the laws that 'determine the form of its being.'"[32]

The advent of socialism would replace competition with altruism. For W. H. Miller, socialism moved evolution beyond the "muscular plan," that is, the animalistic struggle for food and shelter. Evolution was not simply the survival of the fit, argued Miller. The development of consciousness of kind and of altruism separated man from the base animal world. Human evolution, for Miller, was characterized by a movement away from individualism to larger and larger associations. Through the recognition of commonality, mankind cast aside individual existence for the family, then for the tribe, and later for the nation and the class. Each of these associations, as nonsocialist theorists such as Franklin Giddings had demonstrated, was accompanied by antagonisms. Tribal unity was often guaranteed through hostility toward other tribes. In the ever widening circles of association, however, socialism promised the substitution of altruism for antagonism. The sense of collective unity—what Miller termed "a state of *consciousness*"—would make mankind "strong and fit." When mankind stops revering the individual, "we cease to think *antagonistically*, we begin to think *altruistically*. This is the very core of Socialism."[33]

A decrepit and degenerating working class was, in this context, the largest threat to socialism. If workers degenerated toward savagery, they retreated from higher forms of progress. Clifford McMartin published a stinging critique of the acceptance by socialists such as Jack London of the "philosophy of misery." This philosophy was based on a particular understanding of evolutionary progress as grounded in the selection of a few and the misery of the unfit. McMartin declared this an "unnatural philosophy" and a misunderstanding of evolution itself. He derided the concept that workers' "comparative prosperity makes them content, refusing to struggle for further improvement." Progress could not be made by a cringing working class. Instead, slow advancement, like that achieved by the victories of organized labor, sparked the desire for more progress. In this way, "the working class moves in obedience to a law of Nature." While McMartin advocated a specific argument in favor of socialist support of trade unions—"without the trade union the Socialist party in America is a rope of sand"—he also endorsed the belief that evolutionary progress emerged from collective sympathy. The future of socialism lay, not in a deepening misery that might provoke workers' revolution, but in the steady development of altruism. Human evolution depended on the rise of what the Russian anarchist and evolutionary theorist Peter Kropotkin called "mutual aid."

American socialists eagerly adopted Kropotkin's arguments. Arthur Lewis, the popular socialist lecturer, described to Chicago audiences how the theory of mutual aid confirmed the essence of socialism. It was only bourgeois scientists, he insisted, who argued that evolution was based solely on individual struggles for survival. Rather, as Kropotkin eloquently demonstrated, evolution did not favor the most brutal. Rather, evolution was directed toward combination and the avoidance of wasteful competition. As insects such as bees had discovered means of cooperation and as the trusts had eliminated wasteful economic competition, so too was the larger social organism moving toward the mutual aid embodied in socialism. The lesson of nature, Lewis insisted, was "practice mutual aid!"[34]

Because socialism was the product of industrial evolution, even if it was achieved through revolution, its advocates recognized that different races would enter the stage of socialism at different moments. Charles A. Steere imagined the transition to socialism in his novel *When Things Were Doing* (1908). The novel begins with the Christmas imbibing of a former New York state assemblyman, Bill Tempest. After cocktails, highballs, gin rickeys, milk punches, cider, Madeira, Rhine wine, Chianti, claret, Burgandy, sherry, port, sauternes, champagne, brandy, and a nightcap of eggnog, Tempest—understandably—snoozed by the fire. In his alcoholic haze, Tempest dreamed that he had become president after a bloodless revolution that brought socialism to the United States and Europe. After a brief fantasy about socialist organization and military prowess, Steere switched tone to define socialism as simple evolution: "Socialism is here and here to stay. It is scientific and in harmony with the workings of evolution, notwithstanding the dictum of sophists that it conflicts with the law of the survival of the fittest or strongest."[35] The novel traced the success of socialism in advancing industrial evolution through mutual aid. Altruism created new impulses for innovation and invention. Steere cast socialism as a product of evolution and, therefore, as a racial accomplishment. Its realization was uneven across the globe. Domestically, African Americans enjoyed the fruits of socialism, but only under the supervision of "a few hundred experts of the Caucasian race." Blacks built their own schools and factories, but they were permanently segregated in the South. They appeared in this socialist fantasy in a familiar exotic role, complete with cakewalks, voodoo dances, vaudeville artists, and "dusky belles." Their passage to socialism was shrouded in "some far-distant day." Steere, in addition, hinted at African American extinction through "the law of natural selection." In their segregated region, African Americans "work out their racial destiny, whatever that may be and wherever it may finally lead." Socialism was equally distant for inferior races in imperial hinterlands. Steere attacked imperialism as "an alien army of beef-eating swashbucklers." Yet he described socialism as merely a "prophesy…for les miserables of the Orient." He acknowledged that Japan had advanced toward industry "with a meteoric burst of speed that made her Occidental rivals get busy with quirt and spur." Yet even industrializing Japan was not yet ready for socialism.[36]

The Crisis of Masculinity and Femininity

As self-proclaimed dispassionate scholars of society and advocates of science, progressives and socialists sought to help a select few who could be saved from degeneracy. There were confirmed degenerates who rightly should be destined for extinction. Important strains of immigrant stock who faced the brutal struggle of urban life dwelt among these degenerates. As Robert Woods insisted, the settlement house had a responsibility to the neighborhood: "The settlement must continue its work of protecting her goings and comings. The neighborhood must be combed of loafers and degenerates, of coarse men and women."[37] Progressives and socialists drew from the evidence gathered from residency in the slums and vividly described the perils of degeneration that daily faced recent immigrants. Such perils drew them constantly back toward primitivism. These observers described a crisis of degeneration that threatened to engulf immigrant colonies in gendered terms, as a failure of masculinity and femininity.

"We forget that ignorance crowds in upon us from abroad," warned Richard Gilder, one of the University Settlement Society's board members. Ignorance had bred degeneration. He discovered in his work in New York's poorest and most crowded neighborhoods an "amazing tendency downward, the unnumbered evils of congested districts." He mourned that "we are standing on the brink of calamity." Even worse, settlement work had only just begun when he was writing in 1896. Settlements were poorly prepared to confront the degeneration that faced immigrants: "London not only preceded us, but far surpasses us in equipment."[38] He was not alone in his concerned study of the perils of degeneration. Like most college and university settlements, the Andover House provided student fellowships. Fellows lived for two years at the House in order to explore an element of slum life. The House's publicity praised these students from Amherst, the Andover Seminary, Dartmouth, Harvard, and Radcliffe as learned, brave, and well-meaning experts who had transformed the slum into a scientific laboratory. Two residents, Alvan Sanborn and Albert Wolfe, examined nearby lodging houses. Charlotte Price studied women's laundries while Rufus Miles and Everett Goodhue offered more expansive analyses of immigrant character and their local environment. Their studies found ample evidence of the perils of degeneration that lurked in the public and private places of the city's immigrant colonies.[39] Drawing on older models of slum exploration, these students immersed themselves in the sordid culture of the slums. They penetrated what Cleveland's Goodrich Social Settlement called the "waste-places of civilization." When the Andover House changed its name in 1895 to the South End House, it embraced the "unpleasant associations of poverty and vice" that accompanied its new name. "It is because of those unpleasant associations that we are here."[40]

The reports by settlement residents were profoundly self-referential. They focused as much on their position as elite explorers as on the conditions they uncovered.[41] Like other explorations of immigrant, working-class colonies, settlement house studies were dramatic exposés of the reality of degeneration. The South End House, for example, praised Sanborn's studies as "heroic explorations among the cheap lodging houses" that demonstrated the dangers of the tramp problem. The settlement house served as an outpost amid the primitive slums from which educated residents could launch academic expeditions into local "colonies."[42] The evidence of such studies provided ample evidence of the forces that dragged down the immigrant worker into the ranks of the "submerged." There was the constant lure of the primitive in the urban slum. At the same moment that American imperial politics abroad—most notably "benevolent assimilation" in the Philippines—became focused on sweeping public health measures, a parallel domestic settlement house campaign against germs and disease provided a tangible way of understanding how degeneracy might spread from the immigrant colony to the civilized city.[43] Contagion was a powerful idea because of the fears associated with the newly popularized germ theory. Many observers worried about immigrant neighborhoods as the breeding ground for epidemics that might sweep through the rest of the city. The South End House considered its task as no less than the isolation of "social contagion."[44]

The plight of the individual body—its moral and physical breakdown—threatened the social body. Germs might spread in the threads of garments or in bread from immigrant bakeries. Similarly, prostitution and crime, bred in the tenements, could spill over into genteel neighborhoods. The Boston social reformer and tenement investigator B. O. Flower did not limit his understanding of germs to seeing them merely as the source of disease: "So long as the wretched, filthy dens of dirt, vermin, and disease stand as the only shelter for the children of the slums, so long will moral and physical contagion flourish and send forth death-dealing germs; so long will crime and degradation increase."[45]

In typically emotional language, Flower described the perils of contagion in a dream narrative. In his dream, Flower saw "something which resembled smoke, which assumed a thousand fantastic and gruesome forms" rising from the slums of Boston. Drawing on the alarming significance of germ theory as well as the tangible and older concept of miasma, Flower claimed to see contagion itself; the city was cloaked in "plague." The image of the contagion is also illustrative of how degeneracy could effectively be reversed. Degeneracy cast as a disease, rather than as something ingrained, could be cured. Even settlement work and the altruism that it produced might be equally contagious. As James Hamilton, the head worker of the

University Settlement Society, declared, "Tuberculosis, typhoid and scarlet fever and such diseases are not the only things that are contagious. The spirit of social service is also contagious and it may be very spreading." Settlement work was similar to "a great conquering campaign against infectious disease." Facts were gathered through careful study by settlement students to shape "the final vast program of creating a specific, formulated, hygienic gospel, and, with exhaustive system, going out into all the world to make disciples of every creature."[46] However, without a cure, degeneracy, too, might be passed on to future generations.

Children—especially boys—were at special risk of degeneration. In the imperial idiom, the slum street became the unsettled frontier. Street life, so central to the lives of immigrant children, entrenched them in primitivism, not civilization. The street "educates the boy," declared a bulletin from Boston's Lincoln House. The "laws of street environment" acted as a brutal source of natural selection. "The rude struggles which take place in the…street breed a certain sort of rough strength." The street eliminated the "physical coward" and the "modest, retiring, gentle boy." The street selected leaders of boy gangs with characteristics of "unquestionable cruelty, much bullying." In the end, the "right of might, as in certain crude civilizations" was the natural law of the street.[47]

Critics of boys' street life were unwilling to label the street even as a crude civilization. The University Settlement Society's J. H. Chase concluded after his ethnographic study of street gangs and street games that street life was akin to savagery. He quoted from his diary for proof of the brutality of the savage street. In the late fall of 1901, he had observed a confrontation between a stronger and a weaker gang. The stronger gang demanded sticks for a fire from the weaker. The toll was paid without complaint; after all, the strong made the rules in the jungle as in the street. In the winter of the same year, in a scene that retraced an ancient past of rape and wife stealing, he had witnessed a boy "pounding and kicking a girl" as they watched a street fire. "His strength so engrossed him that he forgot to think of the girl's feelings." Chase concluded that children remained in a state of "savagery" where "might makes right."[48] The socialist J. Howard Moore similarly noted savage traits in the child. Like the savage, the child "hates work, and delights in hunting, fishing and fighting.… Savages cry easily, are amused at simple things, love toys and pets." The boy's love of a jackknife was simply "a survival from those ancient days."[49]

The gang and the environment of the street exercised a powerful degenerative pull. "The gang…often trained for crime and graduated through the saloon," concluded the University Settlement Society's John W. Martin.[50]

In his exploration of the "influence of street life," the Society's Frederick King traced a specific example of how the "training and environment" of the "hard gang" led to the degeneration of one fatherless boy of ten. Because his mother went out to work every day, he was left "to lead the life of the streets" where he fell under the influence of a gang. For King, it was a small step from stealing from pushcart peddlers to becoming one of the "frequenters of the lowest type of dance halls, where working girls of 18 and 20 are turned into professional prostitutes and the young men become procurers for the houses which have given such an evil reputation to the neighborhood."[51]

Commentators on street life often turned to "recapitulation theory" to justify the comparison of boys' street life either to savagery or to crude civilization. According to Clark University's G. Stanley Hall, the leading champion of the theory, children "recapitulate" the evolution of the race. The infant was the equivalent of the earliest primates. Childhood, similarly, reflected savage, brutal life.[52] "The 'recapitulation' theory of boy development," concluded the Lincoln House, "has been of service in working out a program for early adolescence." When the Playground Association described a model syllabus for a class on "play," they turned to recapitulation. Childhood was a re-creation of the experience of human evolution. Children's play was akin to savage rituals. "So far as somatic characteristics of play activities are concerned, very close though not perfect correspondence is found between the savage and the child," insisted Lilla Estelle Appleton in a University of Chicago study of children's and adult savages' play that was much indebted to the work of Hall. As the earliest of savage peoples demonstrated the first examples of a sexual division of labor, so too did children's play begin to display sex difference, as girls and boys chose different games. It was in the hands of the settlement house worker and the social reformer to translate the boy's natural and savage thirst for competitive games and the girl's inclination toward games of dolls and nurture them into a progressive evolution toward civilization.[53]

For socialist audiences, Moore offered his "Law of Biogensis" in place of Hall's theory of recapitulation. Civilization was a thin veneer. Children's development revealed the recent past of savagery: "The child is a savage, just as the higher races of men were savages in their infancy." Children's play, he insisted, manifested all the cruelty of savagery. Like the savage, the child was cruel and prone to lying. "Dismembering flies is a favorite amusement," he declared. It was critical to understand the savage past of the civilized in order to divert the savage child toward progress, eventually toward "Industrial Democracy." The savage boy who delights in "doing the cat to death" could be tempted instead by baseball, insisted Mary Marcy. In this way, the socialist could quell the "injurious impulses that survive within us, no longer fitted to the environment of today." The "hoodlum," for example, was merely a degenerate who had not followed the normal passage from savagery to civilization.[54]

Degeneracy manifested itself as a crisis of masculinity and femininity. Whereas a previous generation of social reformers had been concerned with the problem of fallen women, settlement house workers now introduced the dangers of the "boy problem."[55] As King warned, the problem of the fallen woman and the savage boy were intimately linked, as the boy in the gang became the procurer and seducer of desperate girls. The "boy problem" for King was intimately tied to the problem of fallen girls through the process of degeneration.[56] The dirty immigrant home and the perilous slum street were breeding grounds of disease and immorality. Each led to dissolute manhood and the loss of "innocent womanhood."[57] Settlement house workers saw themselves as resurrecting manhood and womanhood, crushed in the environment of the slums.[58] As settlement house worker James Reynolds wrote, "A dirty home in a dark tenement is the explanation of many a case of incorrigible childhood, dissolute youth, and worthless manhood."[59]

Socialists, similarly, understood degeneration in gendered terms. Socialist intellectuals relied on stock images of the degenerate man and woman. The socialist poet Ernest Crosby, for example, described the degeneration of "stalwart country folk" in the slums. They become "bleached and stifled and enervated in the slavery of dull toil."[60] The significance of degeneration for the socialist lay not in the perils of the street, but in the effects of the current stage of industrial evolution that favored the capitalist. According to Murray King, in the capitalist era, the most fit was the "commercial type," which "is as perfect an adaptation to surroundings as was the savage type in the midst of the native forest." By contrast, the worker was the degenerate, the degraded, the least fit.[61]

Working-class degeneration was represented by socialist writers as a breakdown of the stable family and as a crisis of masculinity and femininity. Untermann, for example, described male workers about to start their daily labor with a dinner pail filled with adulterated food. They gazed with "helpless compassion at their invalid wives and their offspring doomed to perpetual drudgery." The "evolution of economic conditions" had destroyed the working-class family and left behind fallen women and degraded, crippled men. Untermann offered the hope of a socialist future to "you young girl with traces of former purity...now degraded and vulgar beyond conception" and "you young toiler…now dwarfed and crippled physically from premature hard work beyond the endurance of his growing body."[62]

If settlement work, like socialism, could be cast as rescuing manhood and womanhood, then the settlement worker and the socialist activist consciously described themselves as representatives of true manhood and womanhood. When Maurice Brent, the director-in-charge of the Lincoln House, died, the settlement mourned him as the perfect settlement worker, "one who combined manliness and strength with humility and gentleness." His memorials illustrate an ideal of the manly settlement worker. If the settlement head worker was cast as strong and manly, female residents were praised as proper models of maternal womanhood, sensitive and nurturing.[63]

Mining the Social Pit

The task of settlement work was akin to mining or dredging. Like exploring, mining, or dredging, it turned up specimens to be studied, rejected, or polished. The settlement was "a sort of shaft sunk down through various strata, whose borings indicate the nature of far-extending veins." The settlement, in short, mined the sedimented working class defined by Booth and his American followers. The upper grades were those that the settlement sought to polish and display. The lower grades, by contrast, were the urban savages who "must be removed as fast as possible…for…the welfare of the community."[64] The settlement houses were not concerned with all the unfortunate residents of surrounding neighborhoods. Rather, they focused on the "strata" at the edge of degeneration. New York's College Settlement assured its board that it was concerned not with the pauper or tramp class— the lowest of the urban degenerates—but only with "hard-working victims of the industrial situation." It was these unfortunates who in the wrong environment of the slum might be forced into the ranks of the unredeemable degenerates. Hartley House's leading donor, J. G. Phelps Stokes, insisted that the settlement set within the tenements could save the unfortunate through a changed environment. The children of the slums could become useful and civilized. "The change could all be effected," he wrote, "by surrounding them with a better, a more wholesome environment." Dramatically, the University Settlement Society promised to dredge for gems and jewels of humanity within the muck of the tenement.[65]

It was in this context that settlement house workers offered a defense of immigrants only as part of the mass that could be saved from degeneration.

They were not advanced, civilized racial stock. In understanding the character of New York's "Russian and Polish Hebrews," Samuel Rosenhohn reminded readers that the settlement worker must "understand their life, their environment, [and] the constructive and destructive forces of the neighborhood." Immigrants must be studied within their slum environment. As James Reynolds reflected on his tenure as the University Settlement Society's head worker, he focused especially on the struggles of newly arrived immigrants within the tenement slum, saying that their "daily struggle of life under the most discouraging conditions has revealed...the noblest type." Jane Addams, similarly, praised the potential of many immigrants, saying that "there is industrial power and skill as well as spiritual energy to be found in almost every foreign colony." The environment of the settlement might realize this potential and avert the degeneration of the immigrant.[66] The reformist defense of immigrant racial stock placed immigrants as part of the middle ranks of the urban poor defined by Charles Booth and his American interpreters. Under the right environment, the immigrant could be lifted to civilization.

Behind this commitment to the primacy of the environment in determining heredity was a lingering belief in neo-Lamarckianism, the theory of the inheritability of at least some acquired traits. It emerged in the 1880s and 1890s as part of a renewed interest in the early nineteenth-century work of Jean-Baptiste Lamarck. Within Europe by the turn of the century, neo-Lamarckianism had been profoundly challenged and, at least in scientific circles, discredited. The most powerful critique of the idea of acquired characteristics came from the German biologist August Weismann. In a series of revolutionary publications, Weismann advanced a theory of "germ plasm." Anticipating the 1900 rediscovery of the work of Gregor Mendel on genes and inheritance, Weismann argued that essential traits were inherited through what he called "germ plasm," an internal substance unaffected by the environment.[67]

In the United States, neo-Lamarckianism had a longer life. As George Stocking has demonstrated, American biologists led by Edward Drinker Cope and sociologists led by Lester Frank Ward launched a spirited defense of the idea of acquired characteristics. Ward, for example, used his 1890 presidential address to the Biological Society of Washington to attack Weismann's theory as inadequate. It could not explain the evolution of higher mammals. Social scientists, from psychologist G. Stanley Hall to anthropologist Frank Baker, echoed Ward's suspicions of theories of germ plasm. In particular, as Stocking notes, Lamarckianism remained crucial to social scientists as they sought to explain the differentiation, emergence, and character of distinct races.[68]

As important as neo-Lamarckianism was to social science, it was equally crucial for the reform and socialist movements. The defense of neo-Lamarckianism obviously was ultimately unsuccessful, but it did preserve it long enough to gain widespread adherence among reform as well as socialist circles. The dedication to the idea of acquired traits among reformers was sometimes explicit, but was most often embedded within defenses of the primacy of environment in determining heredity. As the Hartley House considered its place in the surrounding slums, it turned to the biological importance of environmental factors. "Hartley House stands for a fundamental, human principle…that organisms grow by their exercise of inherent creative faculties…and [grow based] on the nature and character of outside influences in the environment in which the growth takes place." Reynolds drew on his experience as a long-time resident of the slums to argue for the determining importance of the social environment. "The factor of social environment seems," he insisted, "stronger than heredity." The environment, above anything else, would determine the success or the degeneration of new immigrant arrivals.[69]

The importance of reform and settlement work among immigrants lay in its ability to shape the social environment. Reform was not simply the amelioration of conditions for recent arrivals, Reynolds insisted. It halted the degeneration of future generations. "We believe that in grappling with some of these strong problems of human society," declared Reynolds, "we are helping to prevent the continuance of evils whose triumph would mean the pauperizing or degradation of another generation."[70] Even as the social reformer condemned the worst degenerates—the tramp, the feebleminded, and the prostitute—to segregation, they believed that the middle ranks of the immigrant working class could be saved through changes in their environment. Through the settlement house, recent arrivals might acquire and pass on traits of citizenship, civilization, and Americanization to their progeny.

Similarly, socialist intellectuals as well as party activists remained tied to the idea of acquired characteristics. Herman Moellering, for example, used the pages of the *International Socialist Review* to endorse Lamarckianism, at the expense of Weismannism. "We cannot see why Weismann's germ-plasm theory," he argued, "should be presented to the proletariat as favoring socialism." Like so many socialists, Moellering worried about the "inheritance of the perverted and degenerated organisms which breed in the slums." He remained confident about the ability of natural selection to annihilate the most degenerate before their characteristics become fixed in heredity, saying that "natural selection never permits the fixing of degeneracy." He was equally optimistic that a "decent environment" could allow former slum-dwelling children to acquire the characteristics of civilization.[71]

Playing Civilization

Reformers and socialists offered such spirited endorsements of neo-Lamarckianism—both narrowly and broadly defined—because it provided the powerful scientific rationale behind so much of settlement work and socialist activism. The settlement effort to civilize urban primitives was partly about replacing the tribal conditions of the street gang with the civilization of the settlement house club. At the same time, the project of urban uplift focused especially on physically strengthening immigrants. Strengthening their bodies meant more than simply developing muscles. Reformers considered the healthy, strong body as the embodiment of civilization, a state of moral and physical well-being that would be passed on to their progeny. It displayed the success of settlement workers in altering the environment. The preservation of manhood and womanhood could be measured through the examination of the bodies of immigrant children and teens remade—literally re-formed—in the settlement gymnasium and urban playground. Settlement workers and social reformers justified campaigns for playgrounds and gymnasiums based on their promotion of civilization in future generations. Strengthening bodies could arrest the natural "tendency to moral degeneration."[72]

Reformers praised the gymnasiums and the clubs of the settlement house as fundamental transformations of the environment that preserved manhood and womanhood. The settlements' programs enabled the civilizing of the primitive savage street gang. In the sheltering environment of the settlement house gymnasium and meeting rooms, the gang would be transformed into a civilized club. The club was run by rules and morals instead of brute strength. The savage gang could be civilized in the wholesome environment of the settlement. The gang would meet in the settlement rooms under the supervision of a settlement worker who would instruct the gang in rules for games, provide industrial training, and teach dramatic expression.[73]

Most settlements boasted about their success in civilizing the gang. Rarely did they admit the resistance that settlements faced. The Hartley House sadly reported in 1899 that police were forced to be on duty during all club meetings. More typically, Pittsburgh's Kingsley House proclaimed that a process of "natural selection" was occurring as the street gang became a settlement club. The "careless and unambitious boys drop out," leaving behind a vibrant, civilized club.[74] If, as Luther Gulick, the president of the Playground Association of America, declared, the gang depended on a primitive notion of the "strength of the strong," the club taught social control and fostered the civilized learning of rules.[75]

As settlement workers sought to celebrate their success in countering degeneration, they turned to familiar displays of racial exclusion and inclusion. When young club members offered a night's entertainment, they often presented racist vaudeville performances. The Lincoln House in 1898 enjoyed "The Minstrel Show." The next year, the Hartley House celebrated Christmas with "Minstrels," including the stock characters of Whitewash, Washwhite, Sambo, and Tambo. Another Christmas entertainment featured "The Nigger Night School." In 1905, the Lincoln House clubs produced the original operetta "The King of Cannibal Islands" alongside a minstrel show. The next year witnessed a reprise of the popular cannibal plot with "A Cannibal Stew," a musical farce written by club members, and a reprise of "The King of the Cannibal Islands." Often under an ostentatiously displayed American flag—a deliberate claim to racial inclusion and civilization—club members put on performances of primitivism and savagery. A Lincoln House minstrel show included a song entitled "Lincoln Boys" that placed club members in the tradition of American patriotism. Other songs reverted to the traditional minstrel genre and included "Humming Coon" and "Dat Am De Way to Spell Chicken." "Buck and Wing Dancing" was performed by "Mr. 'Sambo' Kaufmann."[76]

It might seem paradoxical for the immigrant Kaufmann to become Sambo, to move from savagery to civilization only to perform primitivism. Yet in putting on elaborate costumes, donning blackface, and acting savagely, club members were signaling their own racial progress. Understanding the settlement house amateur minstrel show depends on looking beyond acting to the very process of putting on a show. The settlement stage presentation hinted at civilized behavior in everything from collective writing to costuming. Every element of the play production had a specific racialized meaning. The popularity of racially themed plays in settlement houses was surely a reflection of the public enthusiasm for vaudeville and the exotic; it was also part of a broader set of claims about the racial effects of settlement house work.

Historians have often argued that professional vaudeville performances generally contained a claim to whiteness. Through the elaborate process of looking black, immigrant performers asserted their whiteness; they were white, and, therefore, had to become black for the stage. The settlement house amateur vaudeville performance was also a claim to a racialized Americanness. Primitivism had become hard work for immigrant performers. In the surroundings of the settlement and in the disciplined context of producing and writing a play, the performance of savagery was central to playing civilization. The minstrel show featured disguise and exaggerated accents from its immigrant actors. At the same time, it demanded precisely the kind of behavior that settlement workers would point to as examples of civilized play: collective cooperation, dramatic expression, and individual creative leadership.[77]

A similar analysis can be offered of the names chosen for clubs. Clubs were generally named after patriotic heroes or Indian tribes, but never after places or people that would recall an immigrant past. Paradoxically, clubs made claims to Americanness, if not whiteness, in names such as the Lincoln Volunteers or Lincoln Associates, but still referenced primitivism in names taken from Indian tribes, such as Mohawks or Mohicans. In effect, clubs offered a vigorous masculinity as their response to degeneration. In patriotic names, clubs celebrated the character of immigrant members. They had learned the lessons of civilization well. The gang might reproduce the primitivism of the savage Indian, but the club was "playing Indian," as Philip Deloria has called it. In the evolution from gang to club, real savagery became deliberate playing and performance. The Eastside House celebrated the transformation of the Mohawks. Once a street gang of "defiant little blackguards," they ascended toward civilization: "With us the 'Mohawks'...are still savages, but of a friendly Indian type. They are older, wiser; they have become civilized in part; are not so dirty in person...even a new *sense* is building."[78] When the Mohawks changed from a gang to a club, they recapitulated the evolution of man from savagery to civilization. In playing Indian the club could lay claim to civilization—they were not real savages, only disguising themselves—and demonstrate the physical strength and raw nobility associated with the almost extinct Indian. In this context, club members reflected an emerging ideal of masculinity that cherished the combination of the moral strength of the civilized and the muscularity of the savage. Even as the Eastside House celebrated the progress of the Mohawks toward civilization, they also mounted an exhibit on Indians at the settlement.[79]

Even the success of the civilizing mission of the club could be read from the physical strength of former gang members, now club participants. Settlement workers argued that club members displayed a better physical type.[80] In the same dream narrative in which Flower described the city suffering from the plague of degeneracy, he also illuminated his imagined cure.

An angel revealed to him the solution to this contagion of class. He beheld "great buildings...and playgrounds where were fountains and many happy children." These "temple-like buildings" were immense gymnasiums.[81] Although Flower's dream of playground temples was obviously never realized, the playground movement did emerge as one of the most successful parts of the larger reform movement. The Playground Association of America, founded in 1906, held well-attended annual meetings. It spread the gospel of structured play. By 1912, the Association's congress hosted 515 delegates from 118 cities in 30 states. The Association's secretary, H. S. Braucher, catalogued the growing consciousness of the importance of play. In 1907, 90 cities maintained playgrounds. By 1909, that number had risen to 336.[82]

Settlement houses, similarly, constructed spacious gymnasiums. Throughout the early years of the century, as settlements launched fund-raising campaigns to construct new and larger buildings, the gymnasium was always a central part of the plan. The role of the settlement in supporting structured play was more than simply providing a safe space for childhood recreation. As Jane Addams of Chicago's Hull House told the Playground Association, the city streets fostered a primitivism that could only be reversed in the supervised gymnasium.[83] The University Settlement boasted that it was "a promoter of cleanliness, both mental and physical" and that its gymnasium "built up...neglected bodies." Just as the sick body revealed a potentially degenerate character, the healthy, vigorous body signaled moral strength.[84]

The success of the drive for playgrounds and gymnasiums was rooted in the argument for reversing racial degeneration through exercise. Moore mourned that while civilized life—in his view, evolving toward socialism—depended on cooperation, children's play taught them lessons about "the vanished life of the savage." Most children's play, especially boy's, featured battles, rather than cooperation. The "*play instinct has never been modernized.*"[85] Structured play, outside of the urban jungle, might teach lessons of civilization. One settlement worker used the example of the gymnasium to explain the overall method of the settlement movement. By making childhood games "clean and profitable" the settlement transformed the play instinct into the work instinct. Without the gymnasium, the savagery of the street would lead to "vice and crime." The games of the gymnasium drew on the insights of recapitulation theory. They were carefully structured to promote all the instincts necessary for survival in "civilization." Young children's games that focused on dolls, pets, and plants fostered the "nurturing instinct" while "games of contest" taught the fighting instinct. As the child grew older, the settlement house designed games to support "the great civilizing instincts of nurture and of construction." In this way, "the young savage has his opportunity to take part in the civilized side of real life." Structured play reinforced the adaptations developed over the course of human evolution. It recapitulated the healthy upbringing of the child and used the instincts of past generations to prepare working-class and immigrant children for the modern city. As the Boston reformer Joseph Lee declared, the group games of the gymnasium and playground were an example of civilization, mutual aid, and the development of altruism. Street games, by contrast, taught only lessons in individual and selfish survival of the strong.[86]

Civilizing was not a subtle and hidden process. Rather, settlement workers and playground advocates constantly reminded children that while they were playing, they revisited each stage of human evolution and developed from primitive savages into civilized "citizens." In 1909 the Playground Association formed the Committee on Storytelling to promote certain kinds of educational storytelling at playgrounds. The committee suggested suitable stories to celebrate such American holidays as Thanksgiving or Washington's birthday. Primitive fiction also appeared in playground storytelling as "social-industrial stories that related the primal activities in the securing of food, clothing, and shelter."[87] The play supervisor would help guide the immigrant child through a complete recapitulation of human racial history.

Reformers and settlement workers pointed to the sound bodies of exercising children as evidence of their success in altering heredity through changes in the environment. "Life—Life—Life for All," declared the Playground Association in 1912. It claimed to have substituted the environment of the playground for the primitivism of the street. The playground reduced "dependency" and "delinquency" and promoted "industrial efficiency" and "morality." "*Social progress,*" it declared, "depends upon the extent to which a people possess the play spirit." Healthy bodies joined by rules, laws, and association meant a healthy social organism. The play movement "in strenuous America...is so essential to the life of our country in shaping our national character."[88]

Underlying settlement house work's focus on manhood and womanhood was the assumption that the degeneration and unfitness of at least some immigrants could be confronted and reversed. Manhood and womanhood, significantly, were cast as traits that could be learned in the classrooms and gymnasiums of settlement houses.[89] The crucial period in determining fitness was childhood. If raised in salubrious surroundings that developed mind, morals, and body, children could become superior examples of men and women. Marcellus T. Hayes, the physical education director of the University Settlement Society, argued that exercise could alter heredity. Without exercise, he declared, "we hand down to posterity ill-shaped, diseased and weak bodies."[90]

For settlement workers, civilization flourished in the present. Their task was to integrate new races and peoples into the existing civilization by consciously combating degeneration through gymnasiums and strenuous play. For socialists, in contrast, civilization lay in the future. Still, the notion of "civilization"—a constant theme of socialist writing—as a stage of social evolution was as important for socialists as for settlement house workers and reformers. Civilization was, for both, the process of race development. For socialists, it was also the ultimate triumph of the working-class movement. Even as Ernest Crosby derided contemporary visions of civilization as merely producing degeneration, he looked forward to a new more perfect "civilization."[91] Thus, according to Murray King, the socialist was a civilizer: "*The socialist is the modern alchemist who handles the moral reagents of the universe as tangible quantities…He is leading the race up the steeps [sic] of achievement.*"[92]

Socialists, like settlement workers, looked to the working-class body as the evidence of advancing civilization—even as they disagreed about whether civilization lay in the future or the present. A "virile, vigorous, well-fed working class," as Clifford McMartin described it, represented the apotheosis of civilization. W. H. Miller similarly equated socialism with the achievement of perfect manhood and womanhood. He compared the degenerate, feeble bodies of the family of an Italian garment worker in Chicago and of crippled former miners with the future civilization of socialism "where women will be free from drudgery and free…to develop such healthy bodies" and where men are "strong and fit." Miller, like settlement workers, was obsessed not simply with the remaking of degenerate proletarian bodies but with future progeny. Healthy mothers meant "*strong, healthy progeny that will be 'fit to survive.'*" Socialism guaranteed the future of the race through healthy children: "Children, instead of being crushed physically and mentally, will be allowed to exercise their inherent right to grow into true men and women, 'strong and fit,' as no preceding race has ever seen or dreamed of."[93]

Better Babies; Better Human Stock

For the socialist, the dream of better progeny lay in the future. "It is yet quite a distance economically," lamented Miller. For the settlement worker, affecting heredity could begin at infancy. If the gymnasium and playground remade the children of newly arrived immigrants, the milk station sought to alter the infant's environment. The better baby contests, in turn, celebrated success in altering the environment and helped to identify the stock that was, sadly, beyond repair and reform.

By the turn of the century, the campaign for pure milk took its place alongside the playground movement as one of the hallmarks of progressive reform. Most major cities began milk stations to provide unadulterated and hygienic milk to poor, often immigrant mothers. They also promoted breast-feeding of newborns. Like the playground movement, milk stations relied on notions of race progress to justify their broader importance. As one Philadelphia milk station declared, "Properly prepared milk fills a most important place in the life-saving agencies of the race."[94] New York's Milk Committee relied on recognized images of degeneration to highlight the importance of fresh milk. In its "Milk. Dirty! Clean!" publicity poster, the Committee linked adulterated and unhygienic milk to poor environments as a source of degeneration. The story of bad milk begins on a filthy farm and continues in haphazard transportation from farm to city. Here, the tramp plays a cameo role, spreading his own "germs" of degeneration as he uses the top of a battered milk jug as a mirror. Later, this already festering milk enters the cramped tenement and is fed to a sick and thin baby. The image of the feeble and stooped mother stands as an emblem of what this poor baby will become—if it survives. On the other side of the poster, hygienic milk takes its place at the center of a healthy environment. The baby fed on this milk and reared in this environment can have the promise of a healthy future and similarly healthy progeny.[95]

The milk station provided much more than comparatively clean and unadulterated milk. Like the settlement house, the milk station sent doctors and nurses to explore the urban slum. In the summer of 1912 the Milk Committee sent six field nurses, a supervising nurse, and a doctor into a "typical Italian colony," among other immigrant districts. After much prior study and "canvassing" of tenement conditions, they penetrated the homes and uncovered the "racial eccentricities" that contributed to the dangerous environment of child rearing. They described different cultural traditions of childbirth and child care as racial traits. Italians and Hebrews favored the ill-trained midwife. "Eastern races" such as Turks constructed an unhygienic bower. They concluded that "among the lower classes of these races, habits of recklessness and intemperance are largely responsible for congenital debility and birth accidents." Environment mattered and the goal of the milk committees in their slum explorations was to create an environment to protect "future citizens."[96]

Figure 6.1. "Milk Dirty! Clean!": This New York Milk Committee poster relies on visual suggestions of degeneration—from the dirty tramp to rodents and flies to the dying baby—to encourage the use of milk stations. (Courtesy: Social Welfare History Archives, University of Minnesota.)

Baby contests accompanied the drive for pure milk. The Milk Committee joined the wave of baby contests in 1912 and 1913. In 1913, the Committee held thirteen contests in the New York area and planned ten more. They judged four thousand babies, offering prizes and printing photographs of the winners. The Committee connected their contests to those begun at the state fairs and stockyard shows of the Midwest that were celebrated by Richardson. The Committee's shows in an urban context and their judging of immigrants' babies were based on the same principle of looking beyond "superficial beauty." Babies were "scored just as cattle are judged according to health and physical development." The contests promised to improve the environment of "children already born and to protect those yet unborn."[97]

If degeneracy was measured in gendered terms, so too were the promises of the better baby contests. The 1912 Philadelphia Baby-Saving Show included a conference on infant hygiene alongside the usual, festive judging of no doubt screaming babies. The University Settlement Society, led by its head worker, Robbins Gilman, hosted the largest of the better baby contests to be organized by a settlement house. Other settlements joined in during 1913—the year of the baby. Among others, the University Settlement Society held its second and larger contest, soon joined by the Hartley House. The goal of "pointing the way to a loftier and more virile manhood" guided the University Settlement Society's contests, according to Gilman. The better baby contests proudly displayed "superfine" immigrant babies and highlighted the effort of the Society to counter the effects of "unhealthy trades,…diseases, conditions which make for sexual immorality, industries that warp the souls of men and incapacitate mothers for their functions of motherhood." The realization and protection of gendered ideals, in short, was the cure for degeneration. "Feeblemindness, tuberculosis, and the social diseases are the dreadful crop society reaps from the seeds of…callousness which are sown in our great cities," the Society declared. "These…we must face and fight if we are…to rear a better race of men."[98] Healthy and chubby babies, the physical specimens of race betterment, displayed by the Society in its publicity material, were "a splendid testimonial to the great war that has been waged…in behalf of the weakest of the race."[99]

The better baby contests, through attention to individual bodies, were medical care for the race. Babies were the malleable new generation of the race and mothers were the keys to its future. The focus on babies was indicative of a widespread faith in the ability of an improved environment—healthy milk, clean tenements, and a bright future of exercise and formative play—to affect heredity. In fact, the shows generally included exhibits about how to alter the environment in order to realize the gendered goals of reform: to forge manly man and motherly women.

As quickly as the contests became popular and widespread, they also died out. The University Settlement Society announced that its contests would be held annually, yet it put on only two contests. In fact, there is no evidence of any settlements holding baby contests after 1913. The fact that the baby contests emerged with such popularity in 1912 and 1913 only to die out is indicative of the paradox of their creation: they represented the height of the faith of reformers in their ability to reshape the environment and, thereby, to better the race. The contests represented a transformation from reform to eugenics. They demonstrated the delicate balance of reform. They posited the positive—eugenic—character of much of working class and immigrant racial stock all the while advocating segregation and extinction for the rest. In their effort to protect and forge virile manhood, the contests sought to separate the truly unfit from those immigrants whose degeneration could be reversed. Through the judging of babies and, more indirectly, their mothers, the contests could separate those who were the victims of poor conditions and those who might be truly unfit. Social reform, Gilman argued, was race purification and race betterment, because it helped to preserve the most desirable element and to maintain gendered order while helping identify "subnormal children." Those congenital degenerates uncovered by the contests were sent to the Clearing House for Mental Defectives, an asylum where their reproduction would be prevented.

The baby contests, behind the din of self-congratulation and the wailing of babies, also revealed the limitations of environmental, neo-Lamarckian thought. As the arguments of Weismann and his followers about germ plasm became American scientific orthodoxy, its implications could not forever be ignored by reformers. By the early 1910s, at the height of the popularity of better baby contests, these dedicated evolutionists were in the awkward position of defending efforts to alter the environment at a time when the accepted biology argued that only the mutation of germ plasm or selective breeding could alter heredity. The contests captured this ambivalence. On the one hand, the baby shows cast babies as tabula rasa; for most, the right environment would determine their racial status. On the other hand, the focus on infancy spoke to the growing importance of breeding. For many, the fitness of the babies on display had already been determined. Critics argued that the judges or their reform sponsors accomplished little to eliminate the unfit.

Not surprisingly, the scientific transformation that caused much consternation within socialist and reform circles helped legitimize an emerging chorus of voices praising the new gospel of eugenics, the careful breeding of the fit and the active elimination of the unfit. At the same moment that the settlements claimed to have discovered jewels of humanity, reformers were voicing a terrible unease about their methods and their science. There, at the University Settlement Society's show, alongside the judging of babies, stood an exhibit on eugenics.

7

OF JUKES AND IMMIGRANTS

Eugenics and the Problem of Race Betterment

We got a queer disease, my sister 'n I
That makes folks allers give us the go-by,
'Taint nothin' like the measles or mumps;
It don't come out in spots, not yet in bumps.
It's cuz o' somethin' 't happened long ago,
When we was home with Ma an' Pa, yer know,
An' Pa was out o'work fer most a year,
An' Ma was sick. Oh my, you'd oughter hear
Her cough! 'Twas fierce. An Pa, he useter say:
'The kids'd do well if he was out the way.'
So then he up an' shot hisself an' Ma.
Folks said 'at he was kinder crazed—poor Pa!
An' somehow 'n other that gave Meg an' Me
This queer disease they call 'heredity.'

"Eugenics," Mary Vida Clark, 1913

In 1911, a young and dedicated eugenicist, Arthur Estabrook, began work that he and his supporters believed would transform American understandings of degeneration. He was going to strike at the very science supporting reform that, he worried, was only helping preserve degenerates to further propagate their dismal strain. He was going to rewrite one of the classic books in American social science and reform: Richard Dugdale's 1877 *The Jukes: A Study in Crime, Pauperism, Disease, and Heredity*.

As the subtitle of the book suggests, the social investigator Dugdale engaged the relationship between environment and heredity more than a decade before the work of August Weismann. In studying a family of paupers and "degenerates" in rural New York that he nicknamed the Jukes, Dugdale reached conclusions that highlighted the role of environment. For a generation of reformers, the Jukes became synonymous with degeneration and the dangers of passing on acquired degenerate traits to new generations. To revise the conclusions of the Jukes would be to strike at the heart of reform. Happily—for the cause, at least—Dugdale's notebooks, in which he recorded the Jukes real family name, were found in 1911. Estabrook could locate his sources. Like urban explorers before him, Estabrook was a tireless field-worker. He would set a model that so many eugenics field-workers would emulate. He followed the trail of degeneration from urban centers to distant mountain valleys to rural dirt farms.

His work enjoyed the backing of Charles B. Davenport, the leading voice of the American eugenics movement. An internationally renowned biologist and animal breeder, Davenport had left a promising position at the University of Chicago to become the director of the fledgling Cold Spring Harbor biological laboratory on Long Island. In 1911, he set up the Eugenics Records Office (ERO), backed by funding from the wealthy Mary Harriman and from the Carnegie Institution. Davenport transformed the laboratory into the leading site for the study and promotion of animal and plant genetics and eugenics. As late as the 1920s, the ERO remained the leading voice for American eugenics. Davenport was stridently conservative and deeply suspicious of immigration. He preferred the order and gentility of the scientific journal and conference to fieldwork. Estabrook would provide the field research and Davenport the science and the funding.

They sought to update Dugdale's study in order to prove that environment mattered little in the fight against degeneracy. The Jukes, even in the best of environments, were congenital defectives. The cost of keeping them alive in jails, poorhouses, and asylums strangled the state. Davenport recognized the significance of the revision of Dugdale's work. He cautioned Estabrook: "Let me urge the desirability of being more discrete than usual on this assignment. The notoriety of the family is great....Be wise as a serpent and harmless as a dove."[1] Their relationship during the four years of the project was tense. Estabrook was frequently at odds with Davenport over issues of pay and reimbursement. Their correspondence lacked the warmth that Davenport shared with other activists, especially Harry Laughlin. Davenport consistently chided Estabrook for unnecessary expenses and suspect reimbursement demands. In return, Estabrook mourned his low salary and reminded his mentor of the difficulty of his labors far from the comforts of civilization and surrounded by degenerates.

Still, Estabrook and Davenport shared a deep concern about the dangers of social reform. The final product, *The Jukes in 1915*, effectively countered the science of reform because it borrowed so liberally from its methods. By 1915, the Jukes had dispersed across the country from their New York locale. Their movement allowed Estabrook to consider Dugdale's claim that poor environment had led to the degeneration of the Jukes. In fact, the Juke descendents continued to display the same degenerate traits of licentiousness, tramping, feeblemindedness, and indolence. Only the poor germ plasm of the Jukes could be at fault. Improved environment for the Jukes achieved little but the preservation of the unfit.

Like social reform, American eugenics was grounded in the migration of European ideas. As settlement house workers looked to English pioneers as their rightful ancestors, so too did American eugenicists admire the work of Europeans, in particular Francis Galton and Karl Pearson. A half cousin of Charles Darwin, Galton had coined the term eugenics in his studies of hereditary genius. Pearson was his intellectual heir. He promoted eugenics research in Britain through a chair endowed by Galton himself. Both American and European eugenicists stressed racial ties over abstract national allegiances. Such ties would be continually refreshed in international eugenics congresses.

Eugenics, no less than social reform, looked to the family as the focus of activism and study. For the eugenicist, however, the family could be mined, not for examples of those who could be saved by reform, but as evidence of the heredity nature of degeneracy. The study of family strains—the Jukes as well as numerous other clans and kin—revealed the evidence of unsolvable degeneracy. By studying degenerate families—whose pseudonyms, such as the Zeros, Kallikaks, or Ishmaelites, became popularly known—eugenicists demonstrated how certain degenerate traits, from feeblemindedness to alcoholism to tramping to pauperism, were ingrained examples of unfitness.

For Laughlin, in particular, the evidence from the study of the Jukes and other degenerate strains could also be cited to combat the immigrant invasion. Laughlin used the findings of Estabrook's study to draw explicit comparisons between the Jukes and strains of immigrant defectives. Eugenics helped turn nativism into science and a campaign for restriction into law. Before taking a position as superintendent of the Cold Spring Harbor laboratory, Laughlin was a little-known professor at an agricultural school in Missouri. Davenport and Laughlin struck up a firm and warm relationship and Laughlin soon moved to Cold Spring and became its most outspoken advocate for eugenics. He expansively defined unfitness and degeneration to favor campaigns for immigration restriction and forced sterilization. His science remained amateur and his research notes were often conspicuously absent from the lab's year-end research reports.

This cast of characters, together with many other supporting actors, would systematically revisit and revise the terms of social reform in order to arrive at a new notion of unfitness as ingrained and immutable. They did so by questioning the degree to which social sciences remained rooted in biology. The biological assumptions of social science, social reform, and socialism were dangerously outdated and outmoded. To revitalize the links between biology and the science of society, they suggested that civilization was imperiled by rising numbers of degenerates, unwise immigration policy, and unregulated reproduction.[2]

Environment and the Problem of Heredity

At the moment that Estabrook began updating *The Jukes,* social reformers as much as socialists anxiously confronted the question of how to argue for the importance of environment when the dominant science was focused on germ plasm. Throughout the 1910s, social reformers and socialists struggled to reconcile their faith in biology with the reality that their own scientific approaches clashed with the scientific mainstream. A few tried to claim the findings of Weismann for their cause. Others sought to advance more refined arguments for environmental approaches that incorporated insights about germ plasm. Others simply recognized the growing importance of the science of eugenics. Recent historians have highlighted the persistence of environmental thought at least until the mid-1910s; even some eugenicists, including Davenport, occasionally paid lip service to the potential of environmental reform. Historians have enumerated the variety of efforts, including home economics, to reconcile environmental reform with the science of germ plasm. Such scholarly work has provided an excellent sense of the larger intellectual shifts of the 1910s. But how was this transition enacted at the grassroots level? How were the methods of progressive investigation co-opted into the practice of eugenics? Movements for reform and even socialism had long been intent on the eradication of the unfit. How did they respond to eugenicists who vastly expanded definitions of unfitness and degeneration?

As the University Settlement Society and kindred social reformers celebrated their success in altering the social environment, they also anticipated the need to regulate the reproduction of the unfit. The University Settlement Society's eugenics exhibit at its better baby contests demonstrates the need for a more nuanced understanding of the relationship between Progressive reform and eugenics. Historians stress the continuity from reform to eugenics. Reformers and eugenicists shared, above all, a faith in the explanatory power of evolution and individuals, most notably the birth control activist Margaret Sanger, crossed fluidly from one campaign to the other. The University Settlement Society spoke for many reformers when it promoted eugenics at its baby show: "Eugenics is fast ceasing to be a topic for academic discussion and is becoming the basis of a science of man-raising that promises much for the future of our civilization." Robbins Gilman applauded the show's eugenics focus: "I look upon the science of eugenics as a great moral movement." Even the Society's founder, Stanton Coit, also spoke of a "religion of eugenics."[3]

Such interest, however, did not translate into a smooth transition from reform to eugenics. Rather, it reflected the ongoing, though expanding, desire of reformers to prevent the reproduction of congenital degenerates. Gilman, for example, used his position as head worker of one of the nation's largest settlements to join eugenicists in advocating a test of heredity for all arriving immigrants—even as he denied that immigrant industrial workers filled the ranks of degenerates. He adopted the vocabulary of eugenics: "I personally would add the eugenic test.... It is an alarming condition—this growth in our country of the potentially dysgenic parenthood.... I hasten to add that I am not one of those who say or even believe that our immigrant stock is causing all our degeneration." Still, he believed that "applying tests of blood and germ-plasm purity seems to me...perfectly proper."[4]

Still, despite Gilman's optimism about the compatibility of eugenics and reform, the notion of germ plasm represented a far-reaching challenge to the scientific basis of settlement houses in particular and of reform in general. Gilman tried to cast reform as a form of eugenics, broadly understood as race betterment. Yet the eugenics he envisioned differed sharply from that promoted by activists such as Estabrook. The "application of eugenic ideals...does not necessarily imply the surgeon's knife or legislation." Nor did Gilman stress the significance of germ plasm as a cause of degeneracy. Instead, he insisted that the criminal and the prostitute were "products of 20th century urban forces" as much as they were examples of ingrained heredity. Gilman supported a eugenics of "conscious, moral action." Even as he talked of "racial poisons," the examples he gave blamed the industrial environment: industrial diseases, political despotism, child labor, tenement housing, and class oppression.[5] Gilman's understanding of eugenics placed him outside the mainstream of organized eugenics that, at least in the 1910s, was centered primarily around Cold Spring Harbor. Historian of science Alexandra Minna-Stern has expanded the history of American eugenics by uncovering its different regional incarnations. Gilman's representation of eugenics is comprehensible in this context—as it stood in stark contrast to the eugenics articulated in the 1910s by those, such as Estabrook and

Davenport, who claimed leadership of eugenics, not simply as a movement and a science but as an institution.

As Mary Vida Clark's poem confirms, the *Survey* was even more concerned about the new scientific consensus that seemed to favor eugenics over reform. As Alice Hamilton, a noted doctor focused on industrial medicine, mourned in its pages, "Heredity has become a word to conjure with, and few of us now dare voice our inner protest when scientists tell us that all characteristics, mental and moral as well as physical, are predetermined in the germ cell, the result of age-long heredity and utterly unaffected by...environment." The eugenicist, she argued, was a latter-day predestinationist, with heredity substituted for the "will of God." She found solace in the work of Edwin Conklin, the Princeton biologist. Conklin fought a rearguard defense of environmental influence. Citing Conklin, she concluded that "society can safely eliminate its worst elements from reproduction, but it cannot wisely go farther at present."[6] Edward Devine, the journal's editor, offered a similar warning about the enthusiasm for eugenics. Like most reformers, he agreed with eugenicists that the "abnormal" should be eliminated. Yet he doubted that eugenicists could truly distinguish the fit from the degenerate. According to Devine, settlement workers were necessary in order to identify and develop the "normal." Only then could the "subnormal" be eliminated. He put his faith in "well-tried eugenic forces as...the instinctive preference of human nature for union with health, strength, beauty and ability, rather than with disease and degeneracy."[7]

Still, the presence of eugenics at the better baby shows suggests that by the "year of the baby" American social reformers and settlement workers were grappling with the new biological orthodoxy: inheritance was shaped by germ plasm. Carl Kelsey admitted, for example, that "we know pretty definitely...that acquired characteristics are not passed on from generation to generation."[8] Some reformers and settlement house workers grudgingly admitted the validity of the science of germ plasm, but tried to identify multiple kinds of heredity. The University of Pennsylvania's Simon Patten, one of the leading evolutionary theorists of reform, issued a 1913 call for the "reorganization of social work," a sharp critique of the effects of unscientific philanthropy. Philanthropy, however well meaning, had actually increased the ranks of the "defective classes." Nonetheless, he defended the potential of reform, training, and philanthropy. "Organic evolution" in the human progressed through three stages: "youthful plasticity, mechanical completeness, and senile degeneration." Germ plasm was responsible only for youth; it was the duty of the social worker to develop that plasticity into adult character. In the end, Patten promoted neither eugenics nor existing philanthropy. Rather, he called for "philogenetics," a conscious "evolution" based not on sympathy or pity, but on the careful molding of character.[9]

Patten defined germ plasm as an ingrained heredity that only shaped essential physical traits. Such an understanding allowed advocates of reform and socialism to limit the scope of germ plasm's importance and to retain the primacy of the environment. Scott Nearing, who in his sociological and economic writings blurred the boundaries between socialist and reformer, described a form of environmental heredity unique to humans. "Social heritage," as Nearing termed it, was the gift of civilization that molded the future of the child. Even the biologist Herbert Conn distinguished between social and organic heredity. The latter was the effect of germ plasm. Alone in the animal world, humans had developed civilization and, with it, exceptional rules of evolution. For Conn, social heredity was the transmission of mental and moral traits, as well as physical property, from generation to generation. Insofar as social heredity separated human acquired traits from the experiments of Weismann, it promised a new—albeit short-lived—lease on life for environmental evolutionary thought.[10]

Other reformers defended the importance of the social environment. They suggested that germ plasm was fragile. Germ plasm evoked a mysterious invisibility. It seemed to linger within the body, but its precise location was poorly understood by biologists. Its almost soul-like relationship to the body lent credence to concern about its fragility. One American advocate of eugenics would even quote the German evolutionary scientist Ernest Haeckel in terming the germ cell a "soul cell."[11] This relationship made moral, especially temperance, arguments easy to adapt to the science of germ plasm. Germ plasm might be damaged by immoral living and hard drink. It seemed, therefore, that transforming the environment might protect germ plasm.[12]

A few socialists accepted the notion of germ plasm and claimed the conclusions of Weismannism for the cause of socialism. Herman Whittaker traced the "Battle of the Darwinians" for socialist readers. He conceded that "Weismann has come out of the fight with flying colors." Weismann, he argued, was "laying a biological foundation for the economic science of the socialist school of philosophy." Weismannism held out great hope for the current generation of the impoverished. For Whittaker, it proved that most slum residents had survived their plight because they boasted good germ plasm. They could, therefore, be raised to the standards of the civilized within a generation. His arguments depended on a position fundamentally at odds with that described by eugenicists in their studies of family strains. He insisted that most "denizens of the slums" were actually fit—but unfortunate. The numbers of confirmed degenerates—those whom Laughlin would come to term the "socially inadequate"—was actually small.[13] The socialist lecturer Arthur Lewis similarly abandoned neo-Lamarckianism for Weismannism, claiming that notions of germ plasm could prove that slum dwellers had not been rendered degenerate by their residence in the slums.

However, Lewis joined Whittaker as lonely voices within socialist ranks. Most socialist intellectuals still defended the significance of the social environment long after neo-Lamarckian biologists had fallen silent. The hostility toward socialists by American eugenicists centered around the ERO was emblematic of the way germ plasm and hereditarian thought undercut the socialist dream of civilization. By World War I, the *International Socialist Review* was fighting two interrelated battles: it struggled against wartime censorship and campaigned against the growing importance of eugenics. In its pages, Eva Trews contrasted the workplace strike with the eugenic support of sterilization. Eugenics, she argued, was merely the primitive law of survival: the elimination of the weak by the strong. The strike, on the other hand, represented a new, more advanced form of selection, based not on brute strength but on conscience and mutual aid. She even targeted Estabrook's revision of the Jukes study. Although she did not deny the degeneracy of the Jukes, she described the descendents of Jonathan Edwards—hailed by Davenport and Estabrook as an example of a eugenic family strain—as social parasites. They benefited only from wealth and privilege. The Edwardses, as part of the "higher classes," might inherit status, but not genius.[14]

Race Betterment

Even after the "year of baby," another event appeared to carry the banner of "race betterment" unfurled at the contests. First held in 1914, the Race Betterment Conferences united the most influential voices in American reform, eugenics, human biology, and immigration policy. The Race Betterment Foundation, centered at the Battle Creek Sanitarium in Battle Creek, Michigan, was the brainchild of the cereal millionaire J. H. Kellogg. The 1914 conference, along with two meetings in 1915 and in 1928, illuminated the eclipse of reform and the growing significance of eugenics at its expense. The first conference was convened at a time that historians often recognize as the apex of the progressive moment. They were also held in the immediate aftermath of the year of the baby. Not surprisingly, the conference held a contest of its own, made a short movie documenting the contest, and published a lavish picture of the winning baby.

In his introductory welcome to the 1914 conference, the foundation's president, Stephen Smith, outlined a familiar history of the evolution of modern philanthropy. He located its origins in religious charity. His narrative would have been familiar to reformers who justified their projects as evolutionary advance. In the midst of his address, though, Smith turned to the "cell" and explained its significance for the prevention of the "birth of degenerates." His focus foretold the conference's direction. The field of "philanthropy," he agreed, was "the world of degenerate humanity." Yet he declared that current work was a failure. It was guided by "sentiment," not by "the immutable truths of science."[15]

Although the disagreement between neo-Lamarckians and advocates of theories of germ plasm had simmered for years in the comparatively private pages of journals, the 1914 conference helped push the conflict into the public arena. Behind the conference's genteel pleasantries—"it has been most gratifying to note the unanimous interest shown in the great purposes of this Conference"—and presentations on better babies, immigration laws, hygiene, and industry, lay a smoldering dispute over the relative significance of environment and germ plasm.[16] Some, such as the reformer Jacob Riis, whose paper was delivered posthumously, fought to defend environmental explanations. "We have heard friends here talk about heredity. The word has rung in my ears until I am sick of it," his paper mourned. His paper was a dead man's defense of a dying theory.[17]

Advocates for reform found themselves face-to-face with eugenicists and both presented competing visions of race betterment. It became clear that the emerging science of the "germ" favored eugenics. In much the same way that reformers and socialists attempted to incorporate notions of germ plasm into their own existing political visions, so too did some conference presenters struggle to articulate a middle ground. The biologist Maynard Metcalf cited Lester Frank Ward in his critique of eugenics. Ward was a lonely voice resisting fears of race suicide. He suggested that the lower classes, though they bred faster than their superiors, were also capable of biological, intellectual, and moral advance. The human race, he reasoned, was replenished from below. The breeder of humans—unlike the breeder of plants and animals—could not truly affect ingrained heredity. Thus, he rejected eugenics in favor of what he termed "euthenics." The term was first used by MIT's Ellen Swallow Richards to describe a domestic science. Euthenics, she wrote, was "*nurture,* pure and simple." Like Ward, Metcalf warned against the rapid application of eugenics. Decades of study would be needed before eugenics could be proved. "Eugenics," he argued, "is, as yet, of little social significance." The sociologist Herbert Adolphus Miller followed Metcalf to the podium to warn that eugenics diverted attention from workable strategies for race betterment. Even if eugenic regulations were in place, the same problems of degeneration would persist. Still, in a gesture of reconciliation, Metcalf advocated at least a reconsideration of marriage, from individual happiness to race betterment. People should marry, he believed, in order to improve the germ plasm of the race. Miller was less compromising: "If Eugenics succeeds in establishing...the tremendous social value of heredity...it will overthrow a mass of valuable work of the last decade which has been pointing the way to a fundamental solution of many of our social problems."[18]

As Miller finished talking, Davenport took the podium to begin that very process of overthrowing. At the conference, Davenport spearheaded the assault against environmental explanations for heredity. In his address, he argued that children removed from bad conditions still displayed the same degenerate traits as their sisters and brothers.[19] Davenport and his allies insisted that reform merely preserved the least fit and allowed them to breed. As Yale's Irving Fisher warned, social reform was keeping "alive to propagate" feeble and unfit children and adolescents: "We prolong the lives of...the defective classes."[20] According to Fisher, while reform might protect a few whose life circumstances prevented the expression of their strong germ plasm, it also increased the "burden on society" by shielding the least fit from extinction.[21]

In a conference where eugenics and euthenics had been promised equal billing, such words echoed loudly. The biologist Leon Cole, from the University of Wisconsin, once an intellectual center of reform, articulated for the conference the gaping divide caused by the new science of germ plasm: "As social reformers were concerned with bettering the environment...the gradual, almost complete acceptance by biologists of Weismann's doctrines...naturally led to a widening of the breach between those who placed their faith in social measures and those who foresaw the direful effects of...the increasingly disproportionate ratio of defective germplasm." It was little wonder, he argued, that with some still believing in the "inheritance of acquired characters" that philanthropy was losing ground in the quest for race betterment. The time had come to recognize that "rational selection" must replace the haphazard selection that had acted upon the early savage. Although Cole, like the ERO's scientists, still afforded a place to philanthropy, he assigned it a secondary role in the struggle against degeneration.[22]

The 1914 Race Betterment Conference afforded pride of place to leading reformers and, indeed, the questions that they raised set much of the agenda of the meeting. Moreover, in featuring papers on smoking, drinking, the lives of children, and prostitution, the conference planners belied an uncertainty about whether immoral living could damage germ plasm. Yet those wary of reform and suspicious of the ability of human philanthropy to reverse degeneracy carried the day. The treatment of the baby contest, the central public event of the conference, was emblematic. The contest was reinterpreted in ways that Gilman might not have anticipated. The contests ignored the positive effects of a good environment. Rather, they highlighted ingrained heredity, that is, the importance of the family strain. The chief judge of the contest warned losing parents that beauty or disposition mattered little. Rather, the contest measured heredity. The picture of the winning baby, given this intellectual transformation, featured sheaths of wheat, themselves symbols of fertility. The viewer learned nothing of the baby's home environment.[23]

The cordiality of the 1914 conference is of little surprise; such were the terms of middle-class debate. What is remarkable was its profound anxiety about strategies of reform. This debate was possible because self-described reformers and eugenicists at the conference spoke a similar language of race betterment and degeneracy. In particular, when they sought to measure the prevalence of degeneration, reformers and eugenicists consciously used the similar method of studying families.[24] The next year saw another conference, this time in San Francisco and connected to the city's world's fair. At this smaller conference, the absence of reformers was notable. Whatever battles had been fought the previous year, reformers did not return for another round. Nor did they return for the final conference, held in 1928 in Battle Creek. The banner of race betterment had been seized by those dedicated to notions of germ plasm.[25] At this conference, attendees looked back with satisfaction at their success in transforming immigration law. Yet the problems of degeneration linked to migrations of the potentially unfit remained unresolved. They focused new attention on African American and Mexican migrants.

The Family Strain

As Estabrook's study of the Jukes demonstrated, the form of eugenics advocated by the ERO revisited and revised the methods of social reform. But it did so not as an outgrowth of Progressive reform but as a conscious effort to transform social and political discussions about degeneration in line with contemporary biological science. By the 1910s, despite the claims of social reformers and socialists to fluency in biological science, a gulf had emerged between scientific orthodoxy and social practice. Eugenics represented a disciplinary resistance to the fragmentation of the academy described by historian Dorothy Ross and others. Eugenics was a synthetic science, especially in the hands of Cold Spring Harbor scientists, its prime advocates in the 1910s. As Laughlin argued in a 1919 article, eugenics united increasingly divergent sciences.[26] Even the oft-cited image of eugenics as the tree of sciences should be understood as more than simply a representation of the intellectual arrogance of those scientists grouped around Cold Spring Harbor, but as a regeneration of ties between biology and the social sciences. The social sciences, with their biological basis restored by their acceptance of the most up-to-date understandings of heredity, were joined with biology to form the roots of the tree of eugenics. In effect, Cold Spring Harbor revitalized the marriage between biology and the social sciences, a relationship that had gone sour with the advent of Weismannism and the rediscovery of Mendel's laws of inheritance. Davenport looked forward to a "new era" in which the sociologist and the legislator consulted the biologist and the biologist "will come out of his laboratory" to engage with society. Together, they would "purify our body politic of the feeble-minded, and the criminalistic and the wayward by using the knowledge of heredity."[27]

The ERO mimicked settlement work. It drew on the same educated, mostly female, college students, who once might have served as settlement residents, to work for summers and for longer periods as field-workers. The laboratory ran a summer program training field-workers. As early as 1916, it boasted of having trained over one hundred active workers.[28] They labored throughout the country and collected data on eugenic and degenerate families. Fieldwork began with the study of an individual, a noteworthy degenerate. The field-worker followed all "trails of the germ plasm she is studying" by tracing the family history as well as tracking down and interviewing living relatives. She used this data to create a larger picture of the inheritance of degeneracy. As settlement house work depended on female residents, Davenport insisted that "women, for the most part, are trained" to study degenerates. "It is believed that, in the long run, feminine tactfulness will prove the most valuable asset for this kind of investigation." Men were trained only for the study of the violent and criminal.[29] The *Eugenical News*, the main organ of American eugenics work, described fieldwork as the "strenuous life"—one suited to manliness or, less obviously, to an updated feminine sense of service.[30] Still, the laboratory refused to jettison negative representations of the female college graduate. It reminded potential field-workers that a desired quality was "no gossip." Not surprisingly,

Figure 7.1. As this image suggests, eugenicists often argued that their science, in the wake of the study of heredity, would revivify links between biological and social sciences. (Courtesy: American Philosophical Society.)

female field-workers often chafed at their lowly title. Ruth Moxey, for example, protested that she was thought of merely as a stenographer, saying "that detracts from one's dignity." She suggested, instead, the title of "Eugenics Investigator."[31]

Her choice of job title illustrates the parallels between eugenics fieldwork and slum exploration. Like slum explorers, self-described "eugenics research workers" cast themselves as courageous explorers of degeneracy. Estabrook regularly detailed in his letters to Davenport the "snubs and unpleasant experiences" faced by field-workers. He complained that he faced "conditions that are very bad," devoid of comfort or contact with "better folk." Fieldwork, no less than settlement residency, was a confrontation with degeneracy. It was the exploration of a hidden, perilous world. One fieldworker described her discovery of the "Timber Rats" degenerate community in dramatic terms. She scrambled down a steep hill to a rocky gully in southeastern Minnesota where she found a colony of defectives living in filth and poverty. She stumbled on so many feebleminded residents that she launched a new study.[32] In similar language, one eugenics researcher described—with obvious pride—to Estabrook how he and his female field-workers braved an epidemic of measles as they conducted their work: "Yesterday seven of the girls and myself ran into the biggest funeral they have had for ten years."[33] The ERO even provided guidelines about how to enter a strange house or community and conduct an interview that included the admonition to "be on the lookout, all the time, for interesting families."[34]

Field-workers submitted visual representations of their explorations. Against rough maps of the regions surrounding Cold Spring Harbor, workers recorded their journeys on the trail of degenerate strains. F. H. Danielson's routes from October to December 1910, for example, took her from the Connecticut coast to the rural hinterlands outside Boston.[35] Field-workers focused especially, but not exclusively, on rural regions, whereas slum explorers had penetrated urban slums. Yet eugenicists' attention to rural areas should not be seen as reflective of a belief that degeneracy was centered outside of cities, only that rural areas provided a simpler context. Because eugenicists denied the significance of environment, rural areas simply made it easier to locate and trace strains of degeneracy. When possible, field-workers tried to uncover urban degenerate families, especially those of immigrants. In fact, the Cold Spring Harbor summer school training for field-workers included in its curriculum a visit to Ellis Island to witness the arrival of degenerate stocks.[36] The field-worker, W. E. Davenport, similarly studied the shiftless family of A____, an immigrant from Sicily. In 1914, the ERO recommended an alliance with a sympathetic settlement worker to study "families with strong criminalistic tendencies who all come from the same small town in Italy." The field-worker Z. E. Udell used the same techniques honed in rural areas to trace the family histories of urban and immigrant degenerates. She traced the heredity of an Italian migrant, Minnie D'Agosta, and uncovered traits of insanity, violence, and alcoholism. Ethel Thayer examined the feeblemindedness and alcoholism of Michael Dooley in New York City and another field-worker studied the heredity of Jacob Markowitz, a Jewish immigrant who lived on New York's Lower East Side.[37]

Moreover, the evidence from rural areas was frequently applied to the urban context. Charles Davenport argued for the rural context of his Nam family study because it made it easier to document how traits appeared in different generations. Still, he warned that the rural context should not make his largely urban readers complacent: "No one should deceive himself by thinking that because this is happening in a far-away rural district it does not affect him. The imbeciles and harlots and criminalistic are bred in the hollow but they do not all stay there." The harlots of the rural areas would become urban prostitutes. "Would you rouse yourself if you learned there were ten cases of bubonic plague at a point not 200 miles away?" Davenport asked his urban audience.[38]

Eugenics fieldwork was a conscious critique of settlement and reform work and its emphasis on the influence of the social environment. Sometimes this meant ridicule. In 1912 and 1913, the ERO performed a satirical play (remarkably, by the otherwise staid Laughlin) entitled "Acquired or Inherited?" The play brought to the stage the same gesture of revision that would guide the study of the Jukes in 1915.[39] More typically, field-workers described philanthropy as misguided help that preserved degenerates. As Mary Storer Kostir declared in tracing the family of "Sam Sixty," the children of degenerates "threaten to *overwhelm the civilization of the future*. Our philanthropy...makes life so easy for them."[40] Davenport and Estabrook used their fieldwork among the Nam family to replicate among human subjects Weismann's studies of heredity and acquired characteristics. Like other eugenics fieldwork studies, they traced family history, examined court and charity records, and interviewed living descendents often spread across a large area. To evaluate reform, they examined the effects of the movement of part of a family to a new locale. Like Weismann's studies of the results of cutting off the tails of mice to see whether tails had disappeared in the next generation, Davenport and Estabrook found that the benefits of changing environment were "frequently discouragingly slight."[41] Reform strategies, they reasoned, were scientifically destined for failure.

The critique of reform in these studies is discernible as well in the changing visual representations of degeneracy. Reformers depended on lavish photographs, drawings, or etchings based on photographs. Such images evoked a sense of pathos from the decrepit bodies of degenerate subjects and slum surroundings. Eugenicists, however, discarded environment in their visual representation of degeneracy, as images from Estabrook's study of the Jukes demonstrate. A photograph presented two Juke descendents perched on a cart. This photograph, from Estabrook's fieldwork, focused almost entirely on the bodies of the individuals in question. They serve as reminders for Estabrook of the symptom of degeneracy he was recording. The final, published study dispensed with photographs and bodies altogether. Instead, Estabrook represented degenerate individuals with symbols. Such imagery reduced degeneracy to family pedigree and read racial inferiority not from bodily clues or environment, but from internal heredity.[42]

Just as there were strains of Jukes, so too were there positive strains of scholars, soldiers, inventors, authors, painters, singers, and lawyers. As

Figure 7.2. Arthur Estabrook collected numerous images like this one of the Jukes. The camera's eye focuses especially on the body of the two women. Degeneracy for Estabrook was grounded in heredity; the surroundings of the Jukes mattered little. (Courtesy: Arthur Estabrook Papers, M. E. Grenander Department of Special Collections and Archives, University at Albany Libraries.)

eugenic fieldwork publicized the dangers embedded in the uncontrolled reproduction of degenerate families, from the Jukes to the Ishmaelites, it also identified eugenic families. The European eugenics tradition highlighted the study of notable families. Such studies provided a counterpoint to the American study of families like the Jukes.[43] Ellsworth Huntington, always willing to use his own experience in the service of science, started the impressively named Research Committee of the Huntington Family Association. He hired a field-worker, Martha Ragsdale, a recent Vanderbilt graduate who was recommended by Laughlin. Together, they distributed detailed questionnaires to help trace the Huntington family strain.[44]

Society needed more eugenic family strains. Thus, Davenport implored a Minneapolis university audience to marry and procreate: "And let me urge the duty of raising fair-sized families—of at least four children, else our best blood will be swamped by the more fertile blood of the less effective strains. At the present rate of reproducing themselves 1000 Harvard graduates will have in 200 years fewer than 100 descendents." From its outset, Cold Spring Harbor, in collaboration with the Race Betterment Foundation, sought to create a registry to encourage eugenic marriages.[45] In the context of the study of family strains, eugenicists extended and eventually replaced the better baby contests with fitter family contests. If the better baby contests once sought to prove that good environment could alter the heredity of at least some babies, the fitter family contests replicated the labors of field-workers and demonstrated the existence of superior family strains. Like the baby contests, fitter family contests had their roots in state fairs, alongside the judging of cattle. As early as 1911, the Iowa State Fair directed its baby contest judges to count heredity as "50 percent" of a baby's total score. In 1920, the Kansas fair launched its eugenics department and broadened its contests to the fitter family. Eugenic babies, they reasoned, were, like a "breed of cattle," merely products of positive family strains. The comparisons of babies and families with livestock had roots in older baby contests; it was also familiar language for eugenicists. Through the work of the ERO and the Race Betterment Foundation, "fitter families" contests moved out of the state fair circuit into the social work and eugenics mainstream. The family contests married older traditions of baby contests with the ERO and the Race Betterment Foundation's desire to maintain records on family heredity to encourage eugenic human pairings. By 1925, the ERO distributed a standardized form to measure family heredity. It recorded everything from health to fruit consumption to political party affiliation to intelligence. The forms were used for fitter family contests in Arkansas, Kansas, Massachusetts, Michigan, Oklahoma, and Texas. The contests distributed trophies and certificates; the ERO stored records and reports as part of its ongoing effort to collect data on American germ plasm.[46]

Framing Congenital Degeneracy

When eugenicists aped the methods of social reformers, they demonstrated how reformers had arrived at erroneous and dangerous conclusions. For Davenport, future generations preserved by "good works" posed a "greater danger." Reform, he declared, provided simply the "veneer of good manners." Their similar methods allowed eugenicists to apply germ plasm theory to the same problems raised by social reformers. They cast degeneration as something fixed. If reformers removed the Nams to new environments they only created new "centers of degeneracy."[47] Reformers and even social-ists had anticipated eugenics insofar as they recognized a category of de-generates who should be isolated and prevented from breeding. Eugenicists, however, uncovered greater numbers of feebleminded, tramps, and moral defectives and argued that most of the degenerates that reformers thought they could save were actually congenitally defective. At the most horrify-ingly personal level, Estabrook confidently told one adoptive mother that her new daughter was condemned to degeneracy. The child's father, he re-vealed, was Fred Bemis of the infamous Nam family. The child, therefore, could never be "intellectual," and likely had inherited "eroticism."[48]

Many of the problems investigated in family studies, such as tramping, feeblemindedness, and sexual immorality, had long been accepted by social reformers and many socialists as manifestations of degeneracy. Davenport, therefore, returned to the tramp in one of his studies of congenital degen-eracy. Studying the family histories of one hundred cases, he concluded that the "wandering impulse" may have been a fundamentally human trait, but it was "typically inhibited in intelligent adults of civilized peoples." Daven-port still understood vagrancy as a problem of men. He declared that the wandering impulse was a male "sex-linked recessive monohybrid trait." He denied environmental explanations for wandering and condemned the tramp as beyond salvation. Following this logic, the *Eugenical News* re-viewed the autobiography of the tramp William Davies, not as literature, but as a case study of a vagrant "feeble in inhibitions."[49]

Davenport's work on nomadism was part of a broad project to expand and link definitions of congenital unfitness. Family studies examined the inheritance of confirmed degenerates housed in institutions, including insane asylums, poorhouses, and colonies for the feebleminded.[50] Henry Goddard's 1912 study of the Kallikaks was an early far-reaching study of confined and segregated degenerates. Goddard studied the inheritance of Deborah Kallikak, a fictional name of a "feeble-minded" women living in the Vineland Home that he oversaw. Goddard traced her degeneracy to Martin Kallikak, a normal Revolutionary War soldier who had a brief af-fair with a feebleminded barmaid. Their "dalliance" resulted in the strain that produced Deborah. His study linked different forms of degeneracy, such as alcoholism, feeblemindedness, sexual immorality, prostitution, and tramping, into an overall category of degenerate or socially inadequate.[51] As family studies proliferated throughout the 1910s, they often focused on a particular set of degenerate traits. When Davenport and Estabrook set out to write a study of the Nam family, they differentiated their work from other family studies. They identified a specific form of degeneracy they hoped to prove was hereditary. If the Jukes demonstrated criminality and the Ishmaelites pauperism, the Nams displayed the inheritance of laziness and alcoholism. Davenport concluded that "indolent strains arise chiefly, if not exclusively, by marriage with indolent persons. Such matings are eugenically unfortunate."[52]

The plethora of family studies produced in the 1910s helped Laughlin synthesize a broader definition of "social inadequacy." In his landmark 1921 article in the *American Journal of Sociology,* Laughlin proposed an understanding of unfitness meant to supersede the older sociological categories of "defective, dependent and delinquent classes." Laughlin began by citing an article by Franklin Giddings that satirized the proliferating number of types labeled as degenerate. Giddings's article, "The Seven Devils," suggested that the "defective, dependent and delinquent classes" had come to include even more Ds, including the dirty, disorderly, deformed, and deranged. Giddings mocked the expanding definition of degeneracy. For Laughlin, however, this was serious inspiration. Laughlin announced a definition of the "socially inadequate" that assimilated a range of degenerate types into a single category. He linked the pauper to the alcoholic, the diseased to the feebleminded, and tramps to the insane. In a conscious effort to update definitions of degeneracy, Laughlin sent his formulation to a host of academics and social workers, including leading progressive champions. He reprinted their comments. Some, such as Edward Ross and Albion Small, found the new formulation a more inclusive and positive substitution for "defective, dependent and delinquent classes." Others, such as Princeton's Frank Fetter and Cornell's Walter Wilcox, condemned Laughlin's broadness.[53] Laughlin's work—and the ire that it generated—was more than an exercise in nomenclature. His progressive critics easily recognized that Laughlin and the ERO created an expansive definition that could easily be used in future campaigns for the exclusion and isolation of degenerates and undesirables. Armed with this definition of degeneracy, Laughlin, in fact, turned the fight against social inadequacy represented by family strains like the Jukes into a struggle to turn back the tide of immigration. The arrival of immigrants, he sought to argue, was really the arrival of new types of Jukes, that is, new strains of the socially inadequate.

The Immigrant Juke

Eugenicists built upon and expanded older concerns about racial degeneration that had emerged from the study of industry. In particular, they remained concerned about race suicide and race substitution. "Today the inferior stock is vastly more productive than the superior stock," warned Davenport.[54] Eugenicists broadened these older fears in the new context of germ plasm: if the best sort reproduced at lower rates than their inferiors, their ingrained genius would become extinct.[55] The degenerate Nams, for example, outbred their betters. As Davenport and Estabrook warned, "The average fecundity [of the Nams] is much greater than that of the most cultured families of the eastern United States. Consequently, they are falling behind even such a degenerate population as this."[56]

The leading American eugenicists viewed the coming of World War I as an opportunity to evaluate the condition of the American stock and measure the relative increase of the unfit. They modeled their measurements and studies after British examinations of soldiers recruited for the Boer War. Such studies had provided ample ammunition for British eugenicists worried about the spread of degeneration. British recruiters were shocked at the condition of the working-class recruits meant to fill the ranks of cannon fodder. Nearly two-fifths were rejected as physically unfit, with weak muscles and underdeveloped chests. A growing fear of degeneracy—as much as the army's struggles on the battlefield—was the alarming outcome of the war. Parliament rushed to appoint a committee to study the deterioration of the urban working class.[57]

Though earlier British observers tended to reinforce the role of the environment in degeneration and made sweeping recommendations for the improvement of working-class health and housing, its focus on anthropometric measurements of recruits appealed to American eugenicists. They were concerned about a similar degeneration among American workers. World War I provided the opportunity to measure American rates of deterioration. During the war, eugenicists and their allies convinced the army to conduct far-reaching tests of recruits in order to determine the overall condition of the new American stock that had been reshaped by industry and by immigration. Estabrook, like many other eugenics experts, postponed his work among degenerate families to offer his skills to the military. Estabrook spent the war years—resplendent in his uniform—not at the western front but at home conducting psychological and physical examinations of drafted men. He paid particular attention to the peculiar condition of conscientious objectors and immigrants.[58] At the same time, the prominent psychologist Robert Yerkes developed intelligence tests to examine more than one and a half million recruits.[59]

In the short run, Yerkes' Alpha and Beta tests and related eugenics examinations determined military roles for recruits. In the long run, their results helped worried observers argue that the emerging American stock was degenerating. As Albert G. Love and Davenport—now in the uniform of a major—declared, the physical and mental tests conducted on soldiers revealed "the inherent failures in man to make complete adaptation to the rapidly advancing requirements of a highly artificial civilization." They argued that 468 defectives were found among every 1,000 recruits. Moreover, Yerkes as well as Davenport and Love agreed that immigrants displayed the lowest mental and physical levels.[60] The army tests helped move debates about eugenics from laboratories to legislative chambers, particularly as lawmakers worried that immigration might soon return to prewar levels.

Alongside wartime testing, the lessons of the Jukes gained new resonance for the study of immigration. As observers considered the results of the tests, they recast immigrants as new generations of Jukes. A study of a rural family that had lived in the United States for generations was used as evidence to restrict the arrivals of new, largely urban immigrants.[61] After all, noted Public Health Service's surgeon E. K. Sprague on the eve of the war, the Jukes once were immigrants as well: "If Max Jukes and others of his ilk had been prevented from entering this country, as was entirely possible, the saving to us is almost beyond comprehension."[62] The nativist atmosphere of the home front combined with the ominous findings of wartime testing to provide new pressure for immigration restriction. Moreover, the end of the war generated fears of an influx of new immigrant arrivals. By 1920, Laughlin had assumed sweeping powers as the eugenical expert of the House Committee on Immigration, led by the restrictionist Congressman Albert Johnson of Washington.[63] Laughlin used his position to launch an exhaustive study to determine whether Europe was exporting its socially inadequate citizens to the United States. His committee work led him across Europe and later to Mexico. Tests of army recruits during the war alongside the study of degenerate family strains remained his principal evidence. Laughlin depended on this evidence and his overseas explorations to argue for a quota of inadequacy that each race could justifiably contribute to the nation's population and still remain a positive element. In his earliest testimony to the House Committee on Immigration, he cited the examples of the Jukes, Ishmaelites, and Kallikaks to remind the congressmen that "there is such a thing as bad stock." The immigrant was a "reproducer"—and a fecund one at that—and was potentially the originator of a new degenerate strain. "The lesson is that immigrants should be examined, and the family stock should be investigated, lest we admit more degenerate 'blood,'" Laughlin told the committee.[64]

In subsequent testimony, Laughlin returned to the notion of "social inadequacy" and the evidence of the army tests. Like earlier twentieth-century efforts to define a racial hierarchy, Laughlin described sixty-five distinct races (up from the forty-two identified by the *Dictionary of Races or Peoples*). Then he examined the number of inadequates that each race contributed to the population. His results were predictable and confirmed his profound suspicions of immigrants. The army tests had already demonstrated that in essentially every category immigrants contributed more than their share to the ranks of degeneracy. Laughlin's testimony came on the heels of wartime and postwar suspicion of immigrants, but his testimony depended on already established ideas of immigration as colonization and invasion. Immigration was an "invasion" of inadequates.[65] Immigration historians have often, in broad terms, noted the importance of eugenics language and of Laughlin as a star witness in the hearings that led to passage of the 1924 Johnson-Reed Act. This law marked a shift in American immigration policy when it restricted European immigration to quotas of 2 percent of the number of people from a particular country that were living in the United States in 1890 and reinforced Asian exclusion. Historian Mae Ngai, in contrast, has argued that the central place assigned to eugenicists, especially Laughlin, in analyzing the act has "obscured from view other racial constructions that took place in the formulation of immigration restriction."[66] Such a complaint, however, makes sense only if eugenics is treated as a new and contained politics and science whose influence was limited in the reformation of the American racial imagination after 1924.

A closer analysis of eugenics in relation to an earlier generation of reform reveals that "hereditarianism" was able to translate restrictionist longings into the political reality of quotas precisely because of its ability to synthesize those other racial constructions into a powerful and expansive notion of unfitness. Laughlin's lengthy testimony—running to the hundreds of pages—did not represent a new and time-bound science; it was an amalgam of older fears of degeneration and race suicide, established theories of immigration as colonization and invasion, and the progressive study of degenerate families. In this context, Laughlin's testimony married the campaign for immigration restriction to the revision of reform, social work, and philanthropy.[67] Laughlin's references to the Jukes was partly a powerful rhetorical device; many, if not most, of the committee members would have been aware of the Jukes. His important position also marked the role of eugenics family studies in connecting concerns about immigration to fears of degeneration first articulated in an earlier progressive age.

As part of his work for the Johnson-Reed Committee, Laughlin took a leave from the Eugenics Records Office in 1923 and spent the year in Europe researching immigration. He set out to expand American fieldwork from native, rural surroundings to foreign shores, especially in places such as Italy that were the wellsprings of migration. At the same time, he conducted a study of the lineage of thoroughbred horses. The two topics were not unrelated. Both were studies related to the conscious breeding of superior specimens of the race. Just as the thoroughbred chart could reveal much about the future success of the racehorse, an effective study of the background of possible migrants could suggest their future condition in the Untied States. Thus, he proposed that applicants for immigration provide a family pedigree chart.[68] For Laughlin, eugenics was a totalizing, synthetic discipline. On the broadest level, it connected the different sciences. It also linked rural family studies to the breeding of racehorses to the restriction of immigration as part of a larger effort at race betterment. Immigrant exclusion was segregation. It was the moral, political, and biological equivalent of the home for the feebleminded or the tramp colony.

The 1924 act was merely a temporary triumph for Laughlin and his allies. He regarded its passage not as a solution to the immigration invasion, but as a foundation for future progress. Even as the bill was on its way to the printers, Laughlin and his allies were already extending their fears of immigration to African American migrants and Mexican immigrants whose arrival initiated a new round of industrial substitution. At the same time, they warned that the Russian Revolution and its worldwide reverberations were frightening evidence of the triumph of the unfit. For eugenicists, fears of degeneration and race suicide, born in the years before World War I, provided the framework for understanding the immigration and radicalism of the interwar generation, indelibly linking prewar nativism to postwar anti-Communism.

American radicals would be virtually silenced during the war. However, as eugenics had competed with reform earlier in the 1910s, radicals had already begun to rethink definitions of fitness and social inadequacy that mocked the manipulations of eugenicists. They revisited and subverted the meanings of savagery and civilization. A generation of prewar radicals recast the industrialist as a degenerate parasite, not as a eugenic strain who might save the race through progeny. At the same time, they heralded the tramp as the consummate class hero. Yet the war accelerated the eclipse of reform and the repression of radicalism. Accusations of wartime disloyalty joined criticisms that the reform impulse was leading to the defeat of the nation, not only on the battle front but also in the racial war fought in the workplace and bedroom.

8

FOLLOWING THE MONKEY

Blond Beasts and the Rising Tide of Color in War and Revolution

Forward-looking minds are coming to realize that social
revolutions are really social *breakdowns,* caused (in the last analysis)
by a dual process of racial impoverishment—the elimination of superior
strains and the multiplication of degenerates and inferiors. Inexorably the
decay of racial values eorrodes [*sic*] the proudest civilization....Said shrewd
old Rivarol, viewing the French Revolution: "The most civilized empires
are as close to barbarism as the most polished steel is to rust;
nations, like metals, shine only on the surface."

Lothrop Stoddard, 1922

Ben Reitman was just a child when he decided to follow the monkey, an organ grinder's pet. It was the first of many times he wandered and ended up in a police station. Looking back on the monkey in his 1925 unpublished autobiography, Reitman saw this incident as "typical of my whole life."[1] Reitman described his wandering after the monkey as the first incident in a career that would take him out of Midwestern immigrant slums and into the "hobohemia" of boxcars and the "jungles," a term tramps used to label outdoor camps. He would claim the title of "king of the hoboes," even as he began to pursue his medical degree. Reitman used his medical training to shift from living among "degenerates" to studying them. As his first step away from the "road"—the tramp life—Reitman joined the syndicalist upsurge of the 1910s. As the anarchist Emma Goldman's lover, Reitman would play a leading role in some of the first public mobilizations of the Industrial Workers of the World (IWW). The IWW was formed in Chicago in 1905 and made its name as a radical syndicalist union that was an alternative to the more conciliatory American Federation of Labor (AFL). The IWW gathered national headlines in 1912 when radicals, including Reitman, descended on San Diego to protest restrictions on soapbox speakers.[2] As the march toward World War I generated a domestic backlash against the radicalism of the IWW and Goldman, Reitman recoiled. He broke with the IWW, which had opposed the war, and split from Goldman. Once an engaged radical, Reitman became a sociological and medical observer. In jail during the war for distributing birth control pamphlets (not for opposition to the fighting), he held a "sociological clinic." In this clinic, he diagnosed "types" of prisoners for their particular form of degeneration. Upon his release, Reitman, the "king," became a doctor for degenerates. He set up his clinic in the "outcast" districts of Chicago where he treated both hoboes and prostitutes.

By the 1920s, Reitman was leading educational tours of the district, explaining to his fascinated guests the "factors that make men social outcasts." In his clinics and tours, he echoed concerns about the degeneracy of tramps and prostitutes. But, in an age of ascendant eugenics, he claimed to be able to revitalize the tramps, to turn them into "men." For prostitutes, by contrast, he sought not to preserve womanhood but merely to treat disease. Reitman's work as a doctor was not born-again reform. In fact, he had long been critical of reform and its proponents. He had criticized settlement house workers such as Robert Hunter, as well as progressive theorists such as G. Stanley Hall and Jacob Riis. Reitman was especially scornful of Orlando Lewis, the progressive reformer who had long sought to uncover the cause of tramping. Reitman's varied career, as a tramp, radical, hobo, doctor, and sociologist, provides an entry into radical—socialist and syndicalist— efforts to reconsider savagery in the face of a dominant science of eugenics. Even as he profited from the public fascination with and fear of degeneracy as a doctor, author, and guide, Reitman flirted with radical reinterpretations of degeneration at the end of the period of mass migration from Europe and Asia. His rereading of degeneration meant loving and, at times, heroizing, not reforming, the male degenerate. The tramp became a hero and the robber baron became a new kind of overfed degenerate, rather than a Teutonic icon. Reitman's career also highlights the profound limitations of this rethinking of primitivism. Even as he celebrated the tramp as hero, the prostitute remained an object of his pity and concern.

Although eugenicists in the era of World War I believed that they were recreating the marriage between biology and the social sciences, their success actually measured something very different. The Johnson-Reed Act suggested an ascendant and triumphant eugenics. In fact, eugenics represented rupture, the dissolution of an earlier intellectual consensus in which socialists, social Darwinists, and reformers could agree on the basic definitions of civilization and savagery. Although eugenicists imagined that they were reuniting the natural and social sciences, they did so by deliberately evicting those they considered unscientific and biologically dangerous. Because they had little or no place in conversations about eugenics, radicals reconsidered the meanings of progress and civilization. Their limited efforts to blur the boundaries between civilization and savagery announced the end of an era. But fears of degeneration would persist in new contexts.

Reitman recast the tramp as a hero at a time of rapid political and industrial change. Divisions between radicals, reformers, and eugenicists were growing fiercer than ever before. The cohesiveness of thought and theory of an earlier decade had fractured. It was replaced by the growing anxiety and antagonism of the war and its aftermath. The IWW faced brutal repression during the war years. Its reconsideration of savagery was muted. Instead, eugenicists were able to magnify fears about racial degeneration, race suicide, and immigrant colonization. They worried that the finest racial stock had perished at the front. They bemoaned the Bolshevik Revolution in Russia, the comparative strength of Asian powers, the growing industrial dependence at home on African American workers, and the restiveness of "colored races" still under imperial control. They united these disparate trends as evidence of a global threat of racial degeneration.

Eugenicists regarded even the passage of sweeping immigration restrictions in 1924 as a fleeting victory. As historians have noted, by the interwar years the terms of racial difference in the United States had perceptibly begun to change. The problem of color was emerging as the measure of the racial divide. Even those critical of past European migrants, such as the fiercely nativist and anti-Bolshevik journalist Lothrop Stoddard and the eugenicist Harry Laughlin, grudgingly came to define a unified white race. For Stoddard and Laughlin, this was a white race besieged from abroad by industrializing Asia and at home by new migrant streams of Mexicans and African Americans. In the face of new immigrants and new incarnations of radicalism, fears of degeneration and race substitution were transferred into the interwar years, when race and color finally merged.

Diagnosing the Tramp

Americans delighted in reading about the lives of tramps. Tramp narratives were a popular genre produced by journalists and academics who assumed a tramp disguise, such as Josiah Flynt Willard or Walter Wyckoff, and by former hoboes, such as Jack London or Nels Anderson. Reitman's autobiography, though unpublished, reproduced many elements of the tramp narrative. In the tramp narrative, these disguised or "reformed" hoboes described their seduction by the freedoms of hobohemia and their eventual exit, often into a world of education or politics. Tramp narrators claimed heroism, either as disguised brave outside observers or as survivors of the strenuous trials of tramp life who cast tramping aside in favor of radical politics or intellectual labor. Reitman's autobiography served as both a self-serving narrative and as a medical and sociological diagnosis of "wanderlust."

The acute tramp fears of the 1870s and 1880s gave way by the 1890s and the early decades of the new century to a mix of hostility, repulsion, and curiosity. A market emerged for dramatic narratives of tramp life. For reformers, such as the religiously minded John James McCook, the popularity of tramp narratives in magazines and book form meant an expansive public audience. Tramp narrators became cultural entrepreneurs. They were impresarios who profited from an insatiable public fascination with stories of adventure and survival. McCook started his career as a church rector and professor at Hartford's Trinity College, before beginning his life's work investigating tramping. He produced voluminous studies and surveys of conditions in tramp reformatories and lodging houses as well as of the causes of tramping. He was best known for amply illustrated magazine articles that detailed the adventures of tramps with whom he corresponded. Most notably, McCook's detailed letters with the tramp William Aspinwall provided grist for the nine-part serial "Leaves from the Diary of a Tramp."[3] McCook patrolled the class boundaries between the writer and his informants. In the process, he allowed—consciously or not—his correspondents a voice in their own description. Aspinwall, for example, often chided McCook for his conclusions and used his letters to correct, for example, the public equation of the bum and the tramp. Aspinwall, no less than McCook, capitalized on a public fascination with tramping.[4]

McCook's articles were produced through the comfort of correspondence. By contrast, Josiah Flynt Willard's even more colorful narratives, especially his 1899 *Tramping with Tramps,* were based on his adventures disguised as a tramp. He spoke a distinctive tramp language, rode the rails, begged, and faced imprisonment. The plot of his narrative was simultaneously about his escapades of tramping and about his subterfuge. With manly pride, Willard claimed that his disguise as "Cigarette" was undetectable to genuine hoboes and, therefore, was an honest reflection of the submerged tramp life: "with a complete abandonment to his surroundings; no tramp has ever known that 'Cigarette' was not really a tramp."[5]

Former tramps, Jack London in particular, profited from public fascination with the hobo life. London began writing *The Road* more than a decade after he had left hobo life. His stories of life as a "profesh," the highest class of hobo, took their place alongside his popular Klondike adventure narratives.[6] In the essays published in *Cosmopolitan* that he joined to form *The Road*, London cast himself as a hobo "hero" who outdid other hoboes in his feats. He thwarted a complex system of local constables and railway police whose job was to thwart hoboes as they hopped freight trains. The essays chronicled his ascent to the top of tramp hierarchies, first as the most successful of the "road kids," a gang of mere children seduced into the tramping life, and, later, as the only hobo capable of outwitting the most zealous of railway guards. Like his Klondike stories, London's hobo adventures followed a naturalist script of Darwinian competition. As a road kid, London joined a gang trying to hop a train that would go "over the hill," that is, over the Sierras out of California, a critical test for young hoboes. London defended his natural fitness. He boasted that he was the only road kid who avoided being "ditched," thrown off the train by guards. He compared his success with the fate of French Kid, whose legs were crushed by the train when he was ditched.[7]

For London, hobohemia was a world of savage competition for survival. He struggled to the top of a hierarchy from "road kid" to "profesh," repelled unwanted sexual advances, scrambled for food, and battled railroad guards and policemen. To achieve the rank of "profesh," he claimed to have scaled an alternative class and race hierarchy: "And be it known, here and now, that the profesh are the aristocracy of The Road. They are the lords and masters, the aggressive men, the primordial noblemen, the *blond beasts* so beloved of Nietzsche."[8] In a world untouched by philanthropy, a grassroots eugenics had taken hold. When settlement workers and eugenicists declared the tramp as beyond philanthropy or reform, the hobo necessarily fought for survival. Even a night's sleep became an evolutionary quest. London described one epic struggle with zealous train guards. The struggle began when a crowd of hoboes attempted to hop a freight train. As the other hoboes were progressively ditched, London outwitted his pursuers through both physical and mental feats.

Similarly, London proved his fitness during a short stint in jail. The jail, with its brutal guards and meager rations of greasy water and bread, represented the ultimate Darwinian laboratory. London outwitted the guards and smuggled precious tobacco into his cell. With his feat acknowledged by other prisoners, he became a hall-man. Hall-men were the select of the prisoners, charged with keeping order. In his elite position, he engaged in graft to horde ample food and tobacco for himself: "Here were thirteen beasts of us over half a thousand other beasts...and it was up to us thirteen there to rule. It was impossible, considering the nature of the beasts, for us to rule by kindness." Like the capitalists outside, "we were economic masters inside our hall...quite similar to the economic masters of civilization."[9] He cast the hobo life as an adventure in order to subvert the notion of tramping as degeneracy. Yet he admitted—even celebrated—its base savagery. In *The Road*, bruises from battles with police became "savage welts," fellow tramps appear as "bovine and stubborn...brutes," and the Main Stem, a city's hobo quarter, was a "world primeval."[10] This savagery marked hobohemia as a eugenic laboratory—albeit one that was beyond civilization.

The Road is also a tale of seduction. London and other tramp narrators deliberately employed the argot of tramps. The language of tramping, for London, encapsulated the seduction of the road. The use of tramp language gave readers a sense of entering a parallel world of tramping. Words and signs scratched into posts or railroad water tanks provided tramps with covert information about where to collect handouts or about local policemen. Fluency in tramp language, as Willard noted, also marked one's standing in the tramp hierarchy. Willard even provided a lengthy glossary of hobo language.[11] London described his sexualized seduction into hobohemia by relating his fascination with the language used by tramps. In an erotically tinged description, London related swimming with "road kids" and listening to their language: "It was a new vernacular. They were road kids, and with every word they uttered the lure of The Road laid hold of me more imperiously."[12] When he used such language in his writing, London—like so many other tramp narrators—helped his readers vicariously experience the same physical seduction. London subtly induced "wanderlust." Yet while his readers could only wonder how they would fare in the trials of the road, London had thrived.

Wanderlust may have been savage, but its expression plunged the tramp into a desperate struggle for survival. Tramp narrators defined a hierarchy of hobohemia and consciously subverted Lewis's and Edmund Kelly's negative categorization of tramps as degenerates. For Reitman, his elite position was confirmed by his self-proclaimed title as King of the Hoboes.[13] London and his gang of road kids, meanwhile, preyed on the unfit, especially drunken tramps. In a language that evoked the Darwinian world of wolves, London described an attack on a drunken tramp: "No hunting cry is raised, but the pack flings itself forward in quick pursuit."[14] Tramp narrators such as Anderson, London, and Reitman also proved their superiority, paradoxically, by tracing their escape from the road. Anderson, for example, described his arrival at the University of Chicago. After a quick shower, he passed seamlessly from hobo to academic.[15]

When London, Reitman, and Anderson cast aside tramping and, later, described their experiences they became—like disguised investigators— mediators and translators. Anderson used his tramp past in the service of his academic work. He conducted sociological studies of American homelessness. He would later reassume his tramp persona to write a popular and partly fictional memoir of tramping.[16] He mediated between hobohemia and the outside world. For Wyckoff, participant observation helped identify the moral failings of tramps. For Reitman, by contrast, his in-between position allowed him to critique the established strategies of both reformers and eugenicists. Reitman even employed the tramp disguise to evaluate reform. He turned a classic technique of social reform investigation against itself. Reitman—the doctor—donned his tramp disguise to test the benevolence of Chicago's reformers. When he sought work at charitable institutions, he was turned away. Hull House handed him bread and butter, but no job. The Chicago Bureau of Charities sent him to employment agencies that were notorious for their corruption. Graham Taylor, the editor of the Chicago *Commons,* a leading reform journal, directed him to the municipal lodging house, which was known for its horrific conditions. Raymond Robins, a leading advocate for settlement work, simply dismissed him as a "bum."[17]

Even as Reitman sought to undermine reform as ineffectual, he still engaged with ideas about degeneration.[18] Looking back on his own life as a tramp, Reitman was willing to label himself as a "social outcast." He may not have been a "degenerate," but, as he would write from jail to his young son, he certainly dwelt among them: "Little Ben, I wish for you a beautiful, big life. I hope you will love people, all kinds—that you will especially love poor people, the tramps, the criminals and the outcasts—men whom your father devoted most of his life to." He would spend his time in jail substituting sociology and medicine for tramping. When he diagnosed his fellow inmates as specific types of degenerates, Reitman removed himself from their world. He was jailed among confirmed degenerates, but, as the diagnosing observer, he was not one of them.[19] Yet he remained proud of his hobo past. As late as the 1930s, Reitman still defended his title as the "king of the hoboes." At the same time, he led tours of the degenerate quarters of Chicago, especially the Main Stem.

The Hobo Hero

When socialists and syndicalists discovered that their solutions to the problem of degeneration were discredited, they increasingly valorized the hobo as a proletarian hero. The roots of their revision, in fact, could be found in earlier radical, but still condemnatory, explanations for tramping. Among socialist voices in the late nineteenth century and the early years of the twentieth century, the newspaper *Appeal to Reason* was loudest in its denunciation of tramping as a social problem. The socialist Julius Wayland started the popular paper in 1895 in Girard, Kansas, close to the grain fields that seasonally drew itinerant workers riding the rails. The newspaper described the "unmanly" tramp in familiar terms: "The tramp is often dirty, uncombed, and generally unpleasant." The newspaper blamed tramping not on inherited degeneracy but on a "social system [that] has failed." It even repudiated those who sought the colonization of tramps. The rich were ultimately responsible for the problem of tramps: "Find, brother, the seed that produces a tramp and by removing or preventing its germination you can remove the tramp—and by no other means. In analyzing the case you will have to pass through stratas of drunkenness, shiftlessness, etc., until you come to the millionaire."[20]

Unlike Wayland, by the 1910s radicals such as London and Reitman admitted the role of heredity in shaping the hobo. In his "Evolution of the Hobo," for example, Reitman criticized the attempt of "some scientists" to divide tramps into "Atavistic Insane and Degenerate. But while this is a very bad classification it must not be overlooked. Any one who makes a study of the Hobo and overlooks the influence of Heredity…makes a serious mistake."[21] For Reitman and London, their own tramping had been caused by the natural desire to wander. In the idea of "wanderlust," radical observers such as Reitman found an alternative to Charles Davenport's definition of "nomadism" as a congenital defect. For Reitman, wanderlust was little different from the comfortable tourism of the rich. He blurred distinctions between the hobo and the migrant worker and argued that as "animals travel in search of food, water and mates" so too might a casual laborer tramp in search of work. In their own expression of the hereditary "migratory instinct," the wealthy traveled between their different residences in the comfort of their automobiles. Even the wealthy followed the monkey.[22] In this context, the monkey played a key symbolic role. In following the monkey, Reitman reached back across evolutionary time from modern man to prehistory and then to primates in order to define the migratory instinct as natural.

The radical rereading of the hobo depended on the juxtaposition of the tramp and the degenerate boss. The *Appeal* imagined a conversation with "the opponents of Socialism" about "the people who won't work." The newspaper argued that the "man who works" paid for jails for the tramp and for the "fine homes of the "millionaire idler." In a cartoon, the newspaper compared the tramp to the millionaire. Both are corpulent and stand in a similar pose. Readers could easily have identified the signs of degenerate lassitude.[23]

In similar graphic form, the *Appeal* recast the idle tramp as a class victim. It blamed the idle millionaire for the tramp's fate. In a ten-part cartoon series, the newspaper traced the lives of Jim, the tramp, and James, the millionaire. Both were born at the same time, but Jim was forced onto the roads. He perished by the roadside. His body became first a meal for buzzards and then a cadaver for the dissecting theater. James, meanwhile, was idle, alive, and wealthy.[24] The *Appeal* linked the tramp and the millionaire in order to recast the wealthy not as eugenic heroes but as degenerates.

In questioning the fitness of the millionaire, the *Appeal* tapped into a strain of thought that denounced the overfed factory owner and, especially, his pampered wife as degenerates. Thorstein Veblen's 1899 *The Theory of the Leisure Class* combined a sardonic tone with familiar evolutionary logic. Veblen argued that the elite represented a vestige of primitive conquest. As tribal conquerors refused menial work, so too did the industrial elite disdain productive labor. They engaged instead in leisure and consumption. The "leisure class" was a parasite on the "industrial community"— much like the idle tramp. They were atavistic: "the class is in large part a heritage from the past, and embodies much of the habit and ideals of the earlier barbarian period."[25] Veblen's text, though hesitantly saluted by other social scientists, went through multiple printings. It appealed particularly to tramp narrators such as Harry Kemp, who gleefully chided the Newport vacationer: "Don't tilt your nose skyward in superior fashion when…you see the tramp passing by. For he, too, is of the leisure class."[26]

Although Veblen targeted an entire social class, his logic was often employed by others in the critique of elite women. The reform journalist William Hard echoed Veblen's critique of consumption when he bemoaned the elite woman of leisure whose days were filled with shopping and who retreated to her bed at the slightest headache.[27] The special focus on female leisure was critical insofar as it allowed radical and socialist observers to call into question the manhood of the robber baron, all the while recasting the hobo as a manly class hero. The *Appeal*, for example, equated James's idleness with that of his wife. The bloated James luxuriated in a "soft bed." He demonstrated none of the characteristic bourgeois ideals of virile manhood. Corpulence took on gendered political significance, as a visual cue about parasitic leisure and soft manhood.

As socialists and, later, IWW organizers celebrated the hobo as fit and the robber baron as degenerate, they reversed eugenic notions of select family

Figure 8.1. "Hobo Banquet": In 1907, Ben Reitman hosted a hobo banquet in a hotel ballroom, subverting the standards of elite leisure. He boasted that the best-known tramps attended. (Courtesy: University of Illinois Chicago Library.)

strains that followed class lines. This was a reversal that Reitman performed with subversive aplomb during his highly publicized 1907 "hobo banquet." Reitman commandeered a hotel ballroom and transformed elite leisure into a haven for hoboes. Complete with fancy dress and a menu ("Boiled Fresh Halibut, Parsley Sauce," "Pommes Parisienne," "Fromage de Brie and Wafers"), Reitman gathered hoboes from local saloons and the courthouse. He proudly reported the attendance of well-known tramps. He listed their monikers, or road names: Philadelphia Jack, the Dancing Kid, New York Slim, and others. This parody of elite life was completed by "several incidents." Fred the Bum, scheduled to give an after-dinner toast, was found drunk under a table and Shoe-String Chase took advantage of a pause in the program to pass his hat.[28]

While the *Appeal* described the tramp with pity, radicals, such as London, who supported the IWW glorified the tramp as a working-class hero. In 1907, the IWW began its resurrection of the hobo when it claimed Karl Marx himself as a "hobo organizer and tramp agitator. He was a sheeny Dutchman, by name, Karl Marx."[29] London, meanwhile, distinguished between the tramp and the hobo. He demeaned the tramp as a common drunk and celebrated the hobo as a conscious subversive and occasional migrant worker.[30] For the IWW, the hobo was the ideal industrial unionist, the "knight-errant of the social revolution." In his letters to the IWW newspaper, the union's West Coast organizer J. H. Walsh promised to raise a hobo "militant industrial army." His army would ride the rails to Chicago for the IWW convention. It would organize the "Mulligan bunches," as he termed tramps gathered in the jungles. As the journey began, his dispatches took on the tone of a tramp narrative, complete with details of evading "shacks" (railroad police). For Walsh, organizing had been grafted onto the narrative of tramp survival.[31] By 1914, the IWW was marveling that "the nomadic worker of the West embodies the very spirit of the I.W.W. His cheerful cynicism, his frank and outspoken contempt for most of the conventions of bourgeois society…make him an admirable exemplar of the iconoclastic doctrine of revolutionary unionism."[32] Even Reitman, despite his growing distance from the union, recognized that the IWW had become ubiquitous along the "road." The hobo, he argued, had "evolved" into a "powerful…man." "To-day if you will get into a box car," he marveled, "you will almost imagine that you are in the Socialist or I.W.W. meeting." In fact, in 1911 the IWW celebrated the seizure of a train by two hundred hoboes and IWW members. They traveled unmolested from Portland, Oregon, only to be arrested in California.[33]

Where the *Appeal* equated the tramp's degeneracy with the boss's flabby decadence, the IWW and its supporters would cast the hobo as a muscular hero. They still condemned bourgeois degeneracy in order to subvert the meaning of savagery and primitivism. As recent historians of the IWW have noted, the union's artists were pioneers in representing class-conscious workers as muscular, fit men. Such artists compared the muscular union man to the stooped and lean unorganized worker and to the rotund boss. Such images were crucial to the IWW's voluminous and amply illustrated press, available to workers and hoboes in numerous languages. The union man hardened his muscles and manhood when he rejected strenuous labor. IWW members and hoboes may have worked the harvest fields and even in factories, but the union's literature celebrated their unwillingness to endure the draining demands of industry. London described his conversion to socialism during his tramping days as a physical experience. As he set out along the road, he saw himself as a "MAN." He described himself as "raging through life without end like one of Nietzsche's *blond beasts*"—one of London's favorite images—"lustfully roving and conquering by sheer superiority and strength." Yet on the road he saw once "*blond-beastly*" men twisted

THE JUDGE'S ASSOCIATE.

The cave-dweller may be dead, but his voice is still heard daily in our law courts, where learned and cultured judges are employed translating his blood-stained tradition into gentlemanly English.

LABOR WANTS A "PLACE IN THE SUN!"

Capital (deeply shocked at Labor's efforts to emerge): "Back to your abyss, Sir! As it is already there is scarcely enough sun to go round!"

"GIVE US THIS DAY—"

Master Baker: Give us this day workmen pure in heart, meek in spirit, as soft and pliable as the dough they daily punch; and punish the wickedness of the agitators who want them to rise.

"SERMONS IN NUTS."

The Simian Philosopher: No, my child, never speak slightingly of Evolution. It is merely through evolutionary development that we have acquired those higher moral faculties which make it possible for us to luxuriate behind nice iron bars that completely shut out the Capitalists.

Figure 8.2. "Will Dyson's Cartoons": These four cartoons mocked the idea of elite evolutionary fitness, as a caveman appears as a judge's assistant and the capitalist appears as fantastically corpulent. The proletarian appears either as the muscle-bound revolutionary or the clever simian that has separated himself, behind the bars, from the capitalists. *International Socialist Review* (1915).

by the "dignity of labor." His conversion to socialism, then, occurred when he became a profesh, refused hard labor, and protected the "muscles of my body."[34] The IWW similarly celebrated the radicalism of the hobo through evocations of bodily strength, undiminished by factory labor: "Look at our muscles—they tremble with the eagerness of creation." At the same time, the union readily accepted the condemnation of the hobo as primitive: "I am a hobo, a scum-proletarian, a man of the abyss. I am uncivilized, primitive, brutal." The primitive man, though, was a revolutionary: "I believe in direct action."[35]

At the time when it was most sympathetic to the goals of the IWW, the *International Socialist Review* printed four cartoons by the Australian syndicalist Will Dyson. American radicals selected Dyson's four cartoons to question the evolutionary fitness of the elite and to venerate the masculine strength of the revolutionary hero. Dyson's images of muscles, fat, and monkeys would have been familiar to American readers. When he suggested that the judge was merely a "cave-dweller" in robes, Dyson—perhaps deliberately—recalled Veblen's notion of elites as savage survivals. Meanwhile, the monkeys in a zoo, like the "primitive" hobo, claimed a higher evolutionary fitness than their capitalist observers. These cartoons of reversed evolution sandwiched two pictures of fabulously flabby capitalists. In the first of these cartoons, the girth of "Capital" blocked out the sun while he implored a muscular "Labor" to return to the "abyss." Taken together, as the journal intended, the images highlighted the way radicals found themselves condemned by mainstream ideas about heredity. They were free, therefore, to present images of evolution reversed and distorted. The members of the elite were degenerate, the dwellers of the abyss were fit, and monkeys were worth following.[36]

Masculinity in the Tramp Jungle

The IWW union man became a masculine hero through his refusal of the hard labor cherished by bosses, reformist advocates of gymnasiums, and antitramping crusaders. The hobo naturally emerged as the consummate IWW hero. As the IWW declared, hoboes were "the knights of the road." The veneration of the hobo depended on a reinterpretation of manliness, not as a set of gendered moral and chivalrous codes but as masculinity, a combination of muscular physicality and a rejection of capitalist domesticity. The hobo "has too much manhood to be willing to half starve a wife and children."[37] The hobo hero lived outside of the family in a homosocial—and frequently homosexual—world of men. When "Sanny M'Nee" defended the tramp in the *International Socialist Review,* he outlined a proletarian eugenics that questioned family and childbearing. His article imagined a conversation between a tramp and a "decent" ploughman. The ploughman mouthed the elite critique and admonished the tramp to find work. The tramp, meanwhile, questioned raising a family: "You marry a wench and set up a beggarly house." He insisted that the children of the poor were "The Children of the Dead End." When the IWW saluted hoboes as the "guerillas of the revolution—the francs-tireurs of the class struggle," it linked their freedom from wage slavery with their rejection of family: "His mobility is amazing.…He promptly shakes the dust of a locality from his feet whenever…the boss is too exacting.…No wife or family encumber him."[38]

The hobo subverted middle-class chivalrous manhood. The success of a profesh could be measured by his ability to win handouts from unsuspecting women at kitchen doors. The hobo violated female domestic space and converted women from the focus of manly chivalry into victims of masculine deception. Tramp chroniclers regularly boasted about the lies they developed that won them ample food and occasional money. Graft, generally won at the expense of women, was critical enough to the hobo economy and to its standards of masculine success that its different forms were enshrined in the specific hobo language. A "hand-out" was accepted, but a "sit-down"—a hot meal served in a duped woman's home—was cherished. London, for example, began *The Road* with an apology to a woman in Reno who was the victim of his graft. His apology was tinged with pride: "I once lied continuously, consistently, and shamelessly, for the matter of a couple of hours." London recited to his unsuspecting victim a complex story of pathos and domestic tragedy.

In this case, he feasted on bacon, biscuits, and eggs. But graft often meant more than winning food. Tales of graft recited to other tramps were tinged with eroticism. The hobo-turned-academic Nels Anderson recorded one typical story. The hobo narrator claimed that while "scrounging for eats," he won not only a "real sit-down" but also sex. When he departed, the woman told him, "'Don't you ever come back again. If my husband sees you he will shoot you.'" His story was met with disbelief from his audience who claimed to have heard the story three times already. For the storyteller, this was an opportunity to extend the boast: "So you may…and I told it many times. Others are stealing it from me."[39]

As the IWW enshrined hobo manhood, the rejection of domesticity, the subversion of chivalrous manhood, and the sabotage of capitalism merged.

Figure 8.3. "A 'Set-Down' with two maiden ladies…": This domestic tableau demonstrates the expression of tramp masculinity: winning food from unsuspecting women. This image illustrated London's proud claim that he lied regularly to women in order to win a hot meal. Jack London, *The Road* (1907).

In the laws of the road, handouts grafted from women were returned to the male-only space of the hobo camp, known ironically as "jungles." The word jungle had complex meanings in the radical vernacular. It evoked not only the primeval setting of popular novels set in prehistoric or racially backward lands but also Upton Sinclair's muckraking of Chicago's meatpacking slums (first published in serialized form in the *Appeal*). Radical observers mocked dominant notions of racial hierarchy when they memorialized the hobo jungle. Like Dyson's cartoons, the notion of the hobo jungle distorted and reversed evolution. "Did you ever compare the life of the migratory worker…with the life of man in the ages before the dawn of history and note the similarity?" asked the IWW. "The life of the jungle is the life of man, in his earliest stage." It was also the life of the hobo. As prehistoric man sought a cave and battled wild beasts, the hobo flopped in a boxcar and battled with railroad "bulls" (another term for railroad guards). "Workers," it concluded, "still suffer the jungle life."[40] The IWW idealized the savagery of the hobo jungle by comparing it to an Indian camp. The hobo in the jungles lived "the carefree life of the American Indian."[41]

For the IWW, the "jungle" was also a space of mutual aid and homosocial manhood. "It is often a marvel of cooperation," complete with strict
rules and mores. "Usually the procuring of food in such a camp is reduced
to a system such as would interest economists and sociologists." Food was
returned to the camp for the preparation of a "mulligan stew."[42] Reitman
remembered the jungles for their mutuality, but also for an order that
was—as in the tribe—enforced physically. He remembered his first visit
to a jungle near railroad tracks in Lima, Ohio. It was a pleasant shaded
camp, lit by a "cheerful fire." At the order of an older hobo, food and drink
were commonly divided. Younger tramps were sent into the city in search
of food and tobacco to round out the meal. For Reitman, the meal was a
"real banquet. I have wined and dined in many places, in great banquet
halls with scientists and statesmen at my side.... But I do not believe that
any of those meals satisfied my stomach, stimulated my brain and gave me
a better sense of fellowship than the meals I have eaten in the jungles with
my hobo friends." Still, infractions, like holding back money or food, were
met with brutal beatings.[43] The IWW's *Industrial Worker* even introduced
a series of articles that reviewed conditions in different jungles. The newspaper proposed a "Baedeker's" guide to the jungles and provided information on water, food, jails, and the character of local police and judges. The
articles read as a parody of the popular middle-class genre of the tourist
guide. "The next stop of any size," the newspaper declared, "is Ritzville. The
jungles are out at the fairgrounds.... There is a sheriff in Ritzville, but he
is generally out fishing. Chewings are fair in Ritzville."[44] Such a parody recalled the comparison of the tramp and the idle rich and evoked the idea of
wanderlust. The IWW offered an image of the "The Jungles for Workers"
above a photo of a lavish banquet room "for loafers."[45]

The IWW also celebrated the jungles as hotbeds of organizing. Walsh
boasted about the amount of literature his militant army sold in the jungles.
The jungles were, for Walsh, sites of manly solidarity: "All fellow workers
can get a meal at...the jungles—free of charge. Many a poor hungry devil
has been fed by the boys around the camp fires." He concluded one of his
most colorful dispatches with an evocation of the jungle as the epicenter of
IWW organizing: "I must close to join the revolutionary forces on the street
who are now congregating, after a big feed in the jungles."[46]

The hobo jungle and the boxcar were sensual spaces, tinged with the
eroticism of close, physical relations. Tramp narrators and IWW commentators proposed hierarchies of the road. They differentiated between
tramps, bums, and hoboes. Their classifications placed profeshs at the
top with "gay-cats" and "punks" at the bottom. Almost universally, tramp

THE JUNGLES FOR WORKERS.

BANQUET ROOM FOR LOAFERS.

Figure 8.4. "The Jungles for Workers, "Banquet Room for Loafers": The IWW continued to compare tramps and millionaires as part of their touristic review of conditions in hobo jungles. The union celebrated the hobo as a manly proletarian hero. *Industrial Worker* (1909).

narrators recognized the same-sexuality of the hobo world and assigned distinct sexual roles to the different categories of hobo. In particular, they noted the role of "jocker" and "punk." As Reitman noted, the jocker served as an older male mentor for the younger punk, but also relied on the punk for penetrative sex. "The boy does not need to remain long in hobo society to learn of homosexual practices," noted Anderson.[47] London's narrative of the road also captured the erotics of the homosocial space of the box-car. "Never was there such a tangle of humanity. They were all lying in the straw, and over, and under, and around one another," he described one ride. "Eighty-four husky hoboes take up a lot of room....Their bodies heaved under me, like waves of the sea."

Hobo masculinity, especially when expressed to the wider public by tramp narrators, existed within this reality of same-sexuality in contradictory ways.

London defended his claims to a position at the top of the tramp hierarchy by denying that he ever was a jocker's punk: "I did not take kindly to possession."[48] Anderson, meanwhile, turned to Havelock Ellis, the British sexual psychologist, to explain the "types of perverts" among tramps. Citing Ellis, Anderson suggested that some were "different temperamentally and physically from the rest of us." Others "substituted homosexual for heterosexual behavior." Anderson's reliance on the work of a known reformer and social scientist and his dismissal of sexual practices as "perversion" highlighted how he distanced himself from his past hobo life. Tramp narrators expressed manhood and heroism by claiming that they never fell into the role of punk or gay-cat. Reitman proudly recounted that a fight he started in a hobo jungle allowed him to escape unnoticed from his potential jocker. For Reitman, his escape from a sexualized tutelage symbolized his coming into manhood as an independent tramp. In a later essay, Reitman recalled yelling at "a bunch of fagots" that he encountered on a New York bus. "You cheap homos," he screamed, "you haven't any right to parade your sex irregularities." As he sought to categorize "homos," Reitman distinguished between "a distinct biological and glandular type" and the hobo punk and jocker.[49] The 1928 silent film *Beggars of Life*, loosely based on the tramp Jim Tully's 1924 memoir, adapted this same ambiguity for the screen and the wider public. In the film, a male tramp helped a young woman escape after she murdered her sexually abusive adoptive father. When they traveled together, she wore a young boy's outfit. The film captured their budding love when it pictured them sleeping side by side inside a cozy haystack. Because the film does not show her donning the disguise, they appeared more as a jocker and punk then as man and woman. Gendered order was only restored at the film's conclusion when she shed the boy's clothing she had used to disguise herself and put on a dress. Only then, did they proclaim their mutual love.[50]

Radicals revisited the meaning of savagery and degeneration beginning in the 1910s through the expression of proletarian masculinity. Critics of the IWW were denigrated as "ladylike men" and addressed directly as "dearie."[51] In this context, the valorization of the hobo was not mirrored in a rereading of the prostitute. For the IWW, the prostitute remained a symbol of capitalist immorality. The union still denigrated prostitution as the "Social Evil" in a familiar language of degeneration: "Many girls…sink to a lower plane than the animals."[52] Reitman, who remembered happy times as a child spent in the company of prostitutes, articulated distinctly different strategies for tramps and prostitutes. For hoboes, he organized banquets; for prostitutes, he offered his services as a doctor who treated venereal disease. Moreover,

Figure 8.5. This scene from a film based on tramp Jim Tully's 1924 memoir captured the same-sexuality of hobohemia. The male tramp lies next to a young woman, disguised as a young boy. The scene evoked stories of male jockers with their punks. *Beggars of Life* (1928).

the heroic role assigned to hoboes was open to white men, the blond beasts. The cleverness of the white tramp was often contrasted with the foolishness of African Americans, who may have inhabited hobohemia but were not fully admitted into the hallowed circles of the road. If African Americans appeared in tramp narratives at all, it was as victims of hobo graft. The IWW, even as it printed its materials in multiple languages, evoked an unblemished hobo proletarian whiteness. The ethnic background of individual tramps was rarely mentioned in tramp narratives. It was visible only in the occasional tramp moniker, such as Frisco Sheeny or Michigan French. Still, the representation of the class-conscious worker in the form of the muscle-bound hobo would remain important even as the IWW itself drifted toward political irrelevancy in the face of wartime repression.

Une Nation Blessée: War, Reaction, and the Rise of the Underman

Compared to its European allies, the United States suffered relatively few deaths during World War I. Two of its casualties, however, were organized reform and radical trade unionism. During the war and continuing nearly unabated in the years of reaction that followed, reform leaders found themselves silenced. With eugenics ascendant, reformers disappeared from the pages of popular and academic journals alike. Even more dramatically, radical unionists, especially members of the IWW, were targeted in a series

of raids, court trials, and expulsions. The Bolshevik Revolution further increased fears of radicalism. Antiradicalism would continue in the postwar
years with raids on radical groups ordered by Attorney General A. Mitchell
Palmer. The spectacular rise of reaction during the war was the public flow-
ering of a process that undercut the intellectual justification for reform and
had begun more subtly in the early 1910s.

The nation's leading eugenicists, such as Arthur Estabrook and Charles
Davenport, enthusiastically enlisted in the war effort. Wielding eugenics in
time of war kept Estabrook and Davenport far away from the western front.
By contrast, the IWW actively encouraged its members to resist the wartime
draft and the Socialist Party (unlike its European counterparts) remained
opposed to the war even after its leader, Eugene V. Debs, was thrown in jail
and its elected officials refused their seats. As it became clear to American
observers as early as 1914 that the war in Europe would be costly and long,
eugenicists advanced arguments that war was dysgenic. The effects of war
magnified the dangers of immigration and radicalism.

The eugenicist David Starr Jordan was the first to discuss the racial effects of war. He had used his position as president of Stanford University
during the Spanish-American War to warn that war carried off the most
energetic and fit of the race. He resurrected these same arguments again
during the slow march toward U.S. entrance into World War I. For Jordan,
war was a primitive anachronism, a vestige of a barbarian time when "every
man's hand is against the other" and "the life of every man and woman is
a tragedy." Jordan argued that the evolution of civilization was toward cooperation in place of "compulsion." Whereas reformers had used such logic
to defend settlement work and progressive politics, Jordan employed it to
argue that war was a savage remnant. Tribal wars, holy wars, and feuds, he
insisted, had dissipated. What remained was "killing on a large scale…on
the line where great nations meet."[53]

Such mass killing, for Jordan, was a crisis of the present and a burden
on the future. He believed that war carried away young men of "courage,
alertness, dash, and recklessness. The men who are left are…the reverse
of all this and it is they that determine what the future of the nation shall
be." Jordan warned that "if we send forth the best we breed, there is no
way those of the future shall be other than second best." War left a nation
biologically "crippled, *une nation blessée*." Rome, he asserted, fell because
the superior old stock had fallen on the battlefield. The inferior survived to
run and ruin the empire. For a more modern example, Jordan turned to an
emerging power, Japan, which had only recently defeated China and then
Russia. Victorious on the battlefield, its "race-heredity" was still threatened.

Jordan prophesized that in "two hundred and even twenty years" the newly militant Japan was unlikely to remain "virile."[54] On the eve of the Great War, Jordan blamed the persistence of slums and the physical degeneration of its inhabitants on the selection of the unfit that had accompanied past wars. London's expanding slums were proof enough that war was a source of race deterioration.[55]

Jordan compiled his monumental *War and Waste* in 1913 and republished it in 1914. Yet he could not fathom the imminent slaughter of the world war. The Balkan War may have engaged all of the European great powers, but a larger war, he insisted, was impossible: "Europe recoils and will recoil even in the dread stress of spoil-division of the Balkan war."[56] His foresight may have been questioned, but his logic was critical for those who, while patriotically supportive of the nation's war aims, still worried about the war's racial perils. Among reformers, Alice Hamilton, for example, warned that the Great War was a form of race suicide that was carrying off the nation's most fit.[57] American eugenicists readily condemned socialists for their opposition to the war, while in the same breath they echoed Jordan. The biologist Seth K. Humphrey sounded familiar themes when he described the biological disaster of the war: "It is safe to say that among the millions killed will be a million who are carrying *superlatively* effective inheritances—the dependence of the race's future....They are gone forever. The survivors [including, of course, Humphrey himself] are going to reproduce their own less valuable kind."[58] Robert DeCourcey Ward cited Jordan in the pages of the *Eugenics Review* to argue that war "impoverishes the breed." The strongest were slaughtered at the front. Meanwhile, "the weaklings live, marry and continue the race." Ward drew on evidence from past European wars to warn that future American babies would become "an inferior lot of men." For Ward, the inevitable racial deterioration of the postwar nation endowed the restriction of immigration—his principal cause—with greater urgency. The fragile racial stock of the nation could hardly withstand a further wave of immigrant invaders.[59]

By 1918, the *Eugenical News* had identified a new racial threat that had emerged out of the war. The notion of degeneration helped link prewar immigration restrictionism to postwar anti-Communism. For American eugenicists, the ascendance of Vladimir Lenin and the Bolsheviks confirmed fears that war was leading to a selection of the unfit. Within months of the Bolshevik Revolution, eugenicists were struggling to understand the heredity of the Communist. In August 1919, for example, the *Eugenical News* offered a review of Rose Cohen's autobiography, a 1912 book that chronicled the young Jewish immigrant's sweatshop struggles. In the eyes of the eugenicist, her autobiography revealed the heredity of the new radical: "Individualistic, selfish, neurotic, she had all the elements necessary for radicalism and revolution."[60] A few months later, the journal published a front-page analysis of Lenin's heredity. It suggested that he was a "hyperkinetic" who preferred action over thought and logic: "He does not stop to think because the basal part of the brain dominates over the cerebrum." Meanwhile, another eugenicist offered an oft-cited medical analysis of the racial effects of the war. He connected concerns about the slaughter of the fit, older fears about the annihilation of space that made migration too easy, and new worries about revolutionary socialism. Socialistic races from eastern Europe might migrate.[61]

Even as the Bolsheviks in Russia consolidated their control and long before American radicals formed their own domestic Communist parties, American eugenicists claimed to have uncovered the degenerate heredity of the revolutionary radical. For American eugenicists, the Russian Revolution and its worldwide reverberations were part of the inevitable racial retrogression that followed the mass slaughter of the trenches. Lothrop Stoddard offered the loudest warnings about the racial dangers of Communism. Stoddard's rise to prominence and popularity was a journey from reformist journalism to eugenics-tinged criticism of Communism and migration. From an elite eastern background, Stoddard was drawn to world politics. He earned his PhD from Harvard as the war started in 1914. In the same year, he joined the reformist *World's Work* as chief of its foreign affairs department.

During his time covering the war, the Russian Revolution, and the peace process, Stoddard grew increasingly concerned about what he saw as threats to civilization itself. In his frequently shrill, unabashedly elitist, but remarkably popular *The Revolt Against Civilization*—its cover emblazoned with the hammer and sickle—Stoddard characterized the Bolshevik uprising as racial degeneration that threatened a return to savagery. His logic would have been familiar to both reformers and eugenicists. He began his narrative with apes and prehistoric man. He argued that civilization was the result of an evolutionary process toward increased cooperation. The "animality" of the ape-man gave way to savage and later barbarian familial and tribal ties. Through industrial evolution "man had gradually gathered to himself an increasing mass of tools, possessions, and ideas." Some races, including much of the Asian stock, American Indians, and African Negroes, he argued, had not advanced beyond the barbarian stages. To these barbarian races, he added the "lower elements" that persisted within civilization. Civilization "enabled many weak, stupid, and degenerate persons to live and beget children," as the wartime intelligence tests amply demonstrated. Meanwhile, he insisted that the reformist belief in Lamarck's theory of acquired characteristics fostered a sense of complacency that society was racially improving. In fact, the fossil record and ancient history suggested that mankind had made little "racial progress." In what he termed the "iron law of inequality," Stoddard suggested that there were naturally civilized races and classes. The burden of civilization had been carried only by the best races and classes. Yet civilization worked not only to preserve the unfit but also to reduce the birthrate of the select: "The strong *individual* survived even better than before—*but he tended to have fewer children.*"[62]

Civilization faced an inherent threat from those who claimed natural equality. He dismissed the "proletarian ragings" of socialists and syndicalists such as the IWW as "elaborate *pseudo*biological fallacies." Civilization, he warned, was besieged by barbarian colored races from without and by the "underman" from within.[63] The forces of civilization and, especially, the carnage of war weakened the fit, lowered their birthrate, and decimated their stock. The Great War was a "catastrophe.... The racial losses were certainly as grave as the material losses." The war accelerated a process of "racial impoverishment" that led inevitably to the revolt of the barbarian underman.[64] The IWW, by contrast, offered its own conclusions about the meaning of the wartime tests. In the context of the union's opposition to the war, the IWW believed the tests only proved the "overstrain of modern industry."[65]

The Russian Revolution was a manifestation of the same racial tendencies that had produced the IWW domestically. Stoddard warned of a "lure of the primitive," as evident in "expressionist art" as in revolutionary politics, that "seems to lead backward toward the jungle past." Led by inferior Slavic stock and nurtured in the declining civilization of war-torn Russia, the Soviet revolt represented a global racial threat. It was the rising of "the unadaptable, inferior, and degenerate elements, who naturally dislike our civilization and welcome a summons to its overthrow."[66] Stoddard's analysis linked class uprising to racial degeneration. The revolt of what one editorialist deemed the "racially submerged" was class war led by racial inferiors. They were emboldened by the unscientific biological logic of socialists and syndicalists. The Russian Revolution was attractive not only to savage races but also to inferior classes. Bolshevik appeals to class war and their overtures to colonized peoples depended on the idea that the "very existence of superior biologic value is a crime." Class warfare was a revolt against civilization aimed at biological betters and at "white peoples."[67]

Stoddard repeated the eugenics mantra about the Lamarckian failures of reform and about the effects of civilization in preserving the unfit. At the same time, he expanded the political goal of eugenics. It was necessary to preserve superior racial stock as a bulwark against Communist revolution. Not surprisingly, he developed a close relationship with Charles Davenport after their introduction by Henry Fairchild Osborn, the director of New York's Natural History Museum. Davenport had special praise for Stoddard's understanding of the Russian Revolution: "I think you have made out a strong case for your contention that there is world wide revolt against civilization and that we are in danger of a resurgence of the underman which will overwhelm civilization."[68]

The Ebbing Tide: The Chronology of Whiteness

Stoddard first grabbed national headlines in 1920 with the book *The Rising Tide of Color*. His text pictured a white world besieged by colored races. Though he recycled older arguments about the rise of Asia, Stoddard's text reconceptualized the relationship of race to color. Stoddard described World War I as a racial "civil war" that had left colored races confident and untouched. He urged that the ranks of whiteness be closed by warning about the rise of colored races globally and domestically. He deemphasized white racial differences. His portrait of racial conflict in terms of whiteness and color would gain particular significance in the 1920s, with the restriction of European immigration, the weakening of European empires, the prospect of renewed migration from Latin America and the Far East, and the undeniable reality that African Americans were an integral part of domestic industrial civilization.

Stoddard began with a world map that showed the distribution of the white races. His map recognized and mirrored Ellsworth Huntington's earlier map of civilization and the comparison revealed a growing racial pessimism. Whereas Huntington had worried only about the ability of white races to thrive in the tropics, Stoddard feared for the ability of the white races to maintain their grip on Europe and America, let alone to preserve their empires. As in *The Revolt Against Civilization*, Stoddard described civilization as fragile. It could advance; it could also degenerate. Stoddard believed that 1900, when European and the American empires were ascendant and white industrial might was unparalleled, represented the "high-water mark of the white tide." World War I, however, ruptured "white solidarity": "And Europe is the white homeland, the heart of the white world. It is Europe that has suffered all the losses of Armageddon, which may be considered the white civil war."[69] The war carried off the racial elite and emboldened

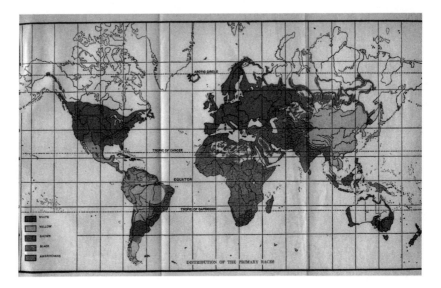

Figure 8.6. "Distribution of the Primary Races": Lothrop Stoddard's map contrasted with maps of European and American empires when it suggested that white races were threatened by a "rising tide of color." In accepting the existence of a unified white race, Stoddard called for white racial solidarity. Lothrop Stoddard, *The Rising Tide of Color* (1920).

nonwhite races: "White solidarity was riven and shattered. And—fear of white power and respect for white civilization together dropped away like garments outworn."[70]

The colored races looked instead to Asia and, potentially, to Bolshevism. In fact, though hardly friendly to Communism, the IWW celebrated the rise of the East. As capitalism transformed Japan, so too was it advancing in China: "Industrial development has to become uniform the world over." The union was confident that the "Chinaman" would soon join European and American "wage slaves" and "hate and fight his economic master."[71] Stoddard resurrected earlier concerns about the Japanese victory in 1904, which, he believed, marked the ebbing of the white tide and the consolidation of colored racial consciousness under the leadership of the yellow races. The Japanese victory, Stoddard argued, "dramatized and clarified ideas which had been germinating half-unconsciously in millions of colored minds, and both Asia and Africa thrilled with joy and hope. Above all, the legend of white invincibility lay, a fallen idol, in the dust." He quoted from aspiring Japanese imperialists to highlight not only antiwhite hostility but also Japanese claims to colored racial leadership. For its excess population, Japan coveted South America and, ultimately, North America. "'North

America alone will support a billion people; that billion shall be Japanese with their slaves'," quoted Stoddard.[72] The colored races admired Japan. They flocked to its universities and reveled in its victories. Since 1911, Stoddard warned, Japanese goods, merchants, and diplomats had flooded Latin America. Anti-American Mexico now considered Japan an ally in its quest to recapture territory seized by the United States in 1848. The brown races of West and South Asia threatened an alliance with yellow races in support of the goal of emancipation from white imperial control. Stoddard recalled the words of a British-educated Afghan to demonstrate brown and yellow racial anger: "You are heaping up material for a Jehad [*sic*], a Pan-Islam, a Pan-Asia Holy War…an invasion of a new Attila."[73]

Stoddard weaved complex conspiracies for a pessimistic postwar public. In an age awash in new forms of migration of African Americans and Mexicans, invigorated anticolonial movements, Communist revolution, and European economic collapse, Stoddard foretold a coming race war. Colored races would be aligned against once-dominant whites. Only white solidarity could combat the rising tide of color. Thus, he downplayed racial differences among the white races. With reverence for the Teuton diminished because of wartime propaganda, Stoddard could trace a history in which the white Mediterranean and Alpine races, though inferior, had been "nordicized" through their conflict with the superior Nordic race. Even Madison Grant, who had categorized European racial differences in his *Passing of the Great Race*, used his introduction to Stoddard's text to offer a revision of his past theories. Grant traced the historical process of Nordic conquest of other European races. It was not that European races had "become white"—as recent historians have argued—but that, in the view of racial observers such as Stoddard or Grant, they had been nordicized. If the elites were purely Nordic, the proletariat was white, but mongrel.

The Rising Tide of Color was alarmist in tone, turgid in style. Yet it vaulted Stoddard into the ranks of the nation's best known racial commentators. It enjoyed multiple printings and admiring reviews. President Warren G. Harding, for example, in a speech on the Southern race problem, urged a Birmingham audience "to read and ponder Mr. Lothrop Stoddard's book."[74] Stoddard even made a cameo appearance in F. Scott Fitzgerald's *The Great Gatsby*. The wealthy but idle Tom celebrates "'The Rise of the Colored Empires' by this man Goddard," a subtle reference to Stoddard. "'Civilization's going to pieces,' broke out Tom violently, 'I've gotten to be a terrible pessimist about things.'"[75] Stoddard's presence translated Lost Generation lassitude into a fear of degeneration and racial decline. Stoddard's text became a rallying call for white solidarity. It linked postwar global anxiety about imperial decline and Communism to domestic concerns about the movement of African Americans into industrial work. Thus, a typical advertisement for his book included an appeal to "white people."[76]

When President Harding evoked Stoddard, it was part of an argument about the transformation of the "race problem" into a national and transnational challenge. "The World War brought us to full recognition that the race problem is national rather than merely sectional," Harding declared. He cited Stoddard to insist that the American race problem "is only a phase of a race issue that the whole world confronts."[77] As late as 1916, eugenicists had dismissed African Americans as a factor in industrial civilization.[78] Confident about Negro mental and industrial marginality, many American eugenicists considered the relationship between race and industry as a problem of immigration and multiple European races. The Southern "race problem" remained—for the moment—a distinctly regional issue. As the war slowed the pace of European immigration while increasing demand for African American laborers, eugenicists came to view race in terms of color.

For Stoddard, industrial evolution contributed to the rising tide of color globally as well as domestically. As late as the end of the nineteenth century, Stoddard argued, "white supremacy was as absolute in industry as it was in politics and war." Even the civilized elite of the brown and yellow races had little role in industry. They were mired, instead, in the "house-industry" stage of industrial evolution. However, when Japanese guns flashed across the "murky waters of Port Arthur" harbor, Asia entered a new industrial stage and emulated Occidental industrial processes.[79] The industrial awakening of Asia ignited the commercial rivalry with white races, stimulated population growth, and, consequently, further immigration. Immigration brought colored workers into direct competition with their white racial superiors. Repeating older concerns, Stoddard insisted Asian workers could "underlive," if not outwork, their white counterparts.[80]

The war accelerated the process of colored industrial advance. In the factories and the merchant marine, colored workers replaced white workers who had perished in the trenches. According to Stoddard, by the war's end, white British dockworkers protested the "Chinese Invasion" of the docks. They insisted that there were two Chinese workers to every white man. Meanwhile, a large Chinese restaurant "had just opened down the West India Dock Road." France, like the United States, witnessed "race-riots" against colored workers. By the mid-1920s, an American observer described the fate of the British and American "sturdy white seamen" who lost their place to Filipino, Chinese, Puerto Rican, and African American sailors. He turned to Stoddard to describe a "rising tide of color."[81]

Stoddard repeated familiar concerns about the racial perils of immigration. Like a generation of nativists before him, Stoddard distinguished between industrial and preindustrial migration. Industrial migration had become too easy. It permitted the unfit to migrate in search of work. Stoddard echoed, as well, the concerns of his mentor Grant about the "Alpine" and "Mediterranean" races that composed the bulk of migrants to the United States in the nineteenth and early twentieth centuries. Increasingly, though, Stoddard distinguished only between white and colored migrants. White immigration may have been of "inferior kindred stocks," but the "influx of wholly alien stocks is infinitely worse." In his later *Re-Forging America,* Stoddard would develop his belief that white immigration was distinct from colored immigration. European immigrants, however inferior, were part of "the white racial group." Chinese, Japanese, Mexicans, or African Americans were "emphatically" so distant as to make "assimilation" impossible. Colored migration, therefore, "is a *universal* peril, menacing every part of the white world. Nowhere can the white man endure colored competition."[82]

Stoddard linked the entrenchment of immigration restriction to the retreat of white empires. Such a connection was comprehensible only through his evocation of ideas of immigration as a form of colonization and conquest. The restriction of immigration was the final dike against the colored flood. Tropical colonies, the first dike, had already been breached. Stoddard argued that white empires should not buttress their rule, especially in Asia. Resistance would only encourage what he called Pan-Colored and Colored-Bolshevist alliances. Stoddard resurrected concerns about white efforts to live in the tropics when he urged the defense of natural white "race-frontiers" through immigration restriction. In similar tones, the expert on China Putnam Weale feared an "invasion" of colored migrants: "A struggle has begun between the white man and all the other men of the world to decide whether non-white men...may or may not invade the white man's countries in order to gain their livelihood." For Stoddard, once "fast-breeding" colored races had breached the "inner dikes" of immigration restriction, white workers inevitably would be inundated: "Nowhere—*absolutely nowhere*—can white labor compete on equal terms with colored immigrant labor. The grim truth is that there are enough hard-working colored men to swamp the whole white world."[83]

The recognition of African Americans as an integral part of American industrial life, rather than as a race destined for extinction, came as part of this growing concern about the rising global tide of "colored immigration," its effect on white workers, and the end of the white monopoly on industry.

The Great Migration that brought African Americans from rural areas to cities and from the South to the North highlighted the growing dependence of industry on black workers. As James Weldon Johnson, then a field secretary for the National Association for the Advancement of Colored People (NAACP), noted, "The Negro has this chance, the first in his history, to get his hand upon the thing by which men live, to become for the first time a real factor in the world of labor."[84]

By 1917, social workers and social reformers noted the mass arrival of Southern migrants. The *Survey* turned to W. E. B. Du Bois to explain the migration of 250,000 blacks to the North. To understand East St. Louis's race riots, however, the journal relied on explanations about industrial race competition that would have been familiar to Stoddard. Oscar Leonard, a St. Louis charity worker, dismissed racism as the cause of the riots. Illinois even permitted intermarriage, he pointed out. Instead, the reason for the riots was race competition for jobs. The "Negro laborer" could, like other colored migrants, under-live the white worker: "Such an element introduced into the community acts as a whip over the heads of the white workers." The new reality of industrial life helped observers such as Leonard define a new white unity. Leonard cast industrial race competition as a struggle between black and white.[85]

Even African American observers tended to recognize an unbroken whiteness. When Charles S. Johnson of the National Urban League described industrial work in the North before World War I, he listed distinct European "racial groups." After the war, he would contrast African American workers only with their "white" counterparts. Similarly, the National Urban League's T. Arnold Hill mourned African Americans' exclusion from unions as a problem of antipathy from white workers, regardless of their racial background. For Hill, the riots in East St. Louis and Chicago realigned racial conflict into divisions between white unionists and "colored workers." For the African American economist George E. Haynes, the "New Order" was defined by "the relation of the Negro wage-earners to white wage-earners."[86]

There was little question that locking the doors of Ellis Island ruptured the Mason-Dixon line, but observers still wondered about African Americans' ability to withstand industrial labor. African American migrations reopened the debate about the black birthrate, the statistic that a previous generation of observers had relied on to argue that blacks were destined for extinction. The *Forum*'s 1924 debate on the "Negro Migrations" centered on the birthrate and the probability of extinction. On one side, the Southern judge Blanton Fortson combined eugenical fears about the fecundity of migrants with notions of a declining black birthrate. The African American migrant to the North will "gradually die out, for there he seems to lose his fecundity." Fortson argued that the North substituted the declining Negro for the fecund European immigrant. NAACP field secretary William Pickens, in response, attacked the "forlorn hope that the Negro will accommodate a solution of the problem by simply 'dying out'." Instead, he insisted that African American populations in the North had increased at precisely the same rate as whites: "Death awaits the hope that the Negro will die."[87] Edward Eggleston's chillingly titled *The Ultimate Solution of the American Negro Problem* and the work of the noted biologist Raymond Pearl still clung to the hope that African Americans were a dying race. Pearl, for example, looked to the birthrate to argue that the "negro is biologically a less fit animal...than the white."[88] For the sociologist Jerome Dowd, however, the persistence of a black population and its rapid integration into industrial labor suggested that the laws of biology would not find a solution to the "Negro Problem": "Have we such a situation in the presence of the Negro and Caucasian in the United States? Is it possible for both to survive under conditions of competition?"[89]

Dowd proposed, as part of debates about the effects of industrial labor on the African American birthrate, an examination of the character of the black worker—just as an earlier generation of observers sought to determine the suitability of immigrant races for industrial labor. Dowd, typically, cast the black worker as inefficient and difficult to discipline. With such negative receptions voiced alongside the poor showing of African Americans in the Army intelligence tests, many African American leaders admitted that black migrants were imperfect laborers. James Weldon Johnson, however, turned to evolutionary theories of migration to argue that the fittest of the race were arriving in Northern factories: "This process is selective, and is already producing a steady flow northward of the best element of the colored working people."

The transformation of the American workplace in the 1920s reconfigured racial debates about fitness, heredity, and civilization. Eugenicists, in particular, had paid scant attention to African Americans in the 1910s, dismissing them as incapable of withstanding the rigors of the factory. In the aftermath of the war, eugenicists confronted the reality that in some industries African Americans were filling the ranks of unskilled labor once composed of European immigrants. They were forced to admit that the African American was perhaps not destined for extinction in the industrial struggle for survival. Eugenicists expanded racialized debates about industry that had previously been largely about Europeans and Asians. By the mid-1920s, Estabrook, for example, had shifted his attention from European migrants to the "Negro problem."[90] Davenport also adapted his racial framework in the face of African American industrial labor. He relied on the evidence of the Army intelligence tests to examine the mental capacity of different races. His study represents a moment of transition in racial thinking, as Davenport examined differences among white races as well as between blacks and undifferentiated whites.[91] By the end of the decade, Davenport had joined a National Research Council committee to study the "American negro." He proposed a range of studies, from the biology of African Americans to their mental capacity.[92]

Historians have often described the passage of the Johnson-Reed Act as a watershed. It marked the end of a period of comparatively open immigration from Europe and the reiteration of the exclusion of Asian immigration. Such an interpretation of the immigration restriction law is not incorrect, but it differs markedly with the views expressed by the eugenicist architects of the law. Laughlin and his allies in the eugenics movement as well as in Congress viewed the act as merely a step on the road toward a national policy of eugenics.[93] Stoddard compared the act to the Constitution, which, in ending the slave trade, also limited the numbers of inferior Africans who would enter the country. But, for Stoddard, it was only "the beginning of a new epoch of national reconstruction and racial stabilization that should normally culminate in a *re-forged America*."[94] Moreover, new perils of immigrant invasion loomed after 1924. Although Laughlin, Stoddard, and other nativists and eugenicists remained worried about the effects of immigration from southern and eastern Europe, they refocused their concern on new sources of migration, especially from Mexico and the American South.

Whereas earlier texts rarely mentioned Mexican immigration, it now took center stage. The Johnson-Reed Act, largely as a result of the labor demands of Southwestern farmers, did not restrict immigration from across the Rio Grande—an omission that did not escape the wary eyes of eugenicists such as Harry Laughlin. By the late 1920s and early 1930s, eugenicists had begun to explore the potential dangers associated with the unrestricted immigration of Mexicans. Predictably, they identified many older concerns, from race substitution to overbreeding. Laughlin, the Immigration Committee's official expert, returned to Congress to testify about the racial effects of Mexican migration. For Laughlin, some Mexicans had primarily white blood and, like European migrants, could be assimilated. However, he strongly urged the exclusion of Mexicans with Indian and black blood. He warned that the migration of fecund Mexicans was leading to a reconquest of the Southwest. Laughlin urged an extension of the Johnson-Reed

Act to include not only intelligence testing and studies of family stock but also the strict limitation of immigration to white races. For the first time in his many appearances in Congress, he minimized distinctions between European races. Laughlin used the problem of Mexicans to appeal for the migration of "assimilable races...who will also improve our existing hereditary family stock qualities."[95]

There is a profound irony in nativists such as Laughlin or Stoddard using a term like "assimilation." After all, Laughlin's testimony to Congress came just a year after the publication of Henry Pratt Fairchild's *The Melting Pot Mistake,* the noted nativist's most recent salvo in the battle against European migration. Fairchild attacked the basis of assimilationist dreams as fantasies.[96] Laughlin's reengagement with the idea of assimilation highlighted distinctions between white and colored races. He still noted older patterns of conquest through industrial substitution, this time by Mexicans. On the railroads and in the stockyards and steel mills, the Mexican was "displacing the Slav."[97]

The concept of "assimilation" gained new currency as an impassable boundary that divided recent colored immigrants and the white immigrants of the prewar generation. Stoddard, like Laughlin and their congressional allies, maintained the negative evaluation of European migrants but confirmed their essential whiteness because of their potential to assimilate. This potential helped Stoddard to continue to argue for racial divides between the mongrel working class and his beloved Nordic elite. The Mexican was, for Stoddard, utterly alien; he was "*a born communist*" and a "primitive tribesman." Meanwhile, Stoddard grudgingly admired the remarkable rise of Japan, but cited the inability of Japanese to assimilate in order to link the Japanese to "inferior stocks" of Mexicans and African Americans. Stoddard resurrected a notion familiar to an earlier generation of reformers (for whom he retained profound contempt) to place concerns about colored migration into a genealogy of twentieth-century anti-immigrant activism. He balanced past nativism with a dedication to "bi-racialism." His elaborate plan for the separation of whites and blacks and for the exclusion of non-white immigrants married Southern Jim Crow and the Northern tradition of immigration restriction.[98] As two discourses about the "race problem" finally merged, Stoddard announced that the question of the Negro was "the gravest of our national problems."[99]

Where once radicals and reformers and nativists had relied on similar logic and voiced similar fears of racial degeneration, they now stared at each other across profound divides. Ironically, American Communists, Stoddard's bitterest foes, would not have disagreed. Communists also came to view the American "race problem" in terms of a global problem of colonized colored races. By the late 1920s and into the 1930s, African Americans in the North and the South became a primary focus for Communist organizing. And, they turned to an older iconography of working-class fitness and emblazoned their literature with images of muscular, class-conscious, and black workers.

9

FAILING OF ART AND SCIENCE

The Abyss in a New Era

> This little book, alas, fails of being either art or science. But while
> it cannot claim these high avenues of expression, it may serve the more
> modest but no less useful purpose of furnishing a plain description
> of the facts of the city dweller's life, together with some indications
> of the evolutionary process going on at the city's heart.
>
> Mary Simkhovitch, *The City Workers World in America*, 1917

This book began with Edward Ross, but it ends with Lothrop Stoddard.
There were many similarities between the two observers. They shared,
above all, a distaste for immigration and worried about its racial effects.
Ross forged his academic credentials as a sociologist, but he made his public
name as an adventurer. He traveled globally and recorded the changes in
China that accompanied European incursions and rising immigration to
North America. He was also abroad at home. He stood at the vanguard of
a host of urban explorers who produced vivid portraits of the foreign races
and places at the heart of industrial America. He helped provide the vocabulary used to describe the anxieties of the era, in particular the notion of
race suicide. Stoddard inherited Ross's language and helped translate fears
of immigration into law. He provided, as well, a continuity between the
pre–World War I anxiety about tramps, prostitutes, and immigrants and
the postwar concerns about African American migrants, the rise of Asia,
and Communism. Yet, in the end, Ross—six years before his death and five
years before Stoddard's—predicted that Stoddard would have no place in
the post–World War II intellectual world.

Ross wrote at the beginning of the period of American industrial dominance, imperial expansion, and mass immigration. He joined an intellectual consensus in which socialists and reformers shared thoughts and

inspiration. Social scientists and biologists crossed still inchoate borders with abandon. Though socialists and reformers hardly agreed on tactics or goals and often traded insults and attacks, they both contrasted the civilization of industry with the savagery exemplified by tribal life in the tropics or—alarmingly—the lives of degenerates in industrial cities. They worried that civilization was a mere veneer over savagery. Their contrast between civilization and savagery united disparate social processes, in particular industrialization, immigration, and imperialism, in common conversation. It also gave rise to fears of racial degeneration as well as to plans for progress—whether gymnasiums and playgrounds or social revolution.

Stoddard wrote at the beginning of a period of imperial contraction, internal migration, and concern about the rise of the East. His shrill tone suggested that the earlier consensus had irrevocably fractured. The consensus at the turn of the century was beginning to fray even before World War I. Some of the leading lights who shaped the science of industrial evolution had died. Lester Frank Ward, the towering figure who helped forge the relationship between biology and sociology and was an inspiration for a generation of socialist feminists and economic feminists, had died in 1913. His fellow sociologist Franklin Giddings would follow him to the grave seven years later. Others had shifted the focus of their research. Richard Ely, for example, turned from the study of industrial evolution to the science of real estate. He abdicated his position as a sage of progressivism. As an older generation of scholars who crisscrossed fields passed on, there was a hardening of intellectual boundaries. The professionalization of the social sciences (and biological sciences) is well documented by historians. But this hardening occurred as political battle lines were forming in the years just before and during World War I. In the run-up to the war, reformers found the intellectual justification of their politics dismissed as erroneous science. Socialists—most dramatically—had been silenced by wartime repression. It could even be argued that this repression would not have been possible without the expulsion of socialists (and to only a slightly less extent, reformers) from conversations about the future of civilization. Their crusade for civilization had passed especially into the hands of eugenicists and their sympathizers, such as Stoddard.

Stoddard was a prophet of pessimism. He connected the alarming trends of the age. For Stoddard, the war was a white racial civil war, not merely a slaughter. That conflict accelerated not only the rise of Asia but also emboldened the racially inferior through the rise of Bolshevism. With the white races in retreat, empire represented an impossible indulgence of power. Tottering empires only brought the colored races to the edge of the white metropole. The decline of empire made immigration restriction necessary. For Stoddard, as for many of his followers, the passage of the Johnson-Reed immigration restriction act in 1924 was not a triumph of a long struggle for protectionism but a lamented necessary response to the twilight of imperialism. The process of immigration restriction inevitably revisited the idea of assimilation. The reforging of assimilation was the first cousin of immigration restriction in an age when empire was on the retreat. Immigration restriction was the result of the ebbing of Nordic civilization.

Inevitably, as well, many of the structures, institutions, and individuals that had sought to understand an earlier era of industrialization—one interwoven with the growth of empires and the transformation of immigration—attempted to adapt to the new era. Most notably, eugenicists shifted to consider African Americans, who were once dismissed as irrelevant to industry but were now considered essential to its future. However, as the post–World War I years stretched into the Great Depression, the science behind the idea of industrial evolution was critiqued and largely rejected. Its advocates faced accusations of racism and their institutions either changed focus or closed.

Yet the intellectual and cultural legacies of the earlier era persisted without the institutions, individuals, or science that produced them. They were born from a particular moment: the public, political, and academic fascination with evolutionary theory. They coalesced around the idea of industry as the highest stage of civilization. Observers who shared a sense of biological time and history forged them. Such observers understood races as biological categories that could be placed in a hierarchy based on their relation to industry. Industry was racial accomplishment, not simply factories, railroads, or mines. They understood industrialization as a process of development that extended back into the recesses of prehistory, but also produced the problem of degeneration. Some races had completed the long march from savagery to civilization. Others remained stalled in savagery. Others had advanced to different stages of primitivism or barbarism. Still others may have degenerated back toward savagery.

Unquestionably, in a process that began in the late 1920s and accelerated during World War II, race lost much of its basis in biology—so much so that many contemporary thinkers quickly deny that they consider culture, ethnicity, or race grounded in biology. Yet the way that we think about the rise and fall of nations; cultural, ethnic, and religious conflict; and immigration remains profoundly tainted by ideas of those who connected race, industry, empire, and immigration. In the end, the significance of the intellectual world of the turn of the twentieth century lies in the fact that its ideas persisted despite the decline of their institutions, the shaming of their leading lights, and the repudiation of their scientific logic.

Eugenics would continue into the 1930s and beyond. It scars were most noticeable in the thousands of woman forcibly sterilized under eugenics laws between the 1920s and 1970s. As historian Alexandra Minna-Stern has noted, eugenics, especially under the broader guise of hereditarianism, remained a powerful force in policy, biology, and social work into the postwar years.[1] Yet it is also true that the reputations of those who had carried the vanguard of eugenics in the 1910s and 1920s were tarnished by the 1930s. Their science was questioned not only for its motivations but also for its simplistic understanding of the advancing science of genetics. Arthur Estabrook, in particular, faced hostility for his study of race mixing, *Mongrel Virginians*. The book was "really absurd and useless," declared one reviewer. It was hardly the rigorous study of genetics and heredity that it promised. Rather, it was "an exposé of small community moral depravity." It was based on the "most trifling morsels of gossip." The ERO itself would quietly close in 1940 after having halted most of its eugenics research in the mid 1930s.[2] Even Lothrop Stoddard, once a name on the lips of presidents, saw his star waning. Stoddard remained popular through the 1920s, but he faced increased suspicion, especially toward the end of the decade. *Re-Forging America* was hailed as a just call for racial purity when it advocated the extension of segregation as the logical next step after the restriction of immigration. "We disposed of the alien problem by the Johnson Act of 1924," declared the government geographer Carleton P. Barnes. "We still have the problem of the negro adequately to solve." Stoddard's response, he believed, was "sane." Yet Stoddard's racial science faced growing skepticism. The essayist Simeon Strunsky, for example, critiqued Stoddard's evocation of biology. "At a time when the very term 'race' is being scrutinized," he viewed Stoddard's Nordic race as myth, not science. It is important to note that Strunsky placed the word race in quotation marks. When he employed quotation marks, Strunsky split racial classification from science. He had "grown weary of this mobilization of the Sciences…in defense of individual likes and dislikes." Still, he sympathized with Ross and the pain he felt when he observed the "stolid 'moon-faces'" of "swarming" immigrants.[3] By 1926, when Stoddard took to national radio for a debate on "The Eastern Menace to Western Civilization," he was roundly attacked for engendering race hatred. By the beginning of the Great Depression, the *Journal of Negro History* could confidently claim that Stoddard, a longtime nemesis of African American intellectuals, was just a "die-hard in a cause that no reputable scientist any longer espouses." His *Lonely America*, a reiteration of the dangers facing white civilization, was an embarrassment to "heredity-mongers." The journal appealed to those "too young to read his *The Rising Tide of Color*" to remind them of an idea "of the last decade." The journal rejoiced that "many scientists will conclude that Mr. Stoddard's biology is as unsound as his history is inaccurate."[4]

Stoddard's reputation, and, with it, ideas about the biological basis of race, dimmed even further with the rise of Nazism and Fascism. In 1939, the fifty-six-year old Stoddard traveled to Nazi Germany where he was—to the distaste of many American readers—heartily received by the Nazi leadership, including Adolf Hitler himself. Stoddard was given unprecedented access to the institutions of the Nazi state, especially its eugenics courts. His *Into the Darkness* (1940) promised—and to some extent delivered—an objective survey of life in the Third Reich. Yet American readers rejected Stoddard's comfortable engagement with Nazism.[5] Moreover, Stoddard was one of the American intellectuals much admired by the Nazis. It did not help Stoddard when he was publicly praised by the führer.[6]

Appropriately enough, it fell to Edward Ross to write Stoddard's political obituary as World War II neared its conclusion. Ross condemned Stoddard (and Madison Grant), but exonerated his many social scientific admirers: "Thirty-odd years ago we sociologists were highly entertained by the books of Madison Grant...[and] Lothrop Stoddard and other exponents of Nordic superiority; but we never imagined the theory being made a hinge for German aggression to turn on." Ross's prophesies about the postwar climate used Stoddard as a straw man. He cast Stoddard as an outrageous Nordic racist whose ravings overshadowed the work of Ross and others in ordering race into a biological hierarchy. It seemed strange to hear Ross opining that in the postwar era "arrangements of the chief races and peoples in a value hierarchy...are going to be greeted with shivers or jeers, now that everyone sees to what diabolical designs such an arrangement lends itself." Ross, after all, had played a key role in "branding of this or that segment of humanity as 'inferior' or 'subhuman,'" as he accused Stoddard of doing.[7]

Historians suggest that there was a rebirth of the progressive impulse in the New Deal of the 1930s, which, in turn, helped shaped the postwar liberalism predicted by Ross. There was unquestionably a genealogy that linked the progressivism of the 1910s and the New Deal politics of the 1930s, though more in goals than in method. Perhaps the heritage is best exemplified by the fixation with fitness in the 1930s. The iconography of the muscle-bound male worker reappeared in everything from public works art to trade union literature to radical broadsides. Though devoid of its obvious links to biology, physical fitness still manifested a political significance.[8]

The environment that had cultivated fears of degeneration, race suicide, and immigration colonization was changing. So, too, was the science that had linked industry, empire, civilization, and savagery in a single conversation and created an interlocking politics. Attempts to understand the origins and meaning of race were often produced more by anthropologists than by biologists. The science of race faced new scrutiny. In fact, by the late 1930s it was enough to make a connection to Nazis to condemn an earlier racial science. The science journalist Waldemar Kaempffert condemned the writings of Stoddard and Grant as befitting a Nazi medal. It was "good Nordic philosophy, but it is not good science."[9] Instead, he joined leading anthropologists such as Alfred Kroeber, a disciple of Franz Boas, and placed the word race in quotation marks. In the immediate postwar years, the assault on the "race concept" continued. In 1952, UNESCO (the United Nations Educational, Scientific and Cultural Organization) convened a blue-ribbon committee to cast judgment on a generation of race scientists. "Racial doctrine is the outcome of a fundamentally anti-rational system of thought," the authors declared, "and is in glaring conflict with the whole humanist tradition of our civilization." This was a sweeping claim about civilization in the face of a science that had claimed the precise opposite: racial development was key to a civilization that only certain races could enjoy. Perhaps UNESCO was too quick to write the obituary of scientific racism. As historian Michelle Brattain notes, the UNESCO report highlighted the errors and abuses of a generation of theorists. It also sparked a defense of scientific studies of race, especially as it related to intelligence testing. Within a year, UNESCO released a backtracking statement that rehabilitated some of the connections between race and biology.[10] The marriage between race and biology has proved hard to annul completely, as contemporary medical research, for example, race-based studies of disease susceptibility, and intelligence testing suggests.[11]

Still, the era of Lothrop Stoddard had waned. By 1960, the *New York Times* editorial page cited Stoddard to condemn a belief in Nordic and white supremacy: "Forty years ago Lothrop Stoddard wrote a book…'The Rising Tide of Color'. He feared that the white races could not indefinitely continue to run the world. Of course they cannot."[12] At the dawn of the modern Civil Rights Movement and in the midst of the painful process of African decolonization, the condemnation of Stoddard served to repudiate the politics of empire and segregation.

Yet if the distinct moment had passed, it has left enduring legacies that stretch to the present day—even if it is largely taboo to explain racial conflict through biology. Much appears quirky, quaint, even bizarre about the

theory of industrial evolution and the fears of degeneration that it produced. The focus on primitive life so central to understanding modern industry might seem shallow, simplistic, or just racist to the contemporary
reader. Primitive fiction, such as *2001: A Space Odyssey,* Stanley Kubrick's
1968 film that began with a tribe of apes discovering how to use bone tools,
is for entertainment, not science. In the hand of labor historians, industrialization is traced only to eighteenth-century innovations and it is compared
to agriculture, not savagery.

Much might also seem eerily current. Stoddard's warning of jihad and
his fear of the effects of Asian industrialization seem like current politics,
rather than almost century-old shrill racism. Today, Stoddard might need a
new editor and he might employ a different vocabulary, but he would have
an audience. In fact, he retains admirers at the far-right fringe of American politics. *American Renaissance,* an anti-immigrant magazine published
by the white supremacist New Century Foundation, lauds Stoddard as a
"prophet" and offers cheap reprints of his *Rising Tide of Color.*[13] Immigration is once again in the headlines of American newspapers, with talk of
new walls (with Mexico), new laws, and new paramilitary anti-immigrant
groups such as the Minutemen who patrol the border with Mexico armed
with guns. The rise of immigration as a primary political issue might seem
surprising in a country focused on war and terrorism. Yet the perspective
of the turn of the twentieth century—a time of war and empire building—
suggests ties between concerns about immigration and war. Critics of immigration still worry about the sanctity of borders and cast immigration
as a form of invasion and hint at the potential of race conquest. That such
racism was ever expressed is a tragedy of American history. That it has persisted is a tragedy of contemporary American politics.

Across the Internet, critics warn of an "immigrant invasion" and of
Aztlan—the pejorative and appropriated name for the former southwestern
United States conquered through Mexican immigration. Such voices may
represent the far right, but their similarity not only to Stoddard but also to
respected observers of migration, economics, and global politics is striking.
It could have easily been Stoddard announcing in the staid and mainstream
pages of *Foreign Affairs* that the post–cold war world faced a "clash of civilizations." Instead, it was Samuel Huntington, a career defense and foreign
affairs specialist. As American Studies scholar Robert G. Lee has noted,
Huntington's forthright nativism echoes Stoddard's Nordic nationalism.
Even their division of the world is similar. Huntington is quick to deny the
role of biology in his formulation. Yet (ironically) civilization, for Huntington, seems virtually a synonym for biological race. It is equally immutable

and ingrained. Huntington argues that the world is divided into competing civilizations and the West is besieged by the East from without and by inassimilable immigrants from within the borders of the United States. Like Stoddard's warnings of "Pan-Turanism," a movement for unity of Turkic peoples, and of jihad, Huntington fears the rise of alliances born in the East. For Lee, the parallel between Stoddard and Huntington confirms the "monotony" of historical nativism.[14] More broadly, it represents how the fears articulated at the turn of the twentieth century still provide the script for conversations about industry, work, immigration, empire, and development at the beginning of the twenty-first century. The right-wing commentator Patrick Buchanan began the new century with the same warnings that began the last with his *The Death of the West: How Dying Populations and Immigrant Invasions Imperil Our Country and Civilization* (2002).[15]

Stoddard would surely join a cacophony of contemporary voices worried about immigration, wages abroad (especially in China), development, and the "clash of civilizations." These are interwoven conversations and their intertwining is comprehensible because of their long histories. Immigration, in the popular mind, is a story not only of new arrivals but also of competition for jobs, differential birthrates, and changing national character. Moreover, immigration to the United States is seen as part of broader patterns of global industrialization and development. In the contemporary world, it is axiomatic to think about industry in its global context. Historians also are quick to remind us that contemporary globalization has many antecedents. Industry has long been considered in its global and transnational contexts often by contrasting East and West, civilized and savage, as well as developed and underdeveloped.

We still talk of economic development as the progress of a nation and people. This is, in many ways, the legacy of ideas about industrial evolution. The notion of development emerged in the dying throes of imperialism and gained currency in the context of postcolonial economics. Partly, the notion of development aimed to repair the damage of empire. Partly, it was a way of reordering the world in the absence of colony and metropole. Development has retained its biological context and aura, even if this is to the displeasure of some critics. The discomfort of development studies scholars with the "organic metaphor" may betray unease with the turn-of-the-twentieth-century parallels. As early as the 1970s, the director of the United Nations Research Institute for Social Development bemoaned the persistence of the "organic metaphor" of development. The "biological metaphor ... tends to assume the inevitability of the course of development of each country according to an innate genius." This suggested that development would simply unroll regardless of intervention.[16] In fact, development studies, as a field, has generally veered toward the eradication of poverty. Is the contemporary development activist a contemporary incarnation of the urban explorer who equated poverty and underdevelopment?

The notion of developed and underdeveloped can easily reproduce the older imperial opposition of civilization and savagery. It also could re-create the sense of threats generated by a developing East. Stoddard's commentary on the Menace of the East is eerily familiar. Stoddard would join critics who warn of working conditions in China. Like many others, his concern would focus on the perceived ability of China to outcompete the West. He would not be motivated by sympathy for Chinese workers. Scholars of contemporary global industry identify a race to the bottom in living and working conditions that has been accelerated by the demands of industrialists. This race to the bottom has been accelerated by neoliberal structural adjustment, the mandated rollback of welfare protection in the interest of international competition. The details of development policy and multinational lending and aid policies are difficult to understand for those whose jobs are at risk. Yet narratives about job loss are easy to create and circulate. The job that fostered a high standard of living in the United States is lost to Mexico, Vietnam, or China where a worker will labor for less in worse conditions. Concerns about poor working conditions, whether in China or elsewhere in rising Asia, are often paired with images of overpopulation. With the machinations of employers, politicians, and policies virtually hidden, the contemporary worker feels pitted against workers in the less-developed world. In this context, the shipping container evokes the steamship steerage of an earlier era. Both annihilated space and brought workers into competition. One can complain about working conditions abroad—and the clamor about Chinese working conditions is virtually omnipresent—without much sympathy for those laboring in those conditions. Rather, such concerns can echo an earlier concern about racial conflict enacted at the points of production and reproduction.[17]

Such parallels profoundly complicate progressive and workers' rights politics. How does one protest the loss of jobs without resorting to rhetoric that describes one group of workers under-living another? How does one speak of the need for progress without referencing the long heritage of development? How can one protest and picture the plight of female and male workers without resort to images of physical and moral decline that recall an older language of degeneration? How can activists raise the issue of population and birth control without their audience hearing concerns about race suicide?

We are entangled in history. The language we use to describe the fundamental trends and problems of the day is scarred with past histories of race, class, gender, nation, and empire. We can repudiate the intellectual heritage bequeathed by those who surveyed the landscape of industrial America and who developed explanations, defined fears, and suggested solutions that we regard with opprobrium. Yet their legacies, all the more nefarious because their obvious roots are obscured, are harder to escape.

NOTES

Introduction

1. Edward Alsworth Ross, *The Old World in the New: The Significance of the Past and Present Immigration to the American People* (New York: Century Co., 1914), 290.

2. David Montgomery, *Citizen Worker: The Experience of Workers in the United States with Democracy and the Free Market during the Nineteenth Century* (New York: Cambridge University Press, 1993); Montgomery, *Workers' Control in America: Studies in the History of Work, Technology, and Labor Struggles* (New York: Cambridge University Press, 1981); Stephen Meyer, *The Five Dollar Day: Labor Management and Social Control in the Ford Motor Company, 1880–1921* (Albany: State University of New York Press, 1981); Herbert George Gutman, *Power and Culture: Essays on the American Working Class*, ed. Ira Berlin (New York: Pantheon Books, 1987); Walter Licht, *Industrializing America: The Nineteenth Century* (Baltimore: Johns Hopkins University Press, 1995); Jonathan Prude, *The Coming of Industrial Order: Town and Factory Life in Rural Massachusetts, 1810–1860* (Amherst: University of Massachusetts Press, 1999 [1983]).

3. Alexander Trachtenberg, *The Incorporation of America: Culture and Society in the Gilded Age,* 25th anniversary ed. (New York: Hill and Wang, 2007), 38–100; Daniel T. Rodgers, *The Work Ethic in Industrial America, 1850–1920* (Chicago: University of Chicago Press, 1978); T. Jackson Lears, *No Place of Grace: Anti-Modernism and the Transformation of American Culture, 1880–1970* (Chicago: University of Chicago Press, 1994), 83–116; Elspeth Brown, *The Corporate Eye: Photography and the Rationalization of American Commercial Culture, 1884–1929* (Baltimore: Johns Hopkins University Press, 2005), 1–158. Intellectual and cultural histories of the era have frequently stressed questions of modernity and consumption. See T. Jackson Lears, *Fables of Abundance: A Cultural History of Advertising in America* (New York: Basic Books, 1995); Kristin Hoganson, *Consumer's Imperium: The Global Production of American Domesticity, 1865–1920* (Chapel Hill: University of North Carolina Press, 2007); Charles McGovern, *Sold American: Consumption and Citizenship, 1890–1945* (Chapel Hill: University of North Carolina Press, 2006), 1–134.

4. See Donna Haraway, *Primate Visions: Gender, Race, and Nature in the World of Modern Science* (New York: Routledge, 1990), 19–83; Gregg Mitman, *The State of Nature: Ecology, Community, and American Social Thought, 1900–1950* (Chicago: University of Chicago Press, 1992).

5. There has been a new focus among American working-class historians on labor in the empire and in former colonies. See, for example, Julie Greene, "The Labor of Empire: Recent

Scholarship on U.S. History and Imperialism," *Labor: Studies in Working-Class History of the Americas* 1 (Summer 2004): 113–29; Nancy Hewitt, *Southern Discomfort: Women's Activism in Tampa, Florida, 1880s–1920s* (Urbana: University of Illinois Press, 2001); Dorothy B. Fujita-Rony, *American Workers, Colonial Power: Philippine Seattle and the Transpacific West, 1919–1941* (Berkeley: University of California Press, 2003).

6. Matthew Frye Jacobson, *Barbarian Virtues: The United States Encounters Foreign Peoples at Home and Abroad, 1876–1917* (New York: Hill and Wang, 2000), links histories of immigration and empire. Nell Irvin Painter, *Standing at Armageddon: The United States, 1877–1919* (New York: W.W. Norton, 1987), is an important critique of the "new labor history" that places the experience of labor within the context of imperial culture.

7. Gail Bederman, *Manliness and Civilization: A Cultural History of Gender and Race in the United States, 1880–1917* (Chicago: University of Chicago Press, 1995).

8. Adam Kuper, *The Invention of Primitive Society: Transformations of an Illusion* (New York: Routledge, 1988); Leah Dilworth, *Imagining Indians in the Southwest: Persistent Visions of a Primitive Past* (Washington: Smithsonian Institution Press, 1996); Patrick Brantlinger, *Dark Vanishings: Discourse on the Extinction of Primitive Races, 1800–1930* (Ithaca: Cornell University Press, 2003).

9. Vernon J. Williams, *The Social Sciences and Theories of Race* (Urbana: University of Illinois Press, 2006); Michel Verdon, "The World Upside Down: Boas, History, Evolution, and Science," *History and Anthropology* 17 (September 2006): 171–87; Corey S. Sparks and Richard L. Jantz, "Changing Times, Changing Faces: Franz Boas' Immigrant Study in Modern Perspective," *American Anthropologist* 105 (June 2003): 333–37; Carl Degler, *In Search of Human Nature: The Decline and Revival of Darwinism in American Social Thought* (New York: Oxford University Press, 1991), 59–186.

10. Yousun Park and Susan P. Kemp, "'Little Alien Colonies': Representations of Immigrants and Their Neighborhoods in Social Work Discourse, 1875–1924," *Social Science Review* 80 (December 2006): 705–34; Kathleen Neils Conzen, "Immigrants, Immigrant Neighborhoods, and Ethnic Identity: Historical Issues," *Journal of American History* 3 (December 1979): 603–15.

11. For notions of the unfit, see Elof Axel Carlson, *The Unfit: A History of a Bad Idea* (Cold Spring Harbor, N.Y.: Cold Spring Harbor Laboratory Press, 2001), 1–278.

12. Mark Pittenger, "A World of Difference: Constructing the 'Underclass' in Progressive America," *American Quarterly* 49 (March 1997): 26–65; Seth Koven, *Slumming: Sexual and Social Politics in Victorian London* (Princeton: Princeton University Press, 2004); Catherine Cocks, *Doing the Town: The Rise of Urban Tourism in the United States, 1850–1915* (Berkeley: University of California Press, 2001). In reinterpreting urban investigation in the context of the imperial travel narrative, this book extends Mary Louise Pratt's notion of contact zones into the domestic sphere and into the period from 1880 to 1930. See Pratt, *Imperial Eyes: Studies in Travel Writing and Transculturation* (New York: Routledge, 1992). See also Patricio N. Abinales, "Progressive-Machine Conflict in Early Twentieth-Century U.S. Politics and Colonial State Building in the Philippines," in *The American Colonial State in the Philippines: Global Perspectives*, ed. Julian Go and Anne L. Foster (Durham: Duke University Press, 2003), 148–81.

13. Warwick Anderson, *Colonial Pathologies: American Tropical Medicine, Race, and Hygiene in the Philippines* (Durham: Duke University Press, 2006); John Farley, *Bilharzia: A History of Imperial Tropical Medicine* (Cambridge: Cambridge University Press, 2003); Dane Kennedy, "The Perils of the Midday Sun: Climatic Anxieties in the Colonial Tropics," in *Imperialism and the Natural World*, ed. John M. MacKenzie (Manchester: Manchester University Press, 1990), 118–40; David Arnold, *The Problem of Nature: Environment, Culture and European Expansion* (Oxford: Blackwell, 1996); Philip D. Curtin, *Disease and Empire: The Health of European Troops in the Conquest of Africa* (Cambridge: Cambridge University Press,

1998); Eric Jennings, *Curing the Colonizers: Hydrotherapy, Climatology, and French Colonial Spas* (Durham: Duke University Press, 2006).

14. The relationship of African Americans to industry has been approached from a number of social-historical angles. The goal here is to examine the discourse that defined African Americans as incapable of industrial work, especially in the North. See Mark Noon, "'It Ain't Your Color, It's Your Scabbing': Literary Depictions of African American Strikebreakers," *African American Review* 38 (Fall 2004): 429–39; Kimberly L. Phillips, *Alabama North: African-American Migrants, Community, and Working-Class Activism in Cleveland, 1915–1945* (Urbana: University of Illinois Press, 1999); Rick Halpern, *Down on the Killing Floor: Black and White Workers in Chicago's Packinghouses, 1904–1954* (Urbana: University of Illinois Press, 1997); William P. Jones, *Tribe of Black Ulysses: African American Lumber Workers in the Jim Crow South* (Urbana: University of Illinois Press, 2005).

15. Pietro Corsi, *The Age of Lamarck: Evolutionary Theories in France 1790–1830* (Berkeley: University of California Press, 1989); Frederick Wollaston Hutton, *Darwinism and Lamarckism: Old and New* (London: Duckworth, 1899); Alfred Kelly, *The Descent of Darwin: The Popularization of Darwinism in Germany, 1860–1914* (Chapel Hill: University of North Carolina Press, 1981); Stephen G. Alter, *Darwinism and the Linguistic Image: Language, Race, and Natural Theology in the Nineteenth Century* (Baltimore: Johns Hopkins University Press, 1999); Joseph Carroll, *Literary Darwinism: Evolution, Human Nature, and Literature* (London: Routledge, 2004); Peter Morton, *The Vital Science: Biology and the Literary Imagination, 1860–1900* (Boston: Allen and Unwin, 1984); Tim Gray, *The Political Philosophy of Herbert Spencer: Individualism and Organicism* (Aldershot: Avebury, 1996); Alan Grafen, *Evolution and Its Influence* (Oxford: Clarendon Press, 1989).

16. Ronald L. Numbers, *Darwinism Comes to America* (Cambridge: Harvard University Press, 1998); Louis Agassiz, *Agassiz on Evolution* (Bristol: Thoemmes Press, 2003); George C. Fry, *Congregationalists and Evolution: Asa Gray and Louis Agassiz* (Lanham, Md.: University Press of America, 1989).

17. Dorothy Ross, *The Origins of American Social Science* (New York: Cambridge University Press, 1991); Michael A. Bernstein, *A Perilous Progress: Economists and Public Purpose in Twentieth-Century America* (Princeton: Princeton University Press, 2001). On the relationship of theories of evolution to academic anthropology, see George W. Stocking Jr., *Race, Culture, and Evolution: Essays in the History of Anthropology* (Chicago: University of Chicago Press, 1982 [1968]); George W. Stocking Jr., *Victorian Anthropology* (New York: Free Press, 1987).

18. Daniel T. Rodgers, *Atlantic Crossings: Social Politics in a Progressive Age* (Cambridge: Belknap Press, 1998).

19. Albert Galloway Keller, *Societal Evolution: A Study of the Evolutionary Basis of the Science of Society* (New York: Macmillan, 1916), 177–80.

20. Edward A. Ross, "Ethnological Notes," Edward A. Ross Papers, Wisconsin Historical Society, Box 26, Folder 11, 15–17.

21. *Reports of the Industrial Commission on Immigration, Including Testimony, with Review and Digest and Special Reports, and on Education, Including Testimony, with Review and Digest* (Washington, D.C.: GPO, 1901); Ross, *The Old World in the New*; John R. Commons, *Races and Immigrants in America* (New York: Macmillan, 1920 [1907]), 63–106; Daniel Garrison Brinton, *Races and Peoples: Lectures on the Science of Ethnography* (New York: N. D. C. Hodges, 1890); Brinton, *The Basis of Social Relations: A Study of Ethnic Psychology* (New York: Putnam, 1902).

22. W. I. Thomas, "The Mind of Woman and the Lower Races," *American Journal of Sociology* 12 (January 1907): 457; Bert Bender, *The Descent of Love: Darwin and the Theory of Sexual Selection in American Fiction, 1871–1926* (Philadelphia: University of Pennsylvania Press, 1996); Wendy Kline, *Building a Better Race: Gender, Sexuality, and Eugenics from the Turn of the Century to the Baby Boom* (Berkeley: University of California Press, 2005), 7–94.

23. Edward A. Ross, "Western Civilization and the Birth-Rate," *American Journal of Sociology* 12 (March 1907): 607–32, with responses by Frank Fetter and William Bailey; Miriam King and Steven Ruggles, "American Immigration, Fertility, and Race Suicide at the Turn of the Century," *Journal of Interdisciplinary History* 20 (Winter 1990): 347–69; Robert Eldridge Bouwman, "Race Suicide, Some Aspects of Race Paranoia in the Progressive Era." (PhD diss., Emory University, 1975); Carroll Smith-Rosenberg and Charles Rosenberg, "The Female Animal: Medical and Biological Views of Woman and Her Role in Nineteenth Century America," *Journal of American History* 60 (September 1973): 332–56; Laura Briggs, *Reproducing Empire: Race, Sex, Science, and U.S. Imperialism in Puerto Rico* (Berkeley: University of California Press, 2002).

24. Daniel Pick, *Faces of Degeneration: A European Disorder, c.1848–c.1918* (Cambridge: Cambridge University Press, 1989); Sander Gilman and Edward J. Chamberlin, eds., *Degeneration: The Dark Side of Progress* (New York: Columbia University Press, 1985); Louis S. Warren, "Buffalo Bill Meets Dracula: William F. Cody, Bram Stoker, and the Frontiers of Racial Decay," *American Historical Review* 107 (October 2002): 1124–57.

25. Mark Pittenger, *American Socialists and Evolutionary Thought, 1870–1920* (Madison: University of Wisconsin Press, 1993); Mina Julia Carson, *Settlement Folk: Social Thought and the American Settlement Movement, 1885–1930* (Chicago: University of Chicago Press, 1990); Kathryn Kish Sklar, *Florence Kelley and the Nation's Work* (New Haven: Yale University Press, 1995); Daphne Spain, *How Women Saved the City* (Minneapolis: University of Minnesota Press, 2001); Shannon Jackson, *Lines of Activity: Performance, Historiography, Hull-House Domesticity* (Ann Arbor: University of Michigan Press, 2000); Ruth Crocker, *Social Work and Social Order: The Settlement Movement in Two Industrial Cities, 1889–1930* (Urbana: University of Illinois Press, 1992); Alan Dawley, *Changing the World: American Progressives in War and Revolution* (Princeton: Princeton University Press, 2003); Glenda Elizabeth Gilmore, *Who Were the Progressives?* (Boston: Bedford/St. Martin's, 2002); Michael McGerr, *A Fierce Discontent: The Rise and Fall of the Progressive Movement in America, 1870–1920* (New York: Free Press, 2003); Mari Jo Buhle, *Women and American Socialism, 1870–1920* (Urbana: University of Illinois Press, 1981).

26. Lawrence J. Friedman and Mark D. McGarvie, eds., *Charity, Philanthropy, and Civility in American History* (New York: Cambridge University Press, 2004); Alice O'Connor, *Poverty Knowledge: Social Science, Social Policy, and the Poor in Twentieth-Century U.S. History* (Princeton: Princeton University Press, 2002), 23–73.

27. Elof Axel Carlson, "R. L. Dugdale and the Jukes Family: A Historical Injustice Corrected," *BioScience* 30 (August 1980): 535–39.

28. John Higham, *Strangers in the Land: Patterns of American Nativism, 1860–1925* (New Brunswick, N.J.: Rutgers University Press, 2002), 131–86; Donald K. Pickens, *Eugenics and the Progressives* (Nashville: Vanderbilt University Press, 1968); Daniel J. Kevles, *In the Name of Eugenics: Genetics and the Uses of Human Heredity* (Cambridge: Harvard University Press, 1998); Alexandra Minna Stern, *Eugenic Nation: Faults and Frontiers of Better Breeding in Modern America* (Berkeley: University of California Press, 2005).

29. Mae M. Ngai, *Impossible Subjects: Illegal Aliens and the Making of Modern America* (Princeton: Princeton University Press, 2004); David G. Gutierrez, *Walls and Mirrors: Mexican Americans, Mexican Immigrants, and the Politics of Ethnicity* (Berkeley: University of California Press, 1995), 39–116.

30. Lothrop Stoddard, *The Revolt Against Civilization: The Menace of the Under Man* (New York: Charles Scribner's Sons, 1922); Matthew Pratt Guterl, *The Color of Race in America, 1900–1940* (Cambridge: Harvard University Press, 2001), 100–192.

31. Matthew Frye Jacobson, *Whiteness of a Different Color: European Immigrants and the Alchemy of Race* (Cambridge: Harvard University Press, 1998); David R. Roediger, *Working toward Whiteness: How America's Immigrants Became White; The Strange Journey from Ellis*

Island to the Suburbs (New York: Basic Books, 2005); Thomas A. Guglielmo, *White on Arrival: Italians, Race, Color, and Power in Chicago, 1890–1945* (New York: Oxford University Press, 2004).

32. On notions of turn-of-the-last-century anxiety, see Rebecca Edwards, *New Spirits: Americans in the Gilded Age, 1865–1905* (New York: Oxford University Press, 2005); Eric Rauchway, *Murdering McKinley: The Making of Theodore Roosevelt's America* (New York: Hill and Wang, 2004).

33. Jack London, *The People of the Abyss* (London: T. Nelson, 1914 [1902]); London, *The Iron Heel* (London: Everett & Co., 1908); "The People of the Abyss," *Guild Journal* 2 (October 1911): 3, in Records of the University Settlement Society of New York City—Jacob S. Eisinger Collection, Wisconsin Historical Society, Series III, Box 4. See also Joseph McLaughlin, *Writing the Urban Jungle: Reading Empire in London from Doyle to Eliot* (Charlottesville: University Press of Virginia, 2000).

1. Cavemen in the Progressive Era

1. Jack London, "The Strength of the Strong," in London, *The Strength of the Strong* (New York: Leslie-Judge Company, 1914 [1908]), 1–33; William Graham Sumner, *What Social Classes Owe Each Other* (New York: Harper, 1884).

2. On London's interest in prehistory, see Bennett Lovett-Graff, "Prehistory as Posthistory: The Socialist Science Fiction of Jack London," *Jack London Journal* 3 (November 1996): 88–104.

3. Such an evolutionary interpretation of industrialization contrasts sharply with the understandings advanced by contemporary historians. See, for example, Joyce Appleby, "The Vexed Story of Capitalism Told by American Historians," *Journal of the Early Republic* 21 (Spring 2001): 1–18; Steven Hahn and Jonathan Prude, *The Countryside in the Age of Capitalist Transformation: Essays in the Social History of Rural America* (Chapel Hill: University of North Carolina Press, 1985); Prude, *The Coming of Industrial Order: Town and Factory Life in Rural Massachusetts, 1810–1860* (Cambridge: Cambridge University Press, 1983); Alexander Trachtenberg, *The Incorporation of America: Culture and Society in the Gilded Age*, 25th anniversary ed. (New York: Hill and Wang, 2007), 38–139.

4. By contrast, see such influential readings of industrialization as grounded in innovations of technology or management as David Hounshell, *From the American System to Mass Production, 1800–1932: The Development of Manufacturing Technology in the United States* (Baltimore: Johns Hopkins University Press, 1985); David Montgomery, *Workers' Control in America: Studies in the History of Work, Technology, and Labor Struggles* (Cambridge: Cambridge University Press, 1979); Ken Fones-Wolf, *Glass Towns: Industry, Labor, and Political Economy in Appalachia, 1890–1930s* (Urbana: University of Illinois Press, 2007).

5. W. M. Flinders Petrie, "Migrations," *Man* 6 (1906): 170.

6. See, for example, Carl Bücher, *Industrial Evolution*, trans. S. Morley Wickett from the 3rd German ed. (New York: Henry Holt & Company, 1901), 1–82; Franklin H. Giddings, "The Economic Ages," *Political Science Quarterly* 16 (June 1901): 193–221; Simon Patten, *Heredity and Social Progress* (New York: Macmillan, 1903); Carroll D. Wright, *The Industrial Evolution of the United States* (New York: Chautauqua-Century Press, 1895); Charlotte Perkins Gilman, *Women and Economics: A Study of the Economic Relation between Men and Women as a Factor in Social Evolution* (Berkeley: University of California Press, 1998 [1898]), 116. On the origins of the idea of the social organism, see Walter Simon, "Herbert Spencer and the 'Social Organism'," *Journal of the History of Ideas* 21 (April–June 1960): 294–99; "Ward's Dynamic Sociology," *Science* 25 (July 27, 1883): 105–8; Franklin Giddings, "The Sociological Character

of Political Economy," *Publications of the American Economic Association* 3 (March 1888): 29–47; Henry Dyer, *The Evolution of Industry* (London: Macmillan, 1895), 26–60.

7. Henry C. Adams, "Another View of Economic Laws and Methods," *Science* 182 (July 30, 1886): 103–5; William Graham Sumner, "Strikes and the Industrial Organization," typescript, *Popular Science Monthly* (July 1887), in William Graham Sumner Papers, Group Number 291, Yale University, Manuscripts and Archives, Folder 989; John Bates Clark, "The Law of Wages and Interest," *Annals of the American Academy of Political and Social Science* 1 (July 1890): 43–65; William Smart, "The Place of Industry in the Social Organism," *International Journal of Ethics* 3 (July 1893): 437–51. See also Dorothy Ross, *The Origins of American Social Science* (New York: Cambridge University Press, 1991); Jeffrey Sklansky, *The Soul's Economy: Market Society and Selfhood in American Thought, 1820–1920* (Chapel Hill: University of North Carolina Press, 2002), 105–224; Helene Silverberg, "'A Government of Men': Gender, the City, and the New Science of Politics," in *Gender and American Social Science,* ed. Silverberg (Princeton: Princeton University Press, 1998), 156–84; Theodore Porter and Dorothy Ross, eds., *The Cambridge History of Science,* vol. 7, *The Modern Social Sciences* (Cambridge: Cambridge University Press, 2003), 205–404; Eileen Janes Yeo, *The Contest for Social Science: Relations and Representations of Gender and Class* (London: Rivers Oram Press, 1996).

8. Lester Frank Ward, *Dynamic Sociology,* vol. 1 (New York: D. Appleton, 1911 [1883]), 15–20; Edward C. Rafferty, *Apostle of Human Progress: Lester Frank Ward and American Political Thought, 1841–1913* (Lanham, Md.: Rowman and Littlefield, 2003), 141–306.

9. Franklin Henry Giddings, *Principles of Sociology* (New York: Macmillan, 1908), 7; John W. Petras, "Images of Man in Early American Sociology," pts. 1 and 2, *Journal of the History of the Behavioral Sciences* 6 (Summer and Autumn 1970): 231–40, 317–34.

10. A. W. Small, "The Era of Sociology," *American Journal of Sociology* 1 (July 1895): 1–2; J. Howard Moore, "Savage Survivals in Higher Peoples," *International Socialist Review* 16 (October 1915): 217–21.

11. Daniel T. Rodgers, *Atlantic Crossings: Social Politics in a Progressive Age* (Cambridge: Harvard University Press, Belknap Press, 1998), 101–7; Benjamin Rader, *The Academic Mind and Reform: The Influence of Richard T. Ely in American Life* (Lexington: University of Kentucky Press, 1966); Rosanne Currarino, "The Politics of 'More': The Labor Question and the Idea of Economic Liberty in Industrial America," *Journal of American History* 93 (June 2006): 17–36.

12. Thorstein Veblen, "Why Is Economics Not an Evolutionary Science?" *Quarterly Journal of Economics* 12 (July 1898): 373–97; Thorstein Veblen, "Industrial and Pecuniary Employments," *Publications of the American Economic Association* 2 (February 1901): 190–235. For a debate on Veblen's engagement with biology, see Ann Jennings and William Waller, "The Place of Biological Science in Veblen's Economics," *History of Political Economy* 30 (Summer 1998): 189–217; Geoffrey Hobson, *Economics and Evolution: Bringing Life Back into Economics* (Ann Arbor: University of Michigan Press, 1997), 123–38.

13. Richard T. Ely, "Competition: Its Nature, Its Permanency, and Its Beneficence," *Publications of the American Economic Association* 2 (February 1901): 55–70; Ely, *Studies in the Evolution of Industrial Society* (New York: Chautauqua Press, 1903), 154.

14. Mark Pittenger, *American Socialists and Evolutionary Thought, 1870–1920* (Madison: University of Wisconsin Press, 1993); Murray King, "Socialism and Human Nature: Do They Conflict?" *International Socialist Review* 5 (December 1905): 328–29; emphasis in original. For similar socialist engagements with evolutionary theory to naturalize the development of socialism, see E. Untermann, "Evolution or Revolution," *International Socialist Review* 1 (January 1901), 406–12; J. W. Summers, "Socialism and Science," *International Socialist Review* 2 (April 1902): 740–48.

15. See, for example, Daniel G. Brinton, *The Basis of Social Relations: A Study in Ethnic Psychology* (New York: G. P. Putnam's Sons, 1902); Maurice Parmelee, *The Science of Human Behavior* (New York: Macmillan, 1913).

16. Frankenthal Y. Weissenburg, "The Ways of the Ant," *International Socialist Review* 16 (April 1916): 606–8.

17. Veblen, "Why Is Economics Not an Evolutionary Science?" 384; J. W. Powell, "Competition as a Factor in Human Evolution," *American Anthropologist* 1 (October 1888), 297–324; S. J. Holmes, *Studies in Evolution and Eugenics* (New York: Harcourt, Brace and Company, 1923); F. Stuart Chapin, *The Introduction to the Study of Social Evolution, The Prehistoric Period* (New York: Century Co., 1921 [1913]); Alexander Sutherland, *The Origin and Growth of the Moral Instinct, Volumes I and II* (London: Longmans, Green, 1898); Thomas E. Snyder, "Communism Among Insects," *Scientific Monthly* 21 (November 1925): 466–77; Carl Kelsey, *The Physical Basis of Society* (New York: D. Appleton, 1916), 64.

18. Ward, *Dynamic Sociology*, 30–31; A. L. Kroeber, "The Superorganic," *American Anthropologist* 19 (April–June 1917): 163–213; Franklin Giddings, "Lecture Notes," Franklin Giddings Papers, Columbia University, Box 3, Folder: "Lecture Notes, 1907–8."

19. Giddings, *Principles of Sociology*, 7, 206–207.

20. Ward, *Dynamic Sociology*, 35.

21. Benjamin Kidd, *Social Evolution* (London: Macmillan, 1895); Stephen Sanderson, *Social Evolutionism* (Oxford: Basil Blackwell, 1990); David Paul Crook, *Benjamin Kidd: Portrait of a Social Darwinist* (Cambridge: Cambridge University Press, 1984).

22. Giddings, *Principles of Sociology*, 14; William Patten, *The Grand Strategy of Evolution* (Boston: Gotham Press, 1920), 358–76; Charles A. Tuttle, "The Wealth Concept: A Study in Economic Theory," *Annals of the American Academy of Political and Social Science* 1 (April 1891): 626.

23. Bücher, *Industrial Evolution*, 6–7.

24. Ely, *Evolution of Industrial Society*, 13–66.

25. "C.L. James to J.B. Andrews" (November 27, 1907), John R. Commons Papers, Wisconsin Historical Society, Box 1, Folder 4; John R. Commons, "A Sketch of Industrial Evolution," *Quarterly Journal of Economics* 24 (November 1909): 39–84.

26. John R. Commons et al., *The History of Labour in the United States* (New York: Macmillan, 1918), 25–31.

27. Emile Vandervelde, *Collectivism and Industrial Evolution*, trans. Charles H. Kerr (Chicago: Charles Kerr, 1904); Jack London, "The Dream of Debs," in London, *Strength of the Strong*, 134–76; E. Backus, "Industrial Evolution and Socialist Tactics," *International Socialist Review* 7 (January 1907): 411–14; King, "Socialism and Human Nature," 330–33.

28. Ely, *Studies in the Evolution of Industrial Society*, 13–18; Charles R. Flint, James J. Hill, James H. Bridge, S. C. T. Dodd, and Francis B. Thurber, *The Trust: Its Book, Being a Presentation of the Several Aspects of the Latest Form of Industrial Evolution* (New York: Doubleday, Page & Company, 1902), xiii–1; E. Dana Durand et al., "Recent Trust Decisions and Business Discussion," *American Economical Review* 4 (March 1914): 173–95.

29. Too often scholars have understood industrialization merely as process, rather than as a term with an intellectual history of its own. See Maury Klein, *The Genesis of Industrial America, 1870–1920* (New York: Cambridge University Press, 2007); David R. Meyer, *The Roots of American Industrialization* (Baltimore: Johns Hopkins University Press, 2003).

30. Bücher, *Industrial Evolution*, 1–82; Veblen, "Why Is Economics Not an Evolutionary Science?" 379–380.

31. "Industrial Evolution" (Lecture, April 2, 1912), Richard Ely Papers, Wisconsin Historical Society, Box 28, Folder 12; "Lecture notes," Summer School of the South (June 28–August 5, 1904), William Jewell College (October 1904), Drew Theological Seminary (April 1906), and "Abstract for an Address on the Evolution of Industrial Society," all in Richard Ely Papers, Box 47, Folder 8.

32. Adam Kuper, *The Reinvention of Primitive Society: Transformations of an Illusion* (New York: Routledge, 1988), 119; Curtis Hinsley, *Savages and Scientists: The Smithsonian Institution and the Development of American Anthropology, 1846–1910* (Washington, D.C.: Smithsonian

Institution Press, 1981); Franz Boas, "The Occurrence of Similar Inventions in Areas Widely Apart" (1911), in *A Franz Boas Reader,* ed. George W. Stocking Jr. (Chicago: University of Chicago Press, 1974), 61–67.

33. Katharine Coman, *The Industrial History of the United States* (New York: Macmillan, 1918 [1905]), 7–8; William James Ghent, *Mass and Class: A Survey of Social Divisions* (New York: Macmillan, 1904), 1–8.

34. Hubert Howe Bancroft, *The Book of the Fair,* vol. 1 (New York: Bounty Books, 1894), 136–41.

35. Chapin, *Social Evolution,* xv–xvii.

36. W. I. Thomas, *Source Book for Social Origins* (Chicago: University of Chicago Press, 1909), 3–28; Carroll D. Wright, *The Battles of Labor* (Philadelphia: George W. Jacob & Co., 1906), esp. 9–56, 165–68.

37. Chapin, *Social Evolution,* frontispiece.

38. Katherine Elizabeth Dopp, *The Later Cave-Men* (Chicago: Rand McNally, 1906).

39. Margaret A. McIntyre, *The Cave Boy in the Age of Stone* (New York: D. Appleton, 1907).

40. Franklin Giddings, *Pagan Poems* (New York: Macmillan, 1914), 19–59.

41. Henry W. Haynes, "Progress of American Archeology during the Past Ten Years," *American Journal of Archeology* 4 (January–March 1900): 17–39. See also Cyrus Thomas, *Introduction to the Study of North American Archeology* (New York: AMS Press, 1973 [1898]); Otis Tufton Mason, "Aboriginal American Zootechny," *American Anthropologist* 1 (January 1899): 45–81; Warren Moorehead, *Primitive Man in Ohio* (New York: Putnam, 1892). The Bureau of Ethnology reports were published annually starting in 1884.

42. Franz Boas, *The Mind of Primitive Man* (New York: Macmillan, 1911); Robert Harry Lowie, *Primitive Society* (New York: Bonie and Liveright, 1920); Robert Francis Murphy, *Robert H. Lowie* (New York: Columbia University Press, 1972); Michel Verdon, "The World Upside Down: Boas, History, Evolutionism, and Science," *History and Anthropology* 17 (September 2006): 171–87; Vernon J. Williams Jr., *Rethinking Race: Franz Boas and His Contemporaries* (Lexington: University Press of Kentucky, 1996).

43. Franz Boas, *Anthropology and Modern Life* (New York: W. W. Norton, 1928). For examples of his work on Northwest Coast cultures, see Boas, *Kwakiutl Tales* (New York: Columbia University Press, 1910); Boas, *The Kwakiutl of Vancouver Island* (Leiden: E. J. Brill, 1905). For examples of Lowie's work, see Robert Harry Lowie, *Myths and Traditions of the Crow Indians* (New York: American Museum of Natural History, 1918).

44. McIntyre, *Cave Boy,* chaps. 15–19.

45. A. L. Kroeber, "Ishi, the Last Aborigine: The Effects of Civilization on a Genuine Survivor of Stone Age Barbarism," *World's Work* 24 (July 1912): 304–12; Ira Jacknis, "The First Boasian: Alfred Kroeber and Franz Boas, 1896–1905," *American Anthropologist* 104 (June 2002): 520–32; Rachel Adams, *Sideshow USA: Freaks and the American Cultural Imagination* (Chicago: University of Chicago Press, 2001), 17–59; Orin Starn, *Ishi's Brain: In Search of America's Last "Wild" Indian* (New York: W. W. Norton, 2004).

46. J. W. Powell, "Technology, or the Science of Industries," *American Anthropologist* 1 (April 1899): 319–49.

47. Mary Marcy, "Stories of the Cave People," *International Socialist Review* 9 (April 1909): 765–71; (May 1909): 845–51; (June 1909): 959–64; 10 (July 1909): 23–31; (August 1909): 144–48; (September 1909): 212–19; (October 1909): 327–32; 16 (October 1915): 233–39; (November 1915): 279–83; (December 1915): 353–56; (January 1916): 424–27; (February 1916): 479–83; (March 1916): 532–35; (May 1916): 672–75; (June 1916): 738–41.

48. "The Gorilla's Divine Unrest: How the Ape's Discontent with Economic Conditions Caused the Origin of Man," *International Socialist Review* 15 (July 1914): 48–54.

49. Thorstein Veblen, "The Instinct of Workmanship and the Irksomeness of Labor," *American Journal of Sociology* 4 (September 1898): 167–201.

50. Franklin Giddings, *Elements of Sociology* (New York: Macmillan, 1918), 13–171; Lester Frank Ward, *Pure Sociology* (New York: Macmillan, 1903); George W. Stocking Jr., "The Turn-of-the-Century Concept of Race," *Modernism/Modernity* 1 (January 1, 1994): 4–16.

51. Giddings, *Principles of Sociology,* 19; George E. Winkler, "The Survival of the Fittest," *International Socialist Review* 10 (February 1910): 740; Chapin, *Social Evolution,* 203–33, 268–77.

52. Giddings, *Principles of Sociology,* 153–57; Giddings, *Elements of Sociology,* 254–89; John Ferguson McLennan, *Studies in Ancient History* (New York: Macmillan, 1896), 97.

53. Giddings, *Principles of Sociology,* 157–68.

54. "C.L. James to J.B. Andrews," John R. Commons Papers.

55. Thomas Nixon Carver, *Essays in Social Justice* (Cambridge: Harvard University Press, 1915), 18, 48, 54–56.

56. William Z. Ripley, *The Races of Europe* (New York: D. Appleton, 1899), 1–47; Ellen Churchill Semple, *Influences of Geographic Environment on the Basis of Ratzel's System of Anthro-Geography* (New York: Russell & Russell, 1968 [1911]), 77–88.

57. Giddings, *Principles of Sociology,* 233, 309; Semple, *Influences of Geographic Environment,* 89–98.

58. Edward A. Ross, "The Causes of Race Superiority," *Annals of the American Academy of Political and Social Science* 18 (July 1901): 71–73.

59. Grant Allen, *Common Sense Science* (Boston: Lothrop, 1886), 7–41; Giddings, *Principles of Sociology,* 329.

60. Richard Wellington Husband, "Race Mixture in Early Rome," *Transactions and Proceedings of the American Philological Association* 40 (1909): 63–81; Alexander F. Chamberlain, "'Fairness' in Love and War," *Journal of American Folklore* 20 (January–March 1907): 1–15.

61. Stanley Waterloo, *The Story of Ab* (Chicago: Way and Williams, 1897); Jack London, *Before Adam* (New York: Macmillan, 1907); Ashton Hilliers, *The Master-Girl, A Romance* (New York: G. P. Putnam's Sons, 1910); Charles Abbott, *The Cliff Dweller's Daughter* (New York: Neely, 1899); Austin Bierbower, *From Monkey to Man or, Society in the Tertiary Age: A Story of the Missing Link, Showing the First Steps in Industry* (Chicago: Dibble, 1894).

62. Ripley, *Races of Europe,* 30; Heather Winlow, "Mapping Moral Geographies: W. Z. Ripley's Races of Europe and the United States," *Annals of the Association of American Geographers* 96 (March 2006): 119–41.

63. Charles Edward Woodruff, *Expansion of Races* (New York: Rebman Company, 1909).

64. Francis A. Walker, "The Tide of Economic Thought," *Publications of the American Economic Association* 6 (January–March 1891): 35–37; Ripley, *Races of Europe,* 57.

65. Ripley, *Races of Europe,* 37–38.

66. Ripley, *Races of Europe,* 37–77; Semple, *Influences of Geographic Environment,* 38–40; John Cummings, "Ethnic Factors and the Movement of Population," *Quarterly Journal of Economics* 14 (February 1900): 171–211.

67. Ripley, *Races of Europe,* 180–452.

68. Waterloo, *Story of Ab,* 356.

69. Giddings, *Principles of Sociology,* 316. For a similar argument that rooted the origins of class difference in ancient race conflict, see Thorstein Veblen, *Theory of the Leisure Class: An Economic Study in the Evolution of Institutions* (New York: Macmillan, 1899).

70. United States Immigration Commission, *Dictionary of Races or Peoples* (Washington, D.C.: GPO, 1911), 13–23.

71. King, "Socialism and Human Nature," 330–37; J. Howard Moore, "Savage Survivals in Higher Peoples," *International Socialist Review* 16 (August–December 1915): 105–9, 161–64, 217–21, 295–99, 360–63; Ross, "The Causes of Race Superiority," 78–79; Brooks Adams, *Law of Civilization and Decay* (London: Swan, 1895).

72. Giddings, *Principles of Sociology,* 11; Carver, *Essays in Social Justice,* 32.

73. Herbert L. Osgood, "Scientific Socialism," *Political Science Quarterly* 1 (December 1886): 585.

74. Charles Wellington Furlong, "Some Effects of Environment on the Fuegian Tribes," *Geographical Review* 3 (January 1917): 1–15. For a cephalic readings of Fuegian racial characteristics and ancient history, see Frederic Ward Putnam, "A Problem in American Anthropology," *Science* 10 (August 25, 1899): 225–236; Francisco P. Moreno, "Notes on the Anthropogeography of Argentina," *Geographical Journal* 18 (December 1901): 574–89.

75. Ward, *Dynamic Sociology,* 16; Carver, *Essays in Social Justice,* 88.

76. Ward, *Dynamic Sociology,* 43.

2. Mapping Civilization

1. On Huntington, see Geoffrey Martin, *Ellsworth Huntington: His Life and Thought* (Hamden, Conn.: Archon Books, 1973); James M. Smythe, *Huntington and Bowman: A Comparative Study of Their Geographic Concepts* (MA thesis, University of Toronto, 1952).

2. "List of Contributors to the Map of Civilization," Ellsworth Huntington Papers, Yale University, Manuscripts and Archives, Series V, Box 10, Folder 69.

3. "Draft of instructions, with details about slips," Huntington Papers, Series V, Box 10, Folder 69.

4. "Map of Civilization," Huntington Papers, Series IV, Box 6, Folder 34.

5. "Map of Civilization, Non-Contributors," Huntington Papers, Series IV, Box 6, Folder 34.

6. Ellsworth Huntington, *Civilization and Climate* (New Haven: Yale University Press, 1924 [1915]), 251–53.

7. Huntington, *Civilization and Climate,* 256–74.

8. S. C. GilFillan, "The Coldward Course of Progress," *Political Science Quarterly* 35 (September 1920): 393–410. Within the field of geography, Huntington's theories are the most emphatic articulations of environmental determinism. To place such theories solely within geography is to miss their wider public and social scientific resonance. See Richard Peet, "The Social Origins of Geographic Determinism," *Annals of the Association of American Geographers* 75 (1985): 309–33; Neil Smith, *American Empire: Roosevelt's Geographer and the Prelude to Globalization* (Berkeley: University of California Press, 2003), 31–82; Susan Schulten, *The Geographical Imagination in America* (Chicago: University of Chicago Press, 2001), 69–118.

9. There has been a flurry of works that have attempted to place the far-reaching cultural notion of civilization within the context of American imperial expansion. Interestingly, the notion of labor has effectively been separated from the discourse of civilization. See Gail Bederman, *Manliness and Civilization: A Cultural History of Gender and Race in the United States, 1880–1917* (Chicago: University of Chicago Press, 1995); Jessica Blatt, "'To Bring Out the Best That Is Their Blood': Race, Reform, and Civilization in the *Journal of Race Development* (1910–1919)," *Ethnic and Racial Studies* 27, no. 5 (2004): 691–709. See also Angel Smith and Emma Aurora Davila Cox, eds., *The Crisis of 1898: Colonial Redistribution and Nationalist Mobilization* (New York: St. Martin's Press, 1999) and Joseph M. Henning, *Outposts of Civilization: Race, Religion, and the Formative Years of American-Japanese Relations* (New York: New York University Press, 2000).

10. This analysis joins efforts to place empire at the center of American cultural history. For a concise statement of this project, see Amy Kaplan, "Left Alone with America: The Absence of Empire in the Study of American Culture," in *Cultures of United States Imperialism,* ed. Amy Kaplan and Donald E. Pease (Durham: Duke University Press, 1993), 3–21. For histories of American imperialism in the Philippines, see Paul A. Kramer, *The Blood of*

Government: Race, Empire, the United States, and the Philippines (Chapel Hill: University of North Carolina Press, 2006); Stuart Creighton Miller, *"Benevolent Assimilation": The American Conquest of the Philippines, 1899–1903* (New Haven: Yale University Press, 1982); Angel Velasco Shaw and Luis Francia, eds., *Vestiges of War: The Philippine-American War and the Aftermath of an Imperial Dream, 1899–1999* (New York: New York University Press, 2002).

11. Matthew Frye Jacobson, *Barbarian Virtues: The United States Encounters Foreign Peoples at Home and Abroad, 1876–1917* (New York: Hill and Wang, 2000); Julie Greene, "The Labor of Empire: Recent Scholarship on U.S. History and Imperialism," *Labor: Studies in Working-Class History of the Americas* 1 (Summer 2004): 113–29; Dana Frank, *Bananeras: Women Transforming the Banana Unions of Latin America* (Cambridge: South End Press, 2005); Nancy Hewitt, *Southern Discomfort: Women's Activism in Tampa, Florida, 1880s–1920s* (Urbana: University of Illinois Press, 2001); Lara Putnam, *The Company They Kept: Migrants and the Politics of Gender in Caribbean Costa Rica, 1870–1960* (Chapel Hill: University of North Carolina Press, 2002); Dorothy B. Fujita-Rony, *American Workers, Colonial Power: Philippine Seattle and the Transpacific West, 1919–1941* (Berkeley: University of California Press, 2003). These important new works tend to focus on the social history of workers and American managers in the colonies and former colonies to the exclusion of the intellectual discourse that connected labor and empire.

12. Karen Ordahl Kupperman, "Fear of Hot Climates in the Anglo-American Colonial Experience," *William and Mary Quarterly* 41 (April 1984): 213–40; Dane Kennedy, "The Perils of the Midday Sun: Climatic Anxieties in the Colonial Tropics," in *Imperialism and the Natural World,* ed. John M. MacKenzie (Manchester: Manchester University Press, 1990), 118–40; David Arnold, *The Problem of Nature: Environment, Culture and European Expansion* (Oxford: Blackwell, 1996); Philip D. Curtin, *Disease and Empire: The Health of European Troops in the Conquest of Africa* (Cambridge: Cambridge University Press, 1998).

13. Jacobson describes how American observers worried about the ability of American industry to thrive and grow if the nation did not expand its markets through some version of colonial expansion. Jacobson, *Barbarian Virtues,* 15–58.

14. David Arnold, *Colonizing the Body: State Medicine and Epidemic Disease in Nineteenth-Century India* (Berkeley: University of California Press, 1993); Arnold, "White Colonization and Labour in Nineteenth-Century India," *Journal of Imperial and Commonwealth History* 9 (January 1983): 133–58; David N. Livingstone, "Tropic Climate and Moral Hygiene: The Anatomy of a Victorian Debate," *British Journal for the History of Science* 32 (March 1999): 93–110. Most work, like those above, focuses on British debates. For a non-British example, see Gavin Bowd and Daniel Clayton, "Tropicality, Orientalism, and French Colonialism in Indochina: The Work of Pierre Gourou, 1927–1982," *French Historical Studies* 28 (Spring 2005): 297–327.

15. Patrick Manson, *Tropical Diseases: A Manual of the Diseases of Warm Climates* (London: Cassell, 1898); Sir James Ranald Martin, *Influence of Tropical Climates in Producing the Acute Endemic Diseases of Europeans* (London: John Churchill, 1861); D. H. Cullimore, *The Book of Climates* (London: Bailliere, Tindall, & Cox, 1891). The study of empire and tropical medicine has emerged as an important focus for historians of medicine and public health. See Roy MacLeod and Milton Lewis, eds., *Disease, Medicine and Empire: Perspectives on Western Medicine and the Experience of European Expansion* (London: Routledge, 1988); David Arnold, ed., *Warm Climates and Western Medicine* (Amsterdam: Rodopi, 1996); Philip D. Curtin, *Death by Migration: Europe's Encounter with the Tropical World* (Cambridge: Cambridge University Press, 1989); Mark Harrison, *Climates and Constitutions: Health, Race, Environment and British Imperialism in India, 1600–1850* (Oxford: Oxford University Press, 1999).

16. Lt.-Col. G. M. Giles, *Climate and Health in Hot Countries* (London: John Bale, Sons, 1904); Philip Curtin, "Medical Knowledge and Urban Planning in Tropical Africa," *American Historical Review* 90 (June 1985): 594–613; Dane Kennedy, *The Magic Mountains: Hill*

Stations and the British Raj (Berkeley: University of California Press, 1996); David Arnold, *The Tropics and the Traveling Gaze: India, Landscape, and Science, 1800–1856* (Seattle: University of Washington Press, 2006); Judith T. Kenny, "Climate, Race, and Imperial Authority: The Symbolic Landscape of the British Hill Station in India," *Annals of the Association of American Geographers* 85, no. 4(1995): 694–714; Eric T. Jennings, "From Indochine to Indochic: The Lang Bian/Dalat Palace Hotel and French Colonial Leisure, Power, and Culture," *Modern Asian Studies* 37 (February 2003): 159–94.

17. Ellen Churchill Semple, *Influences of Geographic Environment on the Basis of Ratzel's System of Anthro-Geography* (New York: Russell & Russell, 1968 [1911]); Charles C. Colby, "Ellen Churchill Semple," *Annals of the Association of American Geographers* 23 (December 1933): 229–40; Preston E. James, Wilford A. Bladen, and Pradyumna P. Karan, "Ellen Churchill Semple and the Development of a Research Paradigm," in *The Evolution of Geographic Thought in America: A Kentucky Root*, ed. Wilford A. Bladen and Pradyumna P. Karan (Dubuque, Iowa: Kendall/Hunt, 1983), 28–57.

18. Daniel T. Rodgers, *Atlantic Crossings: Social Politics in a Progressive Age* (Cambridge: Harvard University Press, Belknap Press, 1998), 33–111.

19. Huntington, *Civilization and Climate*, 41.

20. Huntington, *Civilization and Climate*, 30–52.

21. "Is Civilization Determined by Climate?" *William and Mary College Quarterly Historical Magazine* 24 (July 1915): 56–57.

22. Walter S. Tower, "The Climate of the Philippines," *Bulletin of the American Geographical Society* 35, no. 3 (1903): 259; W. D. Matthew, *Climate and Evolution* (New York: New York Academy of Science, 1915); "Health Conditions in the Philippines," *New York Times* (May 13, 1901), 6. See also David N. Livingstone, "Human Acclimatization: Perspectives on a Contested Field of Inquiry in Science, Medicine and Geography," *History of Science* 25 (1987): 359–94.

23. Robert DeCourcey Ward, *Climate Considered Especially in Relation to Man* (New York: G. P. Putnam's Sons, 1918 [1908]); Robert DeC. Ward, "Some Problems of the Tropics," *Bulletin of the American Geographical Society* 40, no. 1(1908): 7–11.

24. "Philippines Broke Health," *New York Times* (August 5, 1903), 7; "Health of Admiral Dewey," *New York Times* (March 10, 1899), 3.

25. "Sickness in the Philippines," *New York Times* (January 29, 1899), 5.

26. Kristin L. Hoganson, *Fighting for American Manhood: How Gender Politics Provoked the Spanish-American and Philippine-American Wars* (New Haven: Yale University Press, 1998), 180–99; "Health Conditions in the Philippines," 6;

27. Huntington, *Civilization and Climate*, 33–34.

28. Huntington, *Civilization and Climate*, 46–50.

29. "Expansion," *New York Times* (December 26, 1898), 6. See also "The Philippine Problem," *New York Times* (December 13, 1898), 6. See also Benjamin Kidd, *The Control of the Tropics* (New York: Macmillan, 1898).

30. Amos K. Fiske, "If the Philippines, Then Hawaii, Of Course, But—," *New York Times* (July 3, 1898), 16.

31. "Expansion," 6.

32. "Expansion," 6; William Z. Ripley, "Acclimitization," *Popular Science Monthly* 48 (April 1896), 779–93; Huntington, *Civilization and Climate*, 46–50; "For and Against Colonies," *New York Times* (October 19, 1898), 5.

33. Huntington, *Civilization and Climate*, 46–50; Theodore H. Boggs, "The Anglo-Saxon in India and the Philippines," *Journal of Race Development* 2 (January 1912): 309–22; Huntington, "The Adaptability of the White Man to Tropical America," *Journal of Race Development* 5 (October 1914): 185–211; S. P. Verner, "The White Race in the Tropics" *World's Work* 16 (September 1908): 10715–10720; D. G. Brinton, "The Beginning of Man and the Age of the Race," *Forum* 16 (December 1893): 452–458; "For and Against Colonies," 5.

34. Ward, *Climate*, 226–260; Huntington, *Civilization and Climate*, 30–43, 52, 65–75.

35. Charles Woodruff, *The Effects of Tropical Light on White Men* (New York: Rebman Company, 1905); Homer C. Stuntz, *The Philippines and the Far East* (New York: Jennings and Pye, 1904).

36. Warwick Anderson, *Colonial Pathologies: American Tropical Medicine, Race, and Hygiene in the Philippines* (Durham: Duke University Press, 2006); "Colonizing the Tropics," *New York Times* (March 20, 1902), 8; Robert R. Reed, *City of Pines: The Origins of Baguio as a Colonial Hill Station and Regional Capital*, Research Monograph 13 (Berkeley: Center for South and Southeast Asia Studies, 1976); Robert Reed, "Remarks on the Colonial Genesis of the Hill Station in Southeast Asia with Particular Reference to the Cities of Buitenzorg (Bogor) and Baguio," *Asian Profile* 4 (December 1976): 545–91.

37. *Reports of the Taft Philippine Commission* (Washington, D.C.: GPO, 1901), 64.

38. *Report of the Philippine Commission to the President, Vol. IV* (Washington, D.C.: GPO, 1901), 113–358; *Reports of the Taft Philippine Commission*, 65–71, 126–45, 162–221. The continuity with Spanish policy is remarkable. It represents, in effect, a private tacit acceptance of the whiteness of the Spanish, at odds with representations of the defeated Spanish as savage and nonwhite. See Maria DeGuzman, *Spain's Long Shadow: The Black Legend, Off-Whiteness, and Anglo-American Empire* (Minneapolis: University of Minnesota Press, 2005).

39. *Reports of the Taft Philippine Commission*, 65–66, 123, 148; Dean C. Worcester, *The Philippine Islands and Their People: A Record of Personal Observations and Experience* (New York: Macmillan, 1899); Rodney J. Sullivan, *Exemplar of Americanism: The Philippine Career of Dean C. Worcester* (Ann Arbor: Center for South and Southeast Asian Studies, 1991); Greg Bankoff, "'The Tree as the Enemy of Man': Changing Attitudes to the Forests of the Philippines 1565–1898," *Philippine Studies* 53, no. 1 (2005): 321–45.

40. Kramer, *Blood of Government*, 35–284.

41. *Reports of the Taft Philippine Commission*, 149–61, 188–90; Robert Rydell, *All the World's a Fair: Visions of Empire at American International Expositions, 1876–1916* (Chicago: University of Chicago Press, 1984), 105–25; Lew Carlson, "Giant Patagonians and Hairy Ainu: Anthropology Days at the 1904 St. Louis Olympics," *Journal of American Culture* 12 (Fall 1989): 19–26; Paul Kramer, "Making Concessions: Race and Empire Revisited at the Philippine Exposition, St. Louis, 1901–1905," *Radical History Review* 73 (Winter 1999): 74–114; W. J. McGee, "Strange Races of Men," *World's Work* 8 (August 1904): 5185–88; Frederick Starr, "Anthropology at the World's Fair," *Popular Science Monthly* 43 (September 1893): 610–21. For a typical example of ideas about racial "losers" retreating to the hills, see William Z. Ripley, *The Races of Europe* (New York: D. Appleton, 1899), 289–91.

42. *Reports of the Taft Philippine Commission*, 205–7, 245.

43. "Will Make Manila a Brand New City," *New York Times* (September 18, 1904), 1; "Philippines' New Capital, Baguio to Supersede Manila for the Summer for Hygienic Reasons," *New York Times* (April 10, 1904), 4. See also David Brody, "Building Empire: Architecture and American Imperialism in the Philippines," *Journal of Asian American Studies* 4 (June 2001): 123–45; Robert W. Taylor, "The Best of Burnham: The Philippine Plans," *Planning History* 16, no. 3 (1994): 14–17; Thomas S. Hines, "The Imperial Façade: Daniel H. Burnham and American Architectural Planning in the Philippines," *Pacific Historical Review* 41 (February 1972): 33–53.

44. Alexander Supan, *Die territoriale Entwicklung der Europäischen Kolonien; mit einem kolonialgeschichtlichen Atlas von 12 Karten und 40 Kärtchen im Text* (Gotha: Perthes, 1906); Supan, *Grundzüge der physischen Erdkunde* (Leipzig: Veit & Comp., 1903); Waldimir Köppen, *Wind und Wetter in den europäischen Gewässern; ein Ratgeber zur Beurteilung der Wetterlage* (Berlin: E. S. Mittler, 1917); Köppen, *Die Klimate der Erde, Grundriss der Klimakunde* (Berlin: Walter de Gruyter, 1923).

45. Ward, *Climate*, 55–60; Robert DeC. Ward, "The Climatic Zones and their Subdivisions," *Bulletin of the American Geographical Society* 37, no. 7 (1905): 385–96; Robert DeC.

Ward, "Notes on Climatology," *Bulletin of the American Geographical Society* 33, no. 1 (1901): 47–51; Robert DeC. Ward, "Notes on Climatology," *Bulletin of the American Geographical Society* 33, no. 2 (1901): 150–53; Robert DeC. Ward, "Notes on Climatology," *Bulletin of the American Geographical Society* 33, no. 5 (1901): 412–15.

46. "Underwood and Underwood to Ellsworth Huntington" (June 6, 1913), Huntington Papers, Series III, Box 31, Folder 625; "Huntington to Underwood and Underwood" (July 28, 1913), Huntington Papers, Series III, Box 31, Folder 625.

47. John R. Commons, "Amalgamation and Assimilation," *Chautauquan* 39 (May 1904): 222; Ward, "Some Problems of the Tropics," 7; Walter S. Tower, "Geographic Influences in the Evolution of Nations," *Popular Science Monthly* 78 (February 1911): 164–83; "Effects of Climate on Character," *Review of Reviews* 35 (March 1907): 377–79.

48. Huntington, "The Adaptability of White Man to Tropical America," 186–87.

49. Ward, *Climate*, 228–29.

50. Alleyne Ireland, *Tropical Colonization: An Introduction to the Study of the Subject* (New York: Macmillan, 1899); Woodruff, *Effects of Tropical Light;* "White Race and Tropics," *Popular Science Monthly* 27 (July 1899): 534–36; Ripley, "Acclimatization," *Popular Science Monthly* (March 1896): 662–75; S. P. Verner, "The White Race in the Tropics," *World's Work* 16 (September 1908): 10715–20.

51. Ward, "Some Problems of the Tropics," 8; Tower, "Climate of the Philippines," 8–9; Geo. G. Groff, "The Conquest of the Tropics," *Popular Science Monthly* (September 1900): 540–45.

52. *Report of the Philippine Commission to the President* (Washington, D.C.: GPO, 1900), 442. The historiography of coolie labor (whether of Indian or Chinese origin) has largely focused on labor within the British Empire. See, for example, Marina Carter, *Voices from Indenture: Experiences of Indian Migrants in the British Empire* (Leicester: Leicester University Press, 1996); Madhavi Kale, *Fragments of Empire: Capital, Slavery, and Indian Indentured Labor Migration in the British Caribbean* (Philadelphia: University of Pennsylvania Press, 1998).

53. *Report of the Philippine Commission to the President*, 158–59; *Report of the Philippine Commission to the President, Vol. II* (Washington, D.C.: GPO, 1900), 18.

54. *Report of the Philippine Commission to the President, Vol. II*, 289, 338–43.

55. *Report of the Philippine Commission to the President, Vol. II*, 160–66, 237–39, 249–54, 432–45.

56. *Report of the Philippine Commission to the President, Vol. II*, 18, 34, 182–88, 209–16, 237–39, 252–54, 444–45.

57. "Northern Men in Tropical Climates," *New York Times* (June 5, 1898), 18; Edwin Wildman, "Stumbling-Blocks to Philippine Pacification and Development," *New York Times* (May 11, 1902; Sunday magazine), 3; Frederick. W. Eddy, "Philippine Labor Problems, What to Do about the Chinese, The Indolence of the Native," *New York Times* (August 18, 1901; Sunday magazine), 20.

58. *Status of Chinese Persons in the Philippine Islands* (Washington, D.C.: GPO, 1900); *Protests of Chinese Government* (Washington, D.C.: GPO, 1902).

59. "Chinese Temporarily Barred: Congress to Act on Permanent Exclusion from the Philippines," *New York Times* (August 25, 1899), 4; "Chinese Exclusion Complete," *New York Times* (March 26, 1902), 3; "Chinese in the Philippines," *New York Times* (August 27, 1902), 8; "The Philippine Bill of July 1, 1902: An Act Temporarily to Provide for the Administration of the Affairs of Civil Government in the Philippine Islands, and for Other Purposes," *U.S. Statutes at Large* 32 (July 1, 1902): 69; Boggs, "The Anglo-Saxon in India and the Philippines," 320–21; Lanny Thompson, "The Imperial Republic: A Comparison of the Insular Territories under U.S. Dominion after 1898," *Pacific Historical Review* 71, no. 4 (2002): 535–74.

60. Quoted in "Need the Chinese, Says Major Seaman," *New York Times* (March 24, 1907), 10.

61. Ward, *Climate*, 272–73.

62. Ward, *Climate*, 274.

63. Ward, "Notes on Climatology" (no. 1, 1901), 50–51.

64. Ward, *Climate*, 321–37.

65. Ripley, *Races of Europe*, 289–91.

66. Ellsworth Huntington, *The Character of Races, As Influenced by Physical Environment, Natural Selection and Historical Development* (New York: Charles Scribner's Sons: 1927), 66–67.

67. Huntington, *Character of Races*, 110–11; Ellsworth Huntington, *The Red Man's Continent* (New Haven: Yale University Press, 1919).

68. Edward Allsworth Ross, *The Changing Chinese: The Conflict of Oriental and Western Cultures in China* (New York: Century Company, 1920 [1911]); Charles Conant Josey, *Race and National Solidarity* (New York: Charles Scribner's Sons, 1923); George F. Seward, *Chinese Immigration in Its Social and Economical Aspects* (New York: Charles Scribner's Sons, 1881); Edward Alsworth Ross, *The Social Trend* (New York: Century Co., 1922), 137–50; Willard B. Farwell, "New Phases in the Chinese Problem," *Popular Science Monthly* 36 (December 1889): 181–91.

69. Ross, *Changing Chinese*, 54–57; K. S. Latourette, *The Development of China* (Boston: Houghton, 1917).

70. Huntington, *Character of Races*, 184–204.

71. F. W. Williams, *A History of China* (New York: Charles Scribner's Sons, 1897).

72. Mabel Ping-hua Lee, *The Economic History of China* (New York: Columbia University Press, 1921).

73. Huntington, *Civilization and Climate*, 220–39; Ellsworth Huntington, *World-Power and Evolution* (New Haven: Yale University Press, 1919), 98–99; Charles J. Kullmer, "The Shift of the Storm Track," in *The Climactic Factor*, ed. Ellsworth Huntington (Washington, D.C.: Carnegie Institute of Washington, 1914), 193–210.

74. Huntington, *Civilization and Climate*, 314.

75. "H.W. Polton to Ellsworth Huntington," February 20, 1912; "Huntington to Polton," February 23, 1912; "Polton to Huntington," September 23, 1912; "Huntington to Polton," October 8, 1912, all in Huntington Papers, Series III, Box 28, Folder 504; "Efficiency of Factory Workers, Stanley Iron Works, New Britain, CT" (1911), Huntington Papers, Series VI, Box 2, Folder 19.

76. "Wages and Humidity at New Britain" (1913), Huntington Papers, Series VI, Box 2, Folder 15.

77. "Efficiency of Factory Workers, Temperature and Climate, J.P. King and Co., Augusta, GA" (1914–15), Huntington Papers, Series VI, Box 2, Folder 14.

78. Huntington's Turkestan is modern-day Turkmenistan. "Directions for Construction of World Map of Efficiency" (c. 1914), Huntington Papers, Series VI, Box 3, Folder 41; "Ellsworth Huntington to Charles Julius Kullmer" (September 20, 1913), Huntington Papers, Series III, Box 30, Folder 566.

79. Edith Commander Breithus, "The Struggle for Existence: The Unequal Distribution of Labor—A Lesson for Children," *Progressive Woman* 3 (December 1909); J. Howard Moore, *Savage Survivals* (London: Watts, 1933 [1916]), 118–21.

80. "First annual report of the Committee of the Atmosphere and Man" (March 17, 1923), Huntington Papers, Series IV, Box 1, Folder 5, 2–5.

81. "Per. H.W. Peel, Ross' Limited, to Rev. J.D. MacLachlan" (January 3, 1920), Huntington Papers, Series IV, Box 4, Folder 20.

82. "Wm. G. Braemer, Universal Humidifying Company to Ellsworth Huntington" (January 6, 1927), Huntington Papers, Series IV, Box 30, Folder 95; "Kenneth G. Carpenter, D'Arcy Advertising Company to Ellsworth Huntington" (February 2, 1921), Huntington Papers,

Series IV, Box 4, Folder 21. For other pamphlets and correspondence with Huntington about humidifying, heating, and cooling systems, see Huntington Papers, Series IV, Box 30, Folders 90 and 95 and Box 4, Folder 21.

83. "Ellsworth Huntington to Wm. G. Braemer" (January 10, 1927), Huntington Papers, Series IV, Box 30, Folder 95.

84. "Ellsworth Huntington to Rev. J. D. MacLachlan" (April 10, 1920), Huntington Papers, Series IV, Box 4, Folder 20.

85. "Temperature Change Linked to Mortality, Yale Professor Tells Heating and Ventilating Engineers They Are Moving Civilization North," *New York Times* (March 6, 1928), 9. For recent histories of air conditioning, see W. Stephen Comstock, B. J. Spanos, eds., *Proclaiming the Truth: An Illustrated History of the American Society of Heating, Refrigerating and Air-Conditioning Engineers Inc.* (Atlanta: American Society of Heating, Refrigerating and Air-Conditioning Engineers, 1995); Marsha E. Ackermann, *Cool Comfort: America's Romance with Air-Conditioning* (Washington, D.C.: Smithsonian Institution Press, 2002).

86. Semple, *Influences of Geographic Environment*, 328–29.

87. GilFillan, "Coldward Course of Progress," 393–410.

88. GilFillan, "Coldward Course of Progress," 399–401.

89. Ripley, *Races of Europe*, 32.

90. Ellsworth Huntington, "Inventions and Racial Character" (n.d.), Huntington Papers, Series V, Box 63, Folder 575.

91. "Ellsworth Huntington to Samuel L. Rogers, Director United States Census Bureau" (January 15, 1918), and "Madison Grant to Ellsworth Huntington" (March 9, 1918), Huntington Papers, Series IV, Box 16, Folder 161.

92. "Samuel L. Rogers to Ellsworth Huntington" (January 19, 1918), Huntington Papers, Series IV, Box 16, Folder 161.

93. "Resolution adopted by the Home Missions Council, January 16, 1918," Huntington Papers, Series IV, Box 16, Folder 162.

94. *Reports of the Industrial Commission on Immigration* (Washington, D.C.: GPO, 1901), lxvi.

95. United States Immigration Commission, *Dictionary of Races or Peoples* (Washington, D.C.: GPO, 1911), 2–5.

96. *Dictionary of Races or Peoples*, 57–58; Ronald T. Takaki, *Pau Hana: Plantation Life and Labor in Hawaii, 1835–1920* (Honolulu: University of Hawaii Press, 1983); Erika Lee, *At America's Gates: Chinese Immigration during the Exclusion Era, 1882–1943* (Chapel Hill: University of North Carolina Press, 2003): 47–110; Fujita-Rony, *American Workers, Colonial Power*, 25–50.

3. The Other Colonies

1. Frank Julian Warne, *The Slav Invasion and the Mine Workers: A Study in Immigration* (Philadelphia: J.B. Lippincott Company, 1904), 143–56; Robert A. Janosov et al., *The Great Strike: Perspectives on the 1902 Anthracite Coal Strike* (Easton, Pa.: Canal History and Technology Press, 2002); Perry K. Blatz, *Democratic Miners: Work and Labor Relations in the Anthracite Coal Industry, 1875–1925* (Albany: State University of New York Press, 1994).

2. Warne, *Slav Invasion*, 66–89; David Roediger, *Working towards Whiteness: How America's Immigrants Became White* (New York: Basic Books, 2005), 84–85.

3. Warne, *Slav Invasion*, 130–33.

4. See, for example, Matthew Frye Jacobson, *Whiteness of a Different Color: European Immigrants and the Alchemy of Race* (Cambridge: Harvard University Press, 1998), 39–170;

David Roediger, *Towards the Abolition of Whiteness: Essays on Race, Politics, and Working Class History* (London: Verso, 1994); Michelle Brattain, *The Politics of Whiteness: Race, Workers, and Culture in the Modern South* (Princeton: Princeton University Press, 2001); Caroline Waldron Merithew, "Making the Italian Other: Blacks, Whites, and the Inbetween in the 1895 Spring Valley, Illinois, Race Riot," in *Are Italians White? How Race Is Made in America,* ed. Jennifer Guglielmo and Salvatore Salerno (New York: Routledge, 2003), 79–97; Thomas Guglielmo, *White on Arrival: Italians, Race, Color, and Power, 1890–1945* (New York: Oxford University Press, 2003), 59–92.

5. This analysis represents a turn from both the experiential/social historical approach to immigration history and histories of structures (especially of public health) that regulated immigration. See Amy L. Fairchild, *Science at the Borders: Immigrant Medical Inspection and the Shaping of the Modern Industrial Labor Force* (Baltimore: Johns Hopkins University Press, 2003), 23–159; Alan M. Kraut, *Silent Travelers: Germs, Genes, and the "Immigrant Menace"* (Baltimore: Johns Hopkins University Press, 1995). It also redefines immigration restriction away from a coherent, extant politics that slowly reshaped policy. John Higham, *Strangers in the Land: Patterns of American Nativism, 1860–1925,* rev. ed. (New Brunswick, N.J.: Rutgers University Press, 2002); Desmond King, *The Liberty of Strangers: Making the American Nation* (New York: Oxford University Press, 2005), 49–107. Immigration, here, is not studied as "American ethnic history" but in its transnational relationship to race conquest, as it was viewed by the leading commentators of the age. Their view allowed for the reframing of industry as the principal site of race conflict—the struggle of races for survival. In linking race, civilization, and industry, they defined race conflict as a process in which not all races could participate. This explains, in large measure, the alarm about European and Asian immigrants while concurrently restricting, if not dampening, concerns about African American workers (albeit temporarily and in general). In effect, industry, seen in this light, led to the segregating of conversations about race and color (framed around the fate of African Americans) and about race and conflict (framed around colonization). For a recent effort to describe a transnational immigration history, see Matthew Frye Jacobson, "More 'Trans-', Less 'National,'" *Journal of American Ethnic History* 25 (Summer 2006): 74–84.

6. W. Jett Lauck, "The Vanishing American Wage Earner," *Atlantic Monthly* 110 (November 1912): 691; William Z. Ripley, "Races in the United States," *Atlantic Monthly* 102 (December 1908): 745–59; J. R. Commons, "Colonial Race Elements," *Chautauquan* 38 (October 1903): 118–25; United States Immigration Commission (1907–10), *Abstracts of Reports of the Immigration Commission (in two volumes: vol 1)* (Washington, D.C: GPO, 1911), 495–96; Henry Pratt Fairchild, *Immigration: A World Movement and Its American Significance* (New York: Macmillan, 1923 [1913]), 2–4.

7. "Seventh Annual Report of the New York Milk Committee" (1913), Wilbur and Elsie P. Phillips Papers SW094, Social Welfare History Archives, University of Minnesota, Box 5, 39–45; N.A. Erickson, "Colonization of Immigrants" (1912), *National Conference on Social Welfare Proceedings,* 254; "South End House, 1920" (January, 1920), United South End Settlements Records SW204, Social Welfare History Archives, Box 4, Folder: "SEH Reports (1917, 1919, 1920, 1921, 1922)," 15; Gino C. Speranza, "The Italians in Congested Districts," *Survey* 20 (April–October 1908): 55–57; Jane Addams, "The Chicago Settlements and Social Unrest," *Survey* 20 (April–October 1908): 155; Vida D. Scudder, "Experiments in Fellowship: Work with Italians in Boston," *Survey* 22 (April 3, 1909): 47–51; John Daniels, "Americanizing Eighty Thousand Poles," *Survey* 24 (June 4, 1910): 373–85; Simon J. Lubin and Christina Krysto, "The Strength of America: The Significance of Modern Migration," *Survey* 43 (January 24, 1920): 461–63.

8. Fairchild, *Immigration: A World Movement,* 1–5; James Davenport Whelpley, *The Problem of the Immigrant* (London: Chapman & Hall, 1905); Richmond Mayo-Smith, *Emigration and Immigration: A Study in Social Science* (New York: Charles Scribner's Sons, 1890),

12–30; Allan McLaughlin, "Immigration, Past and Present," *Popular Science Monthly* 65 (July 1904): 224–27; James Collier, "The Evolution of Colonies," *Popular Science Monthly* 53 (July–October 1898): 289–307, 452–66, 620–33, 806–8, and 54 (November 1898–March 1899): 52–63, 290–99, 577–88; Collier, "The Theory of Colonization," *American Journal of Sociology* 11 (September 1905): 252–65.

9. James Davenport Whelpley, "International Control of Immigration," *World's Work* 8 (September 1904): 5254–59; Ripley, "Races in the United States," 745–59.

10. F. L. Oswald, "Modern Mongols," *Popular Science Monthly* 57 (October 1900): 618–23; Frank Julian Warne, *The Immigrant Invasion* (New York: Dodd, Mead and Company, 1913), 9–11.

11. Whelpley, *Problem of the Immigrant*, 1; Whelpley, "International Control of Immigration," 5259; Edward Alsworth Ross, *The Social Trend* (New York: Century Co., 1922), 14–15.

12. Warne, *Slav Invasion;* Warne, *Immigrant Invasion.* For a useful study of the metaphors of immigration, see Gerald V. O'Brien, "Indigestible Food, Conquering Hordes, and Waste Materials: Metaphors of Immigrants and the Early Immigration Restriction Debate in the United States," *Metaphor and Symbol* 18, no. 1 (2003): 33–47.

13. Daniel Chauncey Brewer, *The Conquest of New England by the Immigrant* (New York: G. P. Putnam's Sons, 1926); Fairchild, *Immigration: A World Movement*, 2–4, 10–19.

14. Brewer, *Conquest of New England,* 265–86.

15. "Extracts from the Report of the Commissioner-General," *Publications of the Immigration Restriction League* 59 (1912): 8.

16. Brewer, *Conquest of New England,* 286; "South End House Association, Eighth Annual Report" (January 1900), United South End Settlements Records, Box 2, Folder: "South End House Annual Reports 1892–1910," 7–8; Robert Hunter, "Immigration, the Annihilator of Our Native Stock," *Commons* 9 (April 1904): 114–17; Hjalmar H. Boyesen, "Dangers of Unrestricted Immigration," *Forum* 3 (July 1887): 532–42; James John Davis, "Selective Immigration," *Forum* 70 (September 1923): 1857–65; Charlotte Perkins Gilman, "Is America Too Hospitable?" *Forum* 70 (October 1923): 1983–89; Harry H. Laughlin, *Immigration and Conquest: A Study of the United States as the Receiver of Old World Emigrants Who Became the Parents of Future-Born Americans* (New York: New York State Chamber of Commerce, 1939); Charles B. Davenport, *Heredity in Relation to Eugenics* (New York: Henry Holt, 1911), 204–24.

17. Charles Edward Woodruff, *Expansion of Races* (New York: Rebman Company, 1909), 110–17; Ellsworth Huntington, *The Character of Races, As Influenced by Physical Environment, Natural Selection and Historical Development* (New York: Charles Scribner's Sons, 1924), 33–36; Edward Alsworth Ross, *The Old World in the New: The Significance of Past and Present Immigration to the American People* (New York: Century, 1914), 3–23.

18. Prescott F. Hall, "Eugenics, Ethics and Immigration," *Publications of the Immigration Restriction League* 51 (n.d.), 7; Edward Lowry, "Americans in the Raw: The High Tide of Immigrants—Their Strange Possessions and Their Meager Wealth—What Becomes of Them," *World's Work* 4 (October 1902): 2644–55; "The Infertility of the 'Civilized'" and "The Burden of the New Immigration," *World's Work* 6 (July 1903): 3601–03, 3611–12.

19. John Mitchell, "Protect the Workman," *Publications of the Immigration Restriction League* 52 (n.d.), 3.

20. Huntington, *Character of Races,* 303.

21. Huntington, *Character of Races,* 303–4; Charles M. Andrews, *The Colonising Activities of the English Puritans* (New Haven: Yale University Press, 1914); Havelock Ellis, *A Study of British Genius* (Philadelphia: F.A. Davis, Co., 1904); Fairchild, *Immigration: A World Movement,* 27–52.

22. Frederick Jackson Turner, *The Frontier in American History* (New York: Henry Holt, 1935), esp. chap. 1; Ellen Churchill Semple, *Influences of Geographic Environment on the Basis of Ratzel's System of Anthro-Geography* (New York: Russell & Russell, 1968 [1911]), 221–33.

23. William Z. Ripley, *The Races of Europe* (New York: D. Appleton, 1899), 540–41; Frank Norris, "The Frontier Gone at Last," *World's Work* 3 (February 1902): 1728–31.

24. Fairchild, *Immigration: A World Movement*, 38; Mayo-Smith, *Emigration and Immigration*, 33–50; John R. Commons, *Races and Immigrants in America* (New York: Macmillan, 1920 [1907]), 22–38; Charles A. Hanna, *The Scotch-Irish, or, The Scot in North Britain, North Ireland, and North America*, Vol. II (New York: Putnam's, 1902), 172–80; Carlos C. Closson, "The Real Opportunity of the So-Called Anglo-Saxon Race," *Journal of Political Economy* 9 (December 1900): 76–97.

25. Fairchild, *Immigration: A World Movement*, 51–52; Huntington, *Character of Races*, 304–6.

26. Turner, *Frontier in American History*, chap. 8; Ellsworth Huntington, *World-Power and Evolution* (New Haven: Yale University Press, 1919), 98–99; Edward C. Towne, "American Climate and Character," *Popular Science Monthly* 20 (November 1881): 109–16; Woodruff, *Expansion of Races*, 87–109.

27. Commons, *Races and Immigrants*, 30–38; John R. Commons et al., *History of Labour in the United States* (New York: Macmillan, 1918).

28. Fairchild, *Immigration: A World Movement*, 130.

29. Fairchild, *Immigration: A World Movement*, 145; Francis A. Walker, "Restriction of Immigration," *Publications of the Immigration Restriction League* 33 (n.d.), 11.

30. Mayo-Smith, *Emigration and Immigration*, 45–47.

31. Walker, "Restriction of Immigration," 10; Whelpley, *Problem of the Immigrant*, 10–18.

32. Rose Cohen, *Out of the Shadow: A Russian Jewish Girlhood on Lower East Side* (Ithaca: Cornell University Press, 1995 [1918]), 61–64; Pamela S. Nadell, "United States Steerage Legislation: The Protection of Emigrants En Route to America," *Immigrants and Minorities* 5, no. 1 (1986): 62–72.

33. See, for example, "Transformation of the Steerage," *Outlook* 74 (August 15, 1903): 919; K. Durland, "Steerage Impositions," *Independent* 61 (August 30, 1906): 499–504.

34. For examples of observers of the migrant ship, most of whom assumed the disguise of immigrants, see Monsignor Count Vay de Vaya and Luskod, "To America in an Emigrant Ship," *Living Age* 252 (January 19, 1907): 173–82; L. E. Macbrayne, "Judgement of the Steerage," *Harper's Monthly Magazine* 117 (September 1908): 489–99; "Steerage to America" *Living Age* 268 (February 18, 1911): 438–41; S. Graham, "With The Poor Emigrants to America," *Harper's Monthly Magazine* 129 (July 1914): 291–301.

35. *Abstracts of Reports of the Immigration Commission*, 13–15.

36. United States Immigration Commission, *Steerage Conditions: Importation and Harboring of Women for Immoral Purposes; Immigrant Homes and Aid Societies; Immigrant banks.* (Washington, D.C.: GPO, 1911), 23.

37. United States Immigration Commission, *Immigration Legislation* (Washington, D.C.: GPO, 1911), 450.

38. United States Immigration Commission, *Immigration Legislation*, 339–504.

39. United States Immigration Commission, *Steerage Conditions*, 24–29.

40. See, for example, Emily Greene Balch, W. F. Wilcox, Jeremiah W. Jenks, Max J. Kohler, "Restriction of Immigration—Discussion," *American Economic Review* 2 (March 1912): 63–78.

41. United States Immigration Commission, *Steerage Conditions*, 37; Whelpley, *Problem of the Immigrant*, 6–9.

42. Arthur Reeve, "International Aspects of Immigration," *Charities and Commons* 17 (October 1906–April 1907): 506; Warne, *Immigrant Invasion*, 7.

43. *Abstracts of Reports of the Immigration Commission*, 34.

44. United States Immigration Commission, *Brief Statement of the Immigration Commission with Conclusions and Recommendations and Views of the Minority* (Washington, D.C.: GPO,

1911), 189–91. The Commission also noted that such agents helped migrants to avoid local European laws and customs officials trying to limit emigration.

45. Robert De C. Ward, "The New Immigration Act," *Charities and Commons* 18 (April–October 1907): 122–24; Robert De C. Ward, "Opinions of U.S. Consuls on Immigration," *Charities and the Commons* 17 (October 1906–April 1907): 504–5; Fairchild, *Immigration: A World Movement*, 148–53.

46. Whelpley, "International Control of Immigration," 5255. The 1907 "Report of the Commissioner-General" similarly insisted the because of the actions of the steamship companies and their agents, "the line is passed between natural and forced immigration." See "Extracts from Reports of the Commissioner-General, Report of 1907," *Publications of the Immigration Restriction League* 49 (n.d.): 3.

47. Ross, *Social Trend*, 4–5; Jesse Frederick Steiner, *The Japanese Invasion: A Study in the Psychology of Inter Racial Contacts* (Chicago: A.C. McClurg, 1917), vii–xvii; J. B. Bishop, "Quality of Our Latest Immigration," *Nation* 52 (February 5, 1891): 108; "The Tide of Immigration," *Public Opinion* 3, 97–336; "Undesirable Immigration," *Public Opinion* 3, 49.

48. Ward, "Opinions of U.S. Consuls on Immigration," 504–5; Ripley, *Races of Europe*, 89. For similar arguments, see also "Some Phases of Immigrant Travel," *Outlook* 74 (May 23, 1903): 231–36; Viola Paradise, "The Jewish Immigrant Girl in Chicago," *Survey* 30 (September 6, 1913): 699–704; Allan McLaughlin, "Immigration," *Popular Science Monthly* 65 (June 1904): 164–69.

49. Hall, "Eugenics, Ethics and Immigration," 5–6.

50. Fairchild, *Immigration: A World Movement*, 385–86; Warne, *Slav Invasion*, 131–33; Kate Holladay Claghorn, "The Changing Character of Immigration," *World's Work* 1 (February 1901): 381–87.

51. "Extracts from the Report of the Commissioner of Immigration at New York," *Publications of the Immigration Restriction League* 59 (n.d.): 8–9; F. P. Sargent, "Need of Closer Inspection and Greater Restriction of Immigrants," *Century* 67 (January 1904): 470–73; "The Selection of Immigrants," *Survey* 25 (February 4, 1911): 715–16.

52. "General Immigration Statistics," *Publications of the Immigration Restriction League* 48 (n.d.): 19–20.

53. *Publications of the Immigration Restriction League* 62 (1914): 3.

54. Commons, *Races and Immigrants*, 107–27; *Abstracts of Reports of the Immigration Commission*, 490; Francis A. Walker, "Immigration and Degradation," *Forum* 11 (August 1891): 634–44.

55. Lauck, "Vanishing American Wage-Earner," 691; Closson, "Opportunity of the Anglo-Saxon Race," 93–97.

56. Commons, *Races and Immigrants*, 69; E. Dana Durand, "Our Immigrants and the Future," *World's Work* 23 (February 1912): 431–43.

57. Commons, *Races and Immigrants*, 127; Arno Dosch, "Our Expensive Cheap Labor," *World's Work* 26 (October 1913): 699–703.

58. Commons, *Races and Immigrants*, 70–127; Claghorn, "The Changing Character of Immigration," 381–387; Henry Pratt Fairchild, ed., *Immigrant Backgrounds* (New York: John Wiley and Sons, 1927).

59. United States Immigration Commission, *Dictionary of Races or Peoples* (Washington, D.C.: GPO, 1911).

60. *Reports of the Industrial Commission on Immigration*, lxxv.

61. Commons, *Races and Immigrants*, 115–19. Commons explained the rise of anti-Chinese sentiment among white California workingmen as an example of the battle over the standard of living. In that case, workingmen were able to protect, if temporarily, their position through legislation. In other cases, workers with a higher standard faced substitution.

62. John R. Commons, *Trade Unionism and Labor Problems, 2nd Series* (New York: A. M. Kelly, 1967 [1905]), xii–348.

63. Warne, *Slav Invasion*, 37–96; Warne, *Immigrant Invasion*, esp. 105–80.

64. *Abstracts of Reports of the Immigration Commission*, 493; Fairchild, *Immigration: A World Movement*, 163–64, 196–97.

65. *Abstracts of Reports of the Immigration Commission*, 503–30.

66. *Abstracts of Reports of the Immigration Commission*, 517–30; Fairchild, *Immigration: A World Movement*, 214–319, 341–53. Appleton Morgan, "What Shall We Do with the 'Dago'?" *Popular Science Monthly* 38 (December 1890): 172–79; Allan McLaughlin, "The Slavic Immigrant," *Popular Science Monthly* 63 (May 1903): 25–32; McLaughlin, "Italian and Other Latin Immigrants," *Popular Science Monthly* 65 (August 1904): 341–49; McLaughlin, "Hebrew, Magyar and Levantine Immigration," *Popular Science Monthly* 65 (September 1904): 432–42.

67. *Abstracts of Reports of the Immigration Commission*, 387.

68. Fairchild, *Immigration: A World Movement*, 342; Robert DeC. Ward, "The Agricultural Distribution of Immigrants," *Popular Science Monthly* 66 (December 1904): 166–75; Allan McLaughlin, "Social and Political Effects of Immigration," *Popular Science Monthly* 66 (January 1905): 243–55; McLaughlin, "The Problem of Immigration," *Popular Science Monthly* 66 (April 1905): 531–37; A. E. Outerbridge, "Pending Problems for Wage-Earners," *Popular Science Monthly* 49 (May 1896): 57–71; Lindley M. Kreasbey, "Cooperation, Coercion, Competition," *Popular Science Monthly* 63 (October 1903): 526–33; Robert A. Woods, "The Clod Stirs," *Survey* 27 (March 16, 1912): 1929–32.

69. See Fairchild, *Immigration: A World Movement*, 21–26.

70. Commons, *Races and Immigrants*, 39–60, 146–47; William Benjamin Smith, *The Color Line, A Brief in Behalf of the Unborn* (New York: McClure, Phillips & Co., 1905), 166–92. For a broader engagement with the idea of the African American as an industrial worker, see William P. Jones, *The Tribe of Black Ulysses: African American Lumber Workers in the Jim Crow South* (Urbana: University of Illinois Press, 2005).

71. Paul S. Reinsch, "The Negro Race and European Civilization," *American Journal of Sociology* 11 (September 1905): 145–67; W. I. Thomas, "Race Psychology: Standpoint and Questionnaire, With Particular Reference to the Immigrant and the Negro," *American Journal of Sociology* 17 (May 1912): 725–75.

72. Commons, *Races and Immigrants*, 136; Joseph Alexander Tillinghast, "The Negro in Africa and America," *Publications of the American Economic Association* 3 (May 1902): 28–45.

73. Frederick L. Hoffman, "The Race Traits and Tendencies of the American Negro," *Publications of the American Economic Association* 11 (January–March–May 1896): 263–65.

74. Tillinghast and Hoffman reached slightly differing conclusions. Hoffman cited some examples of industrialists who were somewhat positive about their black employees. Tillinghast, however, revisited Hoffman's surveys of industrialists to insist that African American workers were praised only when they were completing the most menial of tasks under the strictest of discipline. See Hoffmann, "Race Traits and Tendencies of the American Negro," 270–75, and Tillinghast, "The Negro in Africa and America," 180–93. See also Woodruff, *Expansion of Races*, 242–73; Chas. H. McCord, *The American Negro as a Dependent, Defective and Delinquent* (Nashville: Benson Printing Company, 1914), 7–46.

75. See J. Stahl Patterson, "Increase and Movement of the Colored Population," *Popular Science Monthly* 19 (October 1881): 784–90; E. W. Gilliam, "The African in the United States," *Popular Science Monthly* 22 (February 1883): 433–44; N. S. Shaler, "The Future of the Negro in the Southern States," *Popular Science Monthly* 57 (June 1900): 147–56.

76. See, for example, Smith, *The Color Line*, 166–261.

77. Alfred Holt Stone, "The Economic Future of the Negro: The Factor of White Competition," *Publications of the American Economic Association* 7 (February 1906): 243–94. See also Thomas, "Race Psychology," 739; Hoffman, "The Race Traits and Tendencies of the American Negro," 263–75; Francis A. Walker, "The Colored Race in the United States," *Forum* 11 (July 1891): 501–9.

78. Commons, *Trade Unionism and Labor Problems,* 349–70.

79. William Graham Sumner, "The Indians in 1887," typescript of article published in *Forum* 3 (May 1887), 38, in William Graham Sumner Papers, Yale University, Manuscripts and Archives, Series II, Box 37, Folder 987. Sumner wrote this article as a critique of the just passed Dawes Severalty Act, which allowed for the division of reservation land into individual holdings in an effort to encourage Indians to become farmers, something Sumner criticized as a misguided effort to force Indians into a new stage of civilization.

80. Smith, *Color Line,* 75; McCord, *American Negro as a Dependent,* 76–85.

81. Walter F. Willcox, W. E. B. DuBois, H. T. Newcomb, W. Z. Ripley, A. H. Stone, "The Economic Position of the Negro," *Publications of the American Economic Association* 6 (February 1905): 216–21; Walter Francis Willcox, Allyn Abbott Young, John S. Billings, Joseph A. Hill, and W. E. B. Dubois, *Supplementary Analysis and Derivative Tables: Twelfth Census of the United States, 1900* (Washington, D.C.: GPO, 1906); Alfred Holt Stone, *Studies in the American Race Problem* (New York: Negro Universities Press, 1969 [1908]), esp. 476–531. For a reading of the 1890 census and its place in the construction of racial and color lines, see Martha Hodes, "Fractions and Fictions in the United States Census of 1890," in *Haunted by Empire: Geographies of Intimacy in North American History,* ed. Ann Laura Stoler (Durham: Duke University Press, 2006), 240–70.

82. Warwick Anderson, "Immunities of Empire: Race, Disease, and the New Tropical Medicine, 1900–1920," *Bulletin of the History of Medicine* 70 (Spring 1996): 94–118; Joseph O. Baylen and John Hammond Moore, "Senator John Tyler Morgan and Negro Colonization in the Philippines, 1901–1902," *Phylon* 29 (Spring 1968): 65–75; "Negroes to the Philippines," *Informer* (February 1903); "Negroes and the Philippines," *New York Times* (May 3, 1903): 23; "Negro Colonization Plan," *New York Times* (December 15, 1902): 6; Smith, *The Color Line,* 166–92.

83. William Noyes, "Some Proposed Solutions to the Negro Problem," *International Socialist Review* 2 (December 1901): 401–13; Clarence Darrow, "The Problem of the Negro," *International Socialist Review* 2 (November 1901): 321–35; Eugene V. Debs, "The Negro in the Class Struggle," *International Socialist Review* 4 (November 1903): 257–60; E. F. Andrews, "Socialism and the Negro," *International Socialist Review* 5 (March 1905): 524–26.

84. W. E. B Du Bois, "The Evolution of the Race Problem," in *W. E. B. Du Bois Speaks: Speeches and Addresses, 1890–1919,* ed. Philip S. Foner (New York: Pathfinder, 1970), 202–7; David Levering Lewis, *W. E. B. Du Bois: Biography of a Race* (New York: Henry Holt, 1993); Eric D. Anderson, "Black Responses to Darwinism, 1859–1915," in *Disseminating Darwinism: The Role of Place, Race, Religion, and Gender,* ed. Ronald L. Numbers and John Stenhouse (Cambridge: Cambridge University Press, 1999), 247–66. See also Alexander Francis Chamberlain, "The Contribution of the Negro to Human Civilization," *Journal of Race Development* 1 (October 1910): 482–502.

85. See, for example, Kelly Miller, *A Review of Hoffman's Race Traits and Tendencies of the American Negro* (Washington, D.C.: American Negro Academy, 1897), and W. O. Thompson, "The Negro: The Racial Inferiority Argument in Light of Science and History," *Voice of the Negro* 3 (1906): 507–13.

86. Giles B. Jackson and D. Webster Davis, *The Industrial History of the Negro Race of the United States* (Freeport, N.Y.: Books for Libraries, 1971 [1908]). W. Fitzhugh Brundage interprets the dioramas as broadly representing black "evolution." Yet evolution was depicted by Warrick, and later by Jackson and Davis, less in biological terms and more simply as progress. The dioramas more specifically should be read as direct responses to intellectual discourses that evicted the African American from American industry. See Brundage, "Meta Warrick's 1907 'Negro Tableaux' and (Re)Presenting African American Historical Memory," *Journal of American History* 89 (March 2003): 1368–1400. See also Sarah Howard Watkins, "The Negro Building: African American Representation at the 1907 Jamestown Tercentennial Exposition" (MA thesis, College of William and Mary, 1994); John Thomas Wilkes, "Enough Glory for Us

All: The 'Negro Exhibit' at the Jamestown Tercentennial Exposition, 1907" (MA thesis, University of Richmond, 2003); see also Ida B. Wells, ed., *The Reason Why the Colored American Is Not in the World's Columbian Exposition* (Chicago: Ida B. Wells, 1893).

87. United States Immigration Commission, *Reports of the Immigration Commission, Immigrants in Industries, Part 25, Japanese and Other Immigrant Races in the Pacific Coast and Rocky Mountain States* (3 vols.) (Washington, D.C.: GPO, 1911); Steiner, *Japanese Invasion,* 96–111, 126–29.

88. Alexander Saxton, *The Indispensable Enemy: Labor and the Anti-Chinese Movement in California* (Berkeley: University of California Press, 1971), 92–178.

89. Matthew Frye Jacobson, *Barbarian Virtues: The United States Encounters Foreign Peoples at Home and Abroad, 1876–1917* (New York: Hill and Wang, 2000), 77; Dr. Albert Allemann, "Immigration and the Future American Race," *Popular Science Monthly* 75 (December 1909): 586–96.

90. Steiner, *Japanese Invasion,* 36–42; Willard B. Farwell, "Why the Chinese Must Be Excluded," *Forum* 6 (October 1888): 196–203; Gerrit L. Lansing, "Chinese Immigration: A Sociological Study," *Popular Science Monthly* 20 (April 1882): 721–35; Sidney L. Gulick, "The Problem of Oriental Immigration," *Survey* 31 (March 7, 1914): 720–22, 730–31; Willard B. Farwell, "New Phases in the Chinese Problem," *Popular Science Monthly* 36 (December 1889): 181–91; Allan McLaughlin, "Chinese and Japanese Immigration," *Popular Science Monthly* 66 (December 1904): 117–21; H. A. Millis, "California and the Japanese," *Survey* 30 (June 7, 1913): 332–36; G. S., "Taking the Queue Out of Quota, An Interview with W. W. Husband, Commissioner-General of Immigration," *Survey* 51 (March 15, 1924): 666–69; "Immigration: A Look Ahead," *Survey* 52 (May 15, 1924): 207–12.

91. Dr. Woods Hutchinson, "The Mongolian as a Workingman," *World's Work* 14 (September 1907): 9373, 9376; Woodruff, *Expansion of Races,* 242–73.

92. Erika Lee, *At America's Gates: Chinese Immigration during the Exclusion Era, 1882–1943* (Chapel Hill: University of North Carolina Press, 2003); Evelyn Nakano Glenn, *Unequal Freedom: How Race and Gender Shaped American Citizenship and Gender* (Cambridge: Harvard University Press, 2002), 18–55; Eiichiro Azuma, *Between Two Empires: Race, History, and Transnationalism in Japanese America* (New York: Oxford University Press, 2005).

93. For discussions of the difficulty of restricting the Japanese in the same manner as the Chinese, see David Starr Jordan, "Relations of Japan and the United States," *Journal of Race Development* 2 (1912): 215–23; "The Japanese in California," *World's Work* 13 (March 1907): 8689–93.

94. Alexander Francis Chamberlain, "The Japanese Race," *Journal of Race Development* 3 (October 1912): 176–87; Sidney Lewis Gulick, *The White Peril in the Far East: An Interpretation of the Russo-Japanese War* (New York: Revell, 1905); Steiner, *Japanese Invasion,* 15; Jihei Hashiguchi, "The Rise of Modern Japan," *World's Work* 7 (April 1904): 4626–46; T. Iyenaga, "The War's Disclosure of the Orient: A New Life for Asia and a New Western Interest in It— The Difference between Oriental and Western Ideas and Activities—Can the Gap of Racial Misunderstanding Be Bridged?" *World's Work* 9 (April 1905): 6041–47.

95. William C. Redfield, "The Progress of Japanese Industry," *Journal of Race Development* 2 (April 1912): 362–72.

96. Hashiguchi, "The Rise of Modern Japan."

97. T. Iyenaga, "The War's Disclosure of the Orient," *World's Work* 9 (April 1905): 6041–47.

98. Steiner, *Japanese Invasion,* 36–38; Woodruff, *Expansion of Races,* 325–359; T. H. Huxley, "The Aryan Question and Prehistoric Man," *Popular Science Monthly* 38 (January 1891): 341–55; (February 1891): 502–16.

99. For a close reading of the debate over the whiteness and influence of the Ainu on the formation of the Japanese race, see Chamberlain, "The Japanese Race," 178–82.

100. William Elliot Griffis, *The Japanese Nation in Evolution: Steps in the Progress of a Great People* (New York: Thomas Y. Crowell & Co., 1907); *Abstracts of Reports of the Immigration Commission*, 253; Steiner, *Japanese Invasion*, 195–209; "The Japanese in California," 8689–93; W. I. Thomas, "Is the Human Brain Stationary?" *Forum* 36 (December 1904): 305–20; Jack London, "The Human Drift," *Forum* 45 (January 1911): 1–14.

101. Gulick, *White Peril*, 87–108.

102. Mary Roberts Coolidge, *Chinese Immigration* (New York: H. Holt, 1909); Coolidge, "Chinese Labor Competition on the Pacific Coast," *Annals of the American Academy of Political and Social Science* 34 (September 1909): 120–30.

103. C. H. Roswell, "Orientophobia," *Collier's* (February 6, 1909): 13.

104. Steiner, *Japanese Invasion*, 103.

105. Ross, *Changing Chinese*, 33–47; Ross, *Social Trend*, 9–10.

106. Harold H. Bender, "The Aryan Question," *Forum* 47 (April 1922): 322–30; Steiner, *Japanese Invasion*, 42, 73–76; Thomas Magee, "China's Menace to the World," *Forum* 10 (October 1890): 197–206; Sunyowe Pang, "The Chinese in America," *Forum* 32 (January 1902): 598–607; "The Hindu, the Newest Immigration Problem," *Survey* 25 (October 1, 1910): 2–3; H. A. Millis, "East Indian Immigration to the Pacific Coast," *Survey* 28 (June 1, 1912): 379–86.

107. Ross, *Social Trend*, 14–15; *Abstracts of Reports of the Immigration Commission*, 45–47.

108. F. G. Moorhead, "The Foreign Invasion of the Northwest," *World's Work* 15 (March 1908): 9992–97; B. L. Putnam Weale, "The Conflict of Color," *World's Work* 28–29 (September, October, December 1909): 12023–29, 12111–25, 12327–32; Ross, *Social Trend*, 14–15; Steiner, *Japanese Invasion*, 78–89, 175–194.

109. Cited in Steiner, *Japanese Invasion*, 88–89.

110. "Breaking Up the Home," *Progressive Woman* 4 (March 1911); N. A. Richardson, "Labor, Capital, and China," *International Socialist Review* 2 (April 1902): 735–39; Cameron H. King Jr., "Asiatic Exclusion," *International Socialist Review* 8 (May 1908): 661–69.

111. Jack London, "The Unparalleled Invasion," in London, *The Strength of the Strong* (New York: Leslie-Judge, 1914), 71–100.

4. Cave Girls and Working Women

1. Edgar Rice Burroughs, *The Cave Girl* (Chicago: A. C. McClurg, 1925). It was first serialized as "The Cave Girl" in *All-Story Weekly* beginning in July 1913.

2. Such a reading suggests the role of empire not only in popular culture but also in domestic ways of visual interpretation. See Catherine A. Lutz and Jane L. Collins, *Reading National Geographic* (Chicago: University of Chicago Press, 1993); Laura Wexler, *Tender Violence: Domestic Visions in an Age of U.S. Imperialism* (Chapel Hill: University of North Carolina Press, 2000); Bradley Deane, "Imperial Barbarian: Primitive Masculinity in Lost World Fiction," *Victorian Literature and Culture* 36 (March 2008): 205–25.

3. William Howe Tolman, "The Tenement House Curse," *Arena* 9 (April 1894): 659–84; Dr. A. E. Winship, "Does City Life Enervate?" *World's Work* 1 (November 1900): 109; B. O. Flower, "Society's Exiles," *Arena* 4 (June 1891): 37–54; "The Exhibit of Congestion Interpreted," *Survey* 20 (April–October 1908): 27–39; Gino C. Speranza, "The Italians in Congested Districts," *Survey* 20 (April–October 1908): 55–57; Alexander Irvine, "From the Bottom Up: VI The Battered Hulks of the Bowery," *World's Work* 19 (December 1909): 12365–77; Irvine, "From the Bottom Up: VII Life Among 'The Squatters,'" *World's Work* 19 (January 1910): 12448–57; Joseph Benjamin, "The Comforts and Discomforts of East Side Tenements," "Report for the Year 1897," Records of the University Settlement Society of New York City—Jacob S. Eisinger Collection, Wisconsin Historical Society, Series 3, Box 5, 25–28;

B. O. Flower, *Civilization's Inferno or Studies in the Social Cellar* (Boston: Arena, 1893), 31–43, 63–78.

4. E. R. L. Gould, "The Housing Problem in Great Cities," *Quarterly Review of Economics* 14 (May 1900): 378–93; Lawrence Veiller, "The Tenement-House Exhibition of 1899," *Charities Review* 10 (1900–1901): 19–25; Joseph Lee, "The Integrity of the Family," *National Conference on Social Welfare Proceedings* (1909): 122–24; John Foster Carr, "The Italian in the United States," *World's Work* 8 (October 1904): 5393–5404; Dr. Richard Gottheil, "A Glimpse into the Jewish World," *World's Work* 6 (July 1903): 3689–92.

5. *Preliminary Report of the Factory Investigating Commission, 1912* (Albany: Argus Company, 1912); Department of Factory Inspection, *Annual Report of the Chief State Factory Inspector of Illinois* (Chicago: Illinois Department of Factory Inspection, 1914); Illinois Office of Inspectors of Factories and Workshops, *Annual Report of the Factory Inspectors of Illinois* (Chicago: Office of Inspector of Factories and Workshops, State of Illinois, 1894–1908); Department of Factory Inspection, *Annual Report of the Factory Inspector of Illinois* (Chicago: Illinois Department of Factory Inspection, 1907–13); Dwight Porter, *Report upon a Sanitary Inspection of Certain Tenement-House Districts of Boston* (Boston: Rockwell and Churchill, 1889); Massachusetts Bureau of Statistics of Labor, *A Tenement House Census of Boston* (Boston: Wright & Potter, 1892–93); Edith Abbott, *The Tenements of Chicago, 1908–1935* (New York: Arno Press, 1970 [1936]).

6. Flower, "Society's Exiles," 37–54.

7. Elizabeth C. Watson, "Home Work in the Tenements," *Survey* 25 (February 4, 1911): 772–81.

8. One example of the breakdown of the immigrant family that was often cited was the desertion of families by the failed male breadwinner. See, for example, this story of an immigrant who deserts his family as his wife is giving birth to his seventh child. His other children are already showing criminal tendencies and he is declining in health. Walter E. Weyl, "The Deserter," *Survey* 21 (December 5, 1908): 388–90.

9. William T. Elsing, "Life in New York Tenement-Houses as Seen by a City Missionary," in *The Poor in Great Cities: Their Problems and What Is Being Done to Solve Them* (New York: Charles Scribner's Sons, 1895), 42–117.

10. Edward W. Townsend, *A Daughter of the Tenements* (New York: Lovell, Coryell & Company, 1895), 56–70.

11. Lieutenant-Colonel A. B. Ellis, "Survivals from Marriage by Capture," *Popular Science Monthly* 39 (June 1891): 207–22.

12. E. A. Ross, "Turning Towards Nirvana," *Arena* 4 (November 1891): 739; "Parenthood and Race Culture," *Survey* 23 (February 12, 1910): 731–33.

13. On the evolution of divorce, see J. P. Lichtenberger, "The Instability of the Family," *Annals of the American Academy of Political and Social Science* 34 (July 1909): 97–105. For a particularly vivid representation of early family life, see Thorstein Veblen, "The Barbarian Status of Women," *American Journal of Sociology* 4 (January 1899): 503–14.

14. F. Stuart Chapin, *The Introduction to the Study of Social Evolution, The Prehistoric Period* (New York: Century Co., 1921 [1913]), 280–96. For similar interpretations, see Alexander Sutherland, *The Origin and Growth of the Moral Instinct, Volumes I and II* (London: Longmans, Green, 1898), 195–212; Edward Westermarck, *The History of Human Marriage*, vol. 2 (New York: Allerton, 1922 [1891]), 240–77; Franklin Giddings, *Readings in Descriptive and Historical Sociology* (New York: Macmillan, 1906), 433–73; Giddings, *The Principles of Sociology* (New York: Macmillan, 1908 [1896]), 256–98; Lewis Henry Morgan, *Ancient Society; or, Researches in the Lines of Human Progress from Savagery, through Barbarism to Civilization* (New York: Holt, 1877); John Ferguson McLennan, *The Patriarchal Theory* (London: Macmillan, 1885); W. J. McGee, "The Beginning of Marriage," *American Anthropologist* 9 (November 1896): 371–83.

15. Lester Frank Ward, *Pure Sociology, A Treatise on the Origin and Spontaneous Development of Society* (New York: Macmillan, 1921 [1903]), 346–416.

16. Charlotte Perkins Gilman, "How Home Conditions React Upon the Family," *American Journal of Sociology* 14 (March 1909): 592–605.

17. Arthur J. Todd, "The Family as a Factor in Social Evolution," *National Conference on Social Welfare Proceedings* (1922): 16; Edward Westermarck, *The Origin and Development of the Moral Ideas, Volume I* (London: Macmillan, 1924 [1906]), 629–69.

18. Maurice Parmelee, "The Economic Basis of Feminism," *Annals of the American Academy of Political and Social Science* 56 (November 1914): 18–26; Janet Pearl, "Woman in Society," *Socialist Woman* 1 (October 1907); Lida Parce Robinson, "From Status to Contract," *Socialist Woman* 1 (March 1908).

19. Adolph F. Bandelier, *The Delight Makers* (New York: Dodd, Mead, 1916); Stanley Waterloo, *The Story of Ab* (Chicago: Way and Williams, 1897); Ashton Hilliers, *The Master-Girl, A Romance* (New York: G. P. Putnam's Sons, 1910), 12.

20. Waterloo, *Story of Ab*, 127–39.

21. Hilliers, *The Master-Girl*, 126–34, 222–41.

22. Frederick H. Costello, *Sure-Dart, A Story of Strange Hunters and Stranger Game in the Days of Monsters* (Chicago: A. C. McClurg, 1909).

23. P. B. McCord, *Wolf, The Memoirs of a Cave-Dweller* (New York: B. W. Dodge & Co., 1908), 47–55, 127.

24. W. I. Thomas, "On the Difference in the Metabolism of the Sexes," *American Journal of Sociology* 3 (July 1897): 31–63; Patrick Geddes and J. Arthur Thompson, *Evolution of Sex* (London: Walter Scott, 1889).

25. G. Delauney, "Equality and Inequality in Sex," *Popular Science Monthly* 20 (December 1881): 184–92; F. Swiney, "Evolution of the Male," *Westminster Review* 163 (March–April 1915): 272–78, 449–55; Helen B. Thompson, *The Mental Traits of Sex: An Experimental Investigation of the Normal Mind in Men and Women* (Chicago: University of Chicago Press, 1903), 169–82.

26. Sue Zscoche, "Dr. Clarke Revisited: Science, True Womanhood, and Female Collegiate Education," *History of Education Quarterly* 29 (Winter 1989): 545–69; Louise Newman, ed., *Men's Ideas/Women's Realities: Popular Science, 1870–1915* (New York: Pergamon Press, 1985).

27. William I. Thomas, *Sex and Society: Studies in the Psychology of Sex* (Chicago: University of Chicago Press, 1907), 35–38.

28. Edward Drinker Cope, *The Primary Factors of Organic Evolution* (New York: Open Court Pub. Co., 1904); Cope, *The Origin of the Fittest: Essays on Evolution* (New York: Macmillan, 1887); Cope, *The Relation of the Sexes to Government* (New York: O'Brien, n.d.).

29. Hilliers, *The Master-Girl*, 172.

30. Eliza Burt Gamble, *The Sexes in Science and History: An Inquiry into the Dogma of Woman's Inferiority to Man* (New York: G. P. Putnam's Sons, 1916), v; Penelope Deutscher, "The Descent of Man and the Evolution of Woman," *Hypatia* 19 (Spring 2004): 36–55. On Gamble, see Rosemary Jann, "Revising the Descent of Woman: Eliza Burt Gamble," in *Natural Eloquence: Women Reinscribe Science*, ed. Barbara T. Gates and Ann T. Shteir (Madison: University of Wisconsin Press, 1997). The literature on Gilman is much more expansive. For recent important focused studies on different elements of her canon, see Mark W. Van Wienan, "A Rose by Any Other Name: Charlotte Perkins Stetson (Gilman) and the Case for American Reform Socialism," *American Quarterly* 55 (December 2003): 603–34; Carol Farley Kessler, *Charlotte Perkins Gilman: Her Progress Toward Utopia with Selected Writings* (Syracuse: Syracuse University Press, 1995); Cynthia J. Davis and Denise D. Knight, eds., *Charlotte Perkins Gilman and Her Contemporaries: Literary and Intellectual Contexts* (Tuscaloosa: University of Alabama Press, 2004); Vai Gough and Jill Rudd, eds., *A Very Different Story: Studies on the Fiction of Charlotte Perkins Gilman* (Liverpool: Liverpool University Press, 1998); Joanne B.

Karpinski, ed., *Critical Essays on Charlotte Perkins Gilman* (New York: G. K. Hall, 1992). See also Charlotte Perkins Gilman, *The Living of Charlotte Perkins Gilman: An Autobiography* (Madison: University of Wisconsin Press, 1991).

31. Gamble, *Sexes in Science and History*, 45; Charlotte Perkins Gilman, *Women and Economics: A Study of the Economic Relation between Men and Women as a Factor in Social Evolution* (Berkeley: University of California Press, 1998 [1898]), 130–34; Miss M. A. Hardaker, "Science and the Woman Question," *Popular Science Monthly* 20 (March 1882): 577–84; Nina Morais, "A Reply to Miss Hardaker on the Woman Question," *Popular Science Monthly* 21 (May 1882): 70–78; Grant Allen, "Plain Words on the Woman Question," *Popular Science Monthly* 36 (December 1889): 170–81; William Graham Sumner, "The Status of Women in Chaldea, Egypt, India, Judea and Greece to the Time of Christ," *Forum* 42 (August 1909): 113–36; Edward Westermarck, "The Position of Women in Early Civilization," *American Journal of Sociology* 10 (November 1904): 408–21.

32. Pearl, "Women in Society;" "Woman, The Toiler," *Socialist Woman* 1 (February 1908); Luella Krehbiel, "Woman, The World Is Waiting for You," *Socialist Woman* 1 (May 1908); Lida Parce, "The Examiner's Glass," *Progressive Woman* 3 (November 1910).

33. Delauney, "Equality and Inequality in Sex," 184.

34. Edward C. Rafferty, *Apostle of Human Progress: Lester Frank Ward and American Political Thought, 1841–1913* (New York: Rowman and Littlefield, 2003), 165–282.

35. Barbara Finlay, "Lester Frank Ward as a Sociologist of Gender: A New Look at His Sociological Work," *Gender & Society* 13 (April 1999): 251–65; Cynthia Eller, "Sons of the Mother: Victorian Anthropologists and the Myth of Matriarchal Prehistory," *Gender and History* 18 (August, 2006): 285–310.

36. Paul Topinard, *Éléments d'Anthropologie Générale* (Paris: Delahaye, 1885). For the debate about Ward's theories that filled the pages of the *Forum*, see Lester F. Ward, "Our Better Halves," *Forum* 6 (November 1888): 266–75; Grant Allen, "Woman's Place in Nature," *Forum* 7 (May 1889): 258–63; Allen, "Woman's Intuition," *Forum* 9 (May 1890): 333–40; Ward, "Genius and Woman's Intuition," *Forum* 9 (June 1890): 401–8.

37. Ward, *Pure Sociology*, 302.

38. Ward, *Pure Sociology*, 314.

39. Ward, *Pure Sociology*, 331; "Charlotte Perkins Gilman to Lester Frank Ward" (January 26, 1908), Lester Frank Ward Papers, Brown University, Series I, Box 33, Folder 10.

40. Ward, *Pure Sociology*, 336–53.

41. Charlotte Perkins Gilman, *The Man-Made World or Our Androcentric Culture* (New York: Charlton Company, 1911).

42. "Lester F. Ward to Gilman" (February 11, 1911), Charlotte Perkins Gilman Papers, Schlesinger Library, Radcliffe Institute, Harvard University, Series III, Folder 124.

43. Gilman, *Women and Economics*, 60–61. For a discussion of the way Gilman interpreted the work of Lester Frank Ward, see Cynthia Davis, "His and Herland: Charlotte Perkins Gilman 'Re-presents' Lester F. Ward," in *Evolution and Eugenics in American Literature and Culture*, ed. Davis (Lewisburg, Pa.: Bucknell University Press, 2003).

44. "Woman, The Toiler"; Pearl, "Woman in Society"; May Wood Simons, *Woman and the Social Problem* (Chicago: Charles Kerr, 1899), 8–9; "Gilman to Ward" (March 20, 1907), Ward Papers, Series I, Box 33, Folder 10.

45. Gamble, *Sexes in Science and History*, 83.

46. Sara Kingsbury, "Lady-Like Woman: Her Place in Nature," *Socialist Woman* 2 (August 1908); Theresa Malkiel, "Woman," *Progressive Woman* 3 (March 1910).

47. Gilman, *Women and Economics*.

48. Gilman, *Women and Economics*, 74–75.

49. Lida Parce Robinson, "The Dangers of Exclusive Masculinism," *Socialist Woman* 2 (July 1908).

50. Josephine C. Kaneko, "Sex Subjugation," *Socialist Woman* 1 (August 1907); Kaneko, "Machinery Makes New Woman with New Matrimonial Ideals," *Socialist Woman* 2 (December 1907); "Woman, The Toiler"; Kingsbury, "Lady-Like Woman."

51. Gilman, "How Home Conditions React upon the Family," 593.

52. Elsa Unterman, "The Mitigator," *Progressive Woman* 3 (June 1910); Simons, *Woman and the Social Problem,* 7.

53. Thorstein Veblen, "The Instinct of Workmanship and the Irksomeness of Labor," *American Journal of Sociology* 4 (September 1898): 187–201; Veblen, "Barbarian Status of Women," 503–504.

54. Thomas, *Sex and Society,* 131–33; Todd, "The Family as a Factor in Social Evolution," 16.

55. Gilman, "How Home Conditions React upon the Family," 596–97; Charlotte Perkins Gilman, *Herland* (1915), in Barbara H. Solomon, ed., *Herland and Selected Stories by Charlotte Perkins Gilman* (New York: Signet, 1992), 137; Thomas, *Sex and Society,* 228–47; Charlotte Perkins Gilman, "Economic Waste in the Home," *Women's Era* 1 (April 1910): 157–58, in Gilman Papers, Series IV, Folder 250.

56. Simons, *Woman and the Social Problem,* 7–9; Mari Jo Buhle, *Women and American Socialism, 1870–1920* (Urbana: University of Illinois Press, 1981), 166–213.

57. Gilman, *Herland,* 72. Gilman introduced the idea of an isolated women's land in a letter to Ward: "Gilman to Ward" (May 9, 1908), Ward Papers, Series I, Box 36, Folder 13.

58. Gilman, "How Home Conditions React upon the Family," 595.

59. Theresa Schmid McMahon, "Women and Economic Evolution or the Effects of Industrial Changes Upon the Status of Women," *Bulletin of the University of Wisconsin* (Madison, 1912); Helen Campbell, "Women Wage-Earners: Their Past, Their Present, and Their Future," *Arena* 7 (January 1893): 153–72; (February 1893): 321–36; (March 1893): 433–40; (May 1893):, 668–79; Otis Tufton Mason, *Women's Share in Primitive Culture* (New York: D. Appleton, 1898); Anna Garlin Spencer, "The Primitive Working-Woman," *Forum* 46 (November 1911): 546–58; Spencer, "Pathology of Women's Work," *Forum* 47 (March 1912): 321–36.

60. McMahon, "Women and Economic Evolution," 86–87; Florence Kelley, "The Family and the Woman's Wage," *National Conference on Social Welfare Proceedings* (1909): 118–21; Josephine Goldmark, *Fatigue and Efficiency, A Study in Industry* (New York: Russell Sage Foundation, 1912), 40.

61. Kaneko, "Machinery Makes New Woman"; Simons, *Woman and the Social Problem,* 19–20; "Why Women Should Be Socialists," *Socialist Woman* 1 (June 1907).

62. Spencer, "Primitive Working-Woman," 552–53; Emily Blackwell, "The Industrial Position of Women," *Popular Science Monthly* 23 (July 1883): 388–99; "Woman as an Inventor and Manufacturer," *Popular Science Monthly* 47 (May 1895): 92–103; Clare de Graffenreid, "The 'New Woman' and Her Debts," *Popular Science Monthly* 49 (May 1896): 664–72; D. R. Malcolm Keir, "Women in Industry," *Popular Science Monthly* 83 (October 1913): 375–80; R. E. Francillon, "The Physiology of Authorship," *Gentlemen's Magazine* 14 (March 1875): 334–35; Havelock Ellis, *Man and Woman: A Study of Human Secondary and Tertiary Sexual Characteristics* (London: Walter Scott, 1894).

63. McMahon, "Women and Economic Evolution," 28–48; Spencer, "Pathology of Women's Work," 321–23; Charlotte Perkins Gilman, "On the Persistence of Primitive Tendencies in the Domestic Relation," (n.d.), Gilman Papers, Series IV, Folder 175.

64. McMahon, "Women and Economic Evolution," 42. Such a logic helps explain the emergence and popularity of the movement for protective legislation, led by women such as Florence Kelley, for the regulation and restriction of women's labors. Restricting women's work was for the good of the individual, as well as for the good of the race. See Kathryn Kish Sklar, *Florence Kelley and the Nation's Work* (New Haven: Yale University Press, 1995); Landon Storrs, *Civilizing Capitalism: The National Consumer's League, Women's Activism, and Labor*

Standards in the New Deal Era (Chapel Hill: University of North Carolina Press, 2000); Eileen Boris, *Home to Work: Motherhood and the Politics of Industrial Homework in the United States* (Cambridge: Cambridge University Press, 1994); Louise Michele Newman, *White Women's Rights: The Racial Origins of Feminism in the United States* (New York: Oxford University Press, 1999), 86–101.

65. Simons, *Woman and the Social Problem*, 13.

66. William Hard and Rheta Childe Dorr, "The Women's Invasion," *Everybody's Magazine* 19 (November 1908): 579–91; 19 (December 1908): 798–810; 20 (January 1909): 73–85; 20 (February 1909): 236–48; 20 (March 1909): 372–85; 20 (April 1909): 521–32; Helen Campbell, "The Working-Women of To-Day," *Arena* 4 (June 1891): 329–39; Helen Sumner, *Women in Industry* (New York: Arno, 1974 [1910]); Mary K. Maule, "What Is a Shop-Girl's Life?" *World's Work* 14 (September 1907): 9311–16; Mabel Potter Daggett, "Women: Building a Better Race," *World's Work* 25 (December 1912): 228–34.

67. Hard and Dorr, "The Women's Invasion" (December 1908), 798–800; Elizabeth Beardsley Butler, *Women and the Trades, Pittsburgh, 1907–1908* (New York: Charities Publication Committee, 1909).

68. Hard and Dorr, "The Women's Invasion" (December 1908), 801–2; Wright cited in U. G. Weatherly, "How Does the Access of Women to Industrial Occupations React on the Family?" *American Journal of Sociology* 14 (May 1909): 740–65.

69. United States Bureau of Labor, *Report on Condition of Woman and Child Wage-Earners in the United States in 19 Volumes: Volume 9: History of Women in Industry in the United States* (Washington, D.C.: GPO, 1910), 11.

70. Helen L. Sumner, "The Historical Development of Women's Work in the United States," *Proceedings of the Academy of Political Science in the City of New York* 1 (October 1910): 11–26.

71. B. O. Flower, "Early Environment in Home Life," *Arena* 10 (September 1884): 483–93; David Dudley Field, "The Child and the State," *Forum* 1 (April 1886): 105–13. The commissioner of labor of Ohio, for example, reported that much of his attention was drawn to "the alarming growth of women and child-labor" and the New York Bureau of Labor devoted its second issue to the problem of children's wage work. Commissioner of Labor for Ohio (1887) and the New York Bureau of Labor (1885), cited in William F. Willoughby, "Child Labor," *Publications of the American Economic Association* 5 (March 1890): 27. See also United States Bureau of the Census, *Child Labor in the United States, Based on Unpublished Information Derived from the Schedules of the Twelfth Census* (Washington, D.C.: GPO, 1900).

72. Claire de Graffenried, "Child Labor," *Publications of the American Economic Association* 5 (March 1890): 71–149. On de Graffenried, see LeeAnn Whites, "The De Graffenried Controversy: Class, Race, and Gender in the New South," *Journal of Southern History* 54 (August 1988): 449–78. For similar representations of the physical and moral decline of child laborers as a symptom of racial decline, see Elias Tobenkin, "The Immigrant Girl in Chicago," *Survey* 23 (November 6, 1909): 189–95; Edith Abbott, *Women in Industry: A Study in American Economic History* (New York: D. Appleton, 1913); United States Bureau of Labor, *Report on Conditions of Woman and Child Wage-Earners in the United States in 19 Volumes, Volume 6: The Beginnings of Child Labor Legislation in Certain States, A Cooperative Study*, 73–221, and *Volume VII: Conditions Under Which Children Leave School to Go to Work* (Washington, D.C.: GPO, 1910).

73. Simons, *Woman and the Social Problem*, 18.

74. Stephen Crane, *Maggie, A Girl of the Streets: A Story of New York* (Boston: Bedford/ St. Martin's, 1999 [1893]); Keith Gandal, *The Virtues of the Vicious: Jacob Riis, Stephen Crane, and the Spectacle of the Slum* (New York: Oxford University Press, 1997); Giorgio Mariani, *Spectacular Narratives: Representations of Class and War in Stephen Crane and the American 1890s* (New York: P. Lang, 1992).

75. "Charles Loring Brace to Fred S. Hall, Secretary Child Labor Committee, March 12, 1903," Records of the University Settlement Society of New York City, Series 4, Box 8; Edward Clopper, *Child Labor in the Streets* (New York: Macmillan, 1913), 52–82; David Nasaw, *Children of the City: At Work and at Play* (Garden City: Anchor Press/Doubleday, 1985); Vincent Richard DiGirolamo, "Crying the News: Children, Street Work, and the American Press, 1830s–1920s" (PhD diss., Princeton University, 1997).

76. Josephine Conger Kaneko, "The Horrible Crime of Child Labor in America," *Socialist Woman* I (March 1908).

77. Sutherland, *Origin and Growth of the Moral Instinct*, 1–103; John Fiske, *The Meaning of Infancy* (Boston: Houghton, 1911); Felix Adler, "The Attitude of Society toward the Child as an Index of Civilization," *Annals of the American Academy of Political and Social Science* 29 (January 1907): 135–41; Margaret F. Byington, "The Normal Family," *Annals of the American Academy of Political and Social Science* 77 (May 1918): 13–18.

78. Adler, "Attitude of Society toward the Child," 136; Byington, "The Normal Family," 15.

79. Sutherland, *Origin and Growth of the Moral Instinct*, 150–51; Edwin Grant Conklin, *Heredity and Environment in the Development of Man* (Princeton: Princeton University Press, 1915), 362; Josephine Conger Kaneko, "The Civilizing Process," *Progressive Woman* 5 (November 1911).

80. G. Stanley Hall, *Youth: Its Education, Regimen, and Hygiene* (New York: Appleton, 1912 [1907]); G. Stanley Hall, *Aspects of Child Life and Education* (Boston: Ginn & Company, 1907); G. Stanley Hall, *Life and Confessions of a Psychologist* (New York: D. Appleton, 1924); Alice Minnie Herts Heniger, *The Kingdom of the Child* (New York: Dutton, 1918); Lilla Estelle Appleton, *A Comparative Study of the Play Activities of Adult Savages and Civilized Children, An Investigation of the Scientific Basis of Education* (Chicago: University of Chicago Press, 1910); Dorothy Ross, *G. Stanley Hall: The Psychologist as Prophet* (Chicago: University of Chicago Press, 1972).

81. Adler, "Attitude of Society toward the Child," 139.

82. Adler, "Attitude of Society toward the Child," 139–40.

83. Townsend, *Daughter of the Tenements*, 66.

84. "Report to the Council for Month ending May 19, 1902," Records of the University Settlement Society of New York City, Series 2, Box 3, Folder 6; Willoughby, "Child Labor," 61–62.

85. Sutherland, *Origin and Growth of the Moral Instinct*.

86. Robert Hunter, *Poverty* (New York: Grosset and Dunlap, 1904), 142–43.

87. Edward A. Ross, "The Causes of Race Superiority," *Annals of the American Academy of Political and Social Sciences* 18 (July 1901): 67–89; Allyn A. Young, "The Birth Rate in New Hampshire," *Publications of the American Statistical Association* 9 (September 1905): 263–91; United States Immigration Commission, *Occupations of the First and Second Generations of Immigrants in the United States: Fecundity of Immigrant Women* (Washington, D.C.: GPO, 1911), 810; Edward A. Ross, "Western Civilization and the Birth-Rate," *American Journal of Sociology* 12 (March 1907): 607–32.

88. Annie Marion MacLean, "The Sweat-Shop in Summer," *American Journal of Sociology* 9 (November 1903): 289–309.

89. Frank Fetter comment on Ross, "Western Civilization and the Birth-Rate," 618. For optimistic readings of infant mortality, see S. J. Holmes, *Studies in Evolution and Eugenics* (New York: Harcourt, Brace and Company, 1923), 151–52.

90. Ross, "The Causes of Race Superiority," 88; Ross, "Western Civilization and the Birth-Rate," 612.

91. J. L. Brownell, "The Significance of the Decreasing Birth-Rate," *Annals of the American Academy of Political and Social Science* 5 (July 1894): 48–89; J. S. Billings, "The Diminishing Birth Rate in the United States," *Forum* 15 (June 1893): 467–77; Samuel J. Holmes, *The*

Trend of the Race, A Study of the Present Tendencies in the Biological Development of Civilized Mankind (New York: Harcourt, Brace and Company, 1921), 120–204; C. M. Doud and S. J. Holmes, "The Approaching Extinction of the *Mayflower* Descendents," *Journal of Heredity* 9 (November 1918): 296–300; R. H. Johnson, "Wellesley's Birth Rate," *Journal of Heredity* 6 (June 1915): 250–53; Frank W. Nicolson, "Family Records of Graduates of Wesleyan University," *Science* 36 (July 19, 1912): 74–76; F. S. Crum, "The Decadence of Native American Stock, A Statistical Study of Genealogical Records," *Publications of the American Statistical Association* 14 (September 1914): 215–22; Nellie Seeds Nearing, "Education and Fecundity," *Publications of the American Statistical Association* 14 (June 1914): 156–74; William Hard, *The Women of Tomorrow* (New York: Baker & Taylor Company, 1911), 135–78.

92. Quoted in Leta S. Hollingworth, "Social Devices for Impelling Women to Bear and Rear Children," *American Journal of Sociology* 22 (July 1916): 22.

93. Ross, "Causes of Race Superiority," 82–88.

94. Henry Pratt Fairchild, *Immigration: A World Movement and Its American Significance* (New York: Macmillan, 1923 [1913]), 341–42; F. A. Walker, *Discussions in Economics and Statistics*, vol. 2 (New York: Holt, 1899), 441–42; Annette Fiske, "Where Race Suicide Does Not Prevail," *Survey* 29 (November 23, 1912): 244–46.

95. Joseph A. Hill, "Comparative Fecundity of Women of Native and Foreign Parentage in the United States," *Publications of the American Statistical Association* 13 (December 1913): 583–604; F. L. Hoffmann, *The Significance of a Declining Birth Rate* (Newark, N.J.: Prudential Press, 1914); Walter F. Willcox, "The Nature and Significance of the Changes in the Birth and Death Rates in Recent Years," *Publications of the American Statistical Association* 15 (March 1916): 1–15; Frederick A. Bushee, "The Declining Birth Rate and Its Cause," *Popular Science Monthly* 63 (August 1903): 355–61; Charles Franklin Emerick, "Is the Diminishing Birth Rate Volitional?" *Popular Science Monthly* 78 (January 1911): 71–80; Scott Nearing, "'Race Suicide' vs. Overpopulation," *Popular Science Monthly* 78 (January 1911): 81–83; Willcox, "Fewer Births and Deaths: What Do They Mean?" *Journal of Heredity* 7 (March 1916): 119–27.

96. For Roosevelt's notions of race suicide, see Theodore Roosevelt, "Race Decadence," *Outlook* (April 8, 1911): 768; Roosevelt, "A Premium on Race Suicide," *Outlook* (September 27, 1913): 164; Roosevelt, *Presidential Addresses and State Papers, April 7, 1904–May 9, 1905* (New York: Kraus, 1970), 238; Thomas G. Dyer, *Theodore Roosevelt and the Idea of Race* (Baton Rouge: Louisiana State University Press, 1980), 143–67.

97. James Morton, "Fewer and Better Children," *International Socialist Review* 15 (November 1914): 301–3; Caroline Nelson, "Neo-Malthusianism in America," *International Socialist Review* 15 (November 1914): 300–301.

98. See, for example, Sidney Webb, "Physical Degeneracy or Race Suicide?" *Popular Science Monthly* 69 (December 1906): 512–29.

99. A. J. McKelway, "The Child Labor Problem—A Study in Degeneracy," *Annals of the American Academy of Political and Social Science* 27 (March 1906): 54.

100. Charles L. Dana, "Are We Degenerating?" *Forum* 19 (June 1895): 458–66; John James Stevenson, *Is This a Degenerate Age?* (n.p., 1903); Hudson Maxim, "Genius and Regeneration," *Arena* 23 (April 1900): 425–31; "Is the Race Degenerating in America?" *Survey* 22 (April 24, 1909): 137–38; Andrew Wilson, "Degeneration," *Popular Science Monthly* 19 (June and July, 1881): 219–29; Franklin Smith, "Reversions in Modern Industrial Life," *Popular Science Monthly* 50 (April 1897): 781–96, (May 1897): 34–43; Daniel G. Brinton, *The Basis of Social Relations: A Study in Ethnic Psychology* (New York: G. P. Putnam's Sons, 1902), 82–119; Dr. Chas. T. Nesbitt, "The Health Menace of Alien Races," *World's Work* 27 (November 1913): 74–78; Allan McLaughlin, "Immigration and the Public Health," *Popular Science Monthly* 64 (January 1904): 232–38.

101. D. Colin Wells, "Social Darwinism," *American Journal of Sociology* 12 (March 1907): 695–716.

102. Willoughby, "Child Labor," 62–64; McKelway, "Child Labor Problem," 64; Thomas Burke, "Physical Deterioration in England," *Forum* 36 (March 1905): 449–57; William Sadler, *Race Decadence* (New York: A. C. McClurg, 1922), 12–13; Simon Patten, "The Principles of Economic Interference," *Survey* 22 (April 3, 1909): 14–16.

103. Simons, *Woman and the Social Problem*, 19.

104. Willoughby, "Child Labor," 63.

105. "Three Things to Do in 1910," *Survey* 23 (January 1, 1910): 433.

106. Raphael Buck, "Natural Selection under Socialism," *International Socialist Review* 2 (May 1902): 787.

107. Jack London, "Wanted: A New Law of Development," *International Socialist Review* 3 (August 1902): 65–78; Wells, "Social Darwinism," 701, and Carl Kelsey reply to "Social Darwinism," 711.

5. Exploring the Abyss and Segregating Savagery

1. The texts that made up the survey were Elizabeth Beardsley Butler, *Women and Trades, Pittsburgh, 1907–1908* (Pittsburgh: University of Pittsburgh Press, 1989 [1909]); Margaret Byington, *Homestead: The Households of a Mill Town* (Pittsburgh: University of Pittsburgh Press, 1989 [1910]); Crystal Eastman, *Work-Accidents and the Law* (Pittsburgh: University of Pittsburgh Press, 1989 [1910]); John Fitch, *The Steel Workers* (Pittsburgh: University of Pittsburgh Press, 1989 [1910]); Paul U. Kellogg, ed., *The Pittsburgh District: Civic Frontage* (New York: Arno, 1974 [1914]); Kellogg, ed., *Wage-Earning Pittsburgh* (New York: Arno, 1974 [1914]). On the broader history and significance of the Survey, see Maurine W. Greenwald and Margo Anderson, *Pittsburgh Surveyed: Social Science and Social Reform in the Early Twentieth Century* (Pittsburgh: University of Pittsburgh Press, 1996); Edward Slavishak, *Bodies of Work: Civic Display and Labor in Industrial Pittsburgh* (Durham: Duke University Press, 2008), 149–199.

2. Matthew Frye Jacobson, *Barbarian Virtues: The United States Encounters Foreign Peoples at Home and Abroad* (New York: Hill and Wang, 2000), 105–38, usefully recasts ghetto sketches and studies of immigrants as part of a broader imperial travelogue genre. Mark Pittenger, "A World of Difference: Constructing the 'Underclass' in Progressive America," *American Quarterly* 49 (March 1997): 26–65, has described the ways undercover observers often assumed the guise of the very subjects they studied. In so doing, they came to represent class as a biological divide that separated social observers from their "underclass" subjects. My reading of immigrant study as imperial observation is influenced by Mary Louise Pratt, *Imperial Eyes: Travel Writing and Transculturation* (New York: Routledge, 1992), particularly for its insight that imperial travel writing was not a closed genre, but remained connected, especially through its pretense to science, to other types of writing and representation. See also Kristi Siegel, ed., *Issues in Travel Writing: Empire, Spectacle, and Displacement* (New York: Peter Lang, 2002).

3. Tyler Andbinder, *Five Points: The 19th-Century New York City Neighborhood That Invented Tap Dance, Stole Elections, and Became the World's Most Notorious Slum* (New York: Simon and Schuster, 2001); Seth Koven, *Slumming: Sexual and Social Politics in Victorian London* (Princeton: Princeton University Press, 2004); Angela M. Blake, *How New York Became American, 1890–1924* (Baltimore: Johns Hopkins University Press, 2006), 49–79; Arthur Pember, *The Mysteries and Miseries of the Great Metropolis* (New York: D. Appleton, 1874); Hutchins Hapgood, *Types from the City Streets* (New York: Funk and Wagnalls, 1910). For a study of urban tourism, including slumming, see Catherine Cocks, *Doing the Town: The Rise of Urban Tourism in the United States, 1850–1915* (Berkeley: University of California Press, 2001), 106–203.

4. "Exploring Chicago with a Sociologist" (ca. 1928), Ben Lewis Reitman Papers, University of Illinois-Chicago, File List 426, Series: Topics, #88.

5. Howard Vincent O'Brien, "The City," *Forum* 54 (October 1915): 385–91.

6. *Seven Days in Chicago* (Chicago: J. M. Wing Co., 1878), 51; Cocks, *Doing the Town*, 136–38.

7. Julian Street, *Abroad at Home: American Ramblings, Observations and Adventures* (Garden City, N.Y.: Garden City Publishing Company, 1914), 164–72.

8. W. Joseph Grand, *Illustrated History of the Union Stockyards: Sketch-Book of Familiar Faces and Places at the Yards* (Chicago: W. Jos Grand, 1901).

9. *Hearings before the Committee on Agriculture* (Washington, D.C.: GPO, 1906), 174–75.

10. Upton Sinclair, *The Jungle* (Urbana: University of Illinois Press, 1988 [1906]), 32–41. Sinclair's vivid description of the packinghouse tour was perhaps the most often cited part of the novel. See, for example, "The Inferno of Packingtown Revealed," *Arena* 35 (June 1906): 651–58.

11. Upton Sinclair, "A Home Colony," *Independent* 60 (June 14, 1906): 1401–02.

12. *Report of the Military Governor of Cuba on Civil Affairs* (Washington, D.C.: GPO, 1902); *Bulletin of the United States Geological Survey No. 192 Series F, Geography, 29. A Gazetteer of Cuba* (Washington, D.C.: GPO, 1902); *Bulletin of the United States Geological Survey, No. 183, Series F, Geography, 25. A Gazetteer of Porto Rico* (Washington, D.C.: GPO, 1902).

13. Edgar Rice Burroughs, *The Efficiency Expert* (Kansas City: House of Greystoke, 1966 [1921]).

14. Hapgood, *Types from City Streets*, 9–25.

15. *Chicago's Dark Places: Investigations by a Corps of Specially Appointed Commissioners and Arranged by the Chief Commissioner* (Chicago: Craig Press, 1891), 9.

16. Jacob Riis, *How the Other Half Lives: Studies Among the Tenements of New York* (New York: Scribner, 1917). For a similar kind of study, see Joseph Kirkland, "Among the Poor of Chicago," *Scribner's Magazine* 12 (July 1892): 3–27.

17. Walter Wyckoff, *The Workers, An Experiment in Reality: The West* (New York: Charles Scribner's Sons, 1899), 134–36, and Wyckoff, *The Workers, An Experiment in Reality: The East* (New York: Charles Scribner's Sons, 1899).

18. Jack London, "South of the Slot," in *The Strength of the Strong* (New York: Leslie-Judge Company, 1908), 34–70.

19. Mary Kingsbury Simkhovitch, *The City Workers World in America* (New York: Macmillan, 1917).

20. Jack London, *People of the Abyss* (New York: Macmillan, 1903).

21. B.O. Flower, *Civilization's Inferno or Studies in the Social Cellar* (Boston: Arena, 1893), 114–23; Edward Ross, "Ethnological Notes and Observations," Edward A. Ross Papers, Wisconsin History Society, Box 26, Folders 9–12. Kristin Hoganson, *Consumer's Imperium: The Global Production of American Domesticity, 1865–1920* (Chapel Hill: University of North Carolina Press, 2007), 153–208, highlights the American fascination with tourism. Her focus on armchair travel, as an engagement with the exotic, presents an important contrast with urban exploration, which represented not only a confrontation with the exotic but a personal measure of civilization.

22. Robert A. Woods, "The City Wilderness," *Commons* 3–6 (January–April 1899): 24–25.

23. Mary B. Sayles, "The Work of a Tenement-House Inspector," *Outlook* 75 (September 12, 1903): 116–22; Helen Campbell, *Prisoners of Poverty* (Boston: Roberts Brothers, 1887), 244–52. Louise Michele Newman, *White Women's Rights: The Racial Origins of Feminism in the United States* (New York: Oxford University Press, 1999), 56–101, highlights the female reform movement's use of racialized ideas of civilization. These women defended their presence in the slums largely through claims to civilizing work.

24. Jessie Davis [pseud.], "My Vacation in a Woolen Mill," *Survey* 40 (August 10, 1918): 538–41.

25. *Chicago's Dark Places*, 13–16. See also Jane Addams, *Hull House Maps and Papers by Residents of Hull-House* (New York: Arno, 1970 [1895]).

26. For example, Ishi, the Yahi Indian of California, was put on display in a natural history museum. A. L. Kroeber, "Ishi, The Last Aborigine," *World's Work* 24 (July 1912): 304–12.

27. Graham Taylor, "Social Tendencies of the Industrial Revolution," *Commons* 9 (October 1904): 459–68; John Martin, "The Exhibit of Congestion Interpreted," *Charities and the Commons* 20 (April 4, 1908): 27–39; Jane Addams, "Labor Museum at Hull House," *Commons* 3–6 (June 1900): 1–4.

28. "Annual Report of the Hale House Association, September 30, 1915," United South End Settlements Records SW204, Social Welfare History Archives, University of Minnesota, Box 2, Folder: "Hale House Annual Report 1915," 3–6; "The City Wilderness," *Commons* 3–6 (January–April 1899): 24–25; Royal Loren Melendy, "Social Function of the Saloon in Chicago," *Commons* 3–6 (November 1900): 1–9; "What Rev. George L. McNutt Sees while Exploring the Working-World as a Day Laborer," *Commons* 3–6 (September 1901): 1–3; William H. Matthews, "The Muckers," *Survey* 35 (October 2, 1915): 5–8; Davis, "My Vacation in a Woolen Mill," 538–41.

29. Helen Campbell, *Darkness and Daylight: Lights and Shadows of New York Life* (Hartford: A. D. Worthing & Co., 1892), vii–xi.

30. W. J. McGee, "Strange Races of Men," *World's Work* 7 (August 1904): 5185–88.

31. For examples from *National Geographic*, see Z. F. McSweeny, "Character of Our Immigration," 16 (January 1905): 1–15; "Some of Our Immigrants," 18 (May 1907): 317–34; "Foreign Born of the United States," 26 (September 1914): 265–71; "Our Foreign Born Citizens," 31 (February 1917): 95–130; William Joseph Showalter, "New York—The Metropolis of Mankind," 34 (July 1918): 1–49. For examples from *World's Work*, see Kate Holladay Claghorn, "The Changing Character of Immigration," 1 (February 1901): 381–87; Edward Lowry, "Americans in the Raw: The High Tide of Immigrants—Their Strange Possessions and their Meager Wealth—What Becomes of Them," 4 (October 1902): 2644–55; Richard Gottheil, "A Glimpse into the Jewish World," 6 (July 1903): 3689–92. See also Catherine Lutz and Jane L. Collins, *Reading National Geographic* (Chicago: University of Chicago Press, 1993).

32. Graham Romeyn Taylor, "Ten Thousand at Play," *Survey* 22 (June 5, 1909): 365–73.

33. Alois B. Koukol, "The Slav's A Man for A' That," *Survey* 21 (January 2, 1909): 589–98.

34. Lewis Hine, "Immigrant Types in the Steel District," *Survey* 21 (January 2, 1909): 581–89.

35. I. L. Nascher, *The Wretches of Povertyville* (Chicago: Jos. L. Lanzit, 1909). For a review of this text that clearly reads it as class tourism, see "The Wretches of Povertyville," *Survey* 22 (August 28, 1909): 723–24. See also Hapgood, *Types from City Streets*, esp. 125–333; Alfred Hodder, "Those Who Lose in the Game of Life: Incidents in the Day's Routine in the Office of Mr. Jerome, District Attorney of New York County," *World's Work* 5 (January 1903): 2999–3002; "An Etcher of Henry Street," *Survey* 35 (November 6, 1915): 135–38. See also Kathryn Oberdeck, *The Evangelist and the Impresario: Religion, Entertainment, and Cultural Politics in America, 1884–1914* (Baltimore: Johns Hopkins University Press, 1999), and Todd Depastino, *Citizen Hobo: How a Century of Homelessness Shaped America* (Chicago: University of Chicago, 2003), 72–75.

36. London, *The People of the Abyss*; Koven, *Slumming*, 27, 140–80.

37. Edward A. Ross, "Ethnological Notes and Observations, ca. 1912 (Vol. 1)," Ross Papers, Box 26, Folder 9, 11–13, 83–84. For the most complete linkage of beauty and racial status, see Knight Dunlap, *Personal Beauty and Racial Betterment* (St. Louis: C. V. Mosby Company, 1920), 15–70.

38. Franklin H. Giddings, "Is the Term 'Social Classes' a Scientific Category?" *National Conference on Social Work Proceedings* (1895): 110–16; R. Stetson, "Industrial Classes as Factors in Racial Development," *Arena* 41 (February 1909): 177–89.

39. Benjamin Seebohm Rowntree, *Poverty, A Study of Town Life* (New York: Garland, 1980 [1910]); Charles Booth, *Life and Labour of the People of London* (London: Macmillan, 1902); "Some of the Social Thinkers…," *Charities and the Commons* 17 (October 1906–April 1907): 182–88; Frank Kellogg, "Charles Booth An Appreciation," *Survey* 37 (December 9, 1916): 267–68.

40. Shelby M. Harrison, "The Development of Social Surveys," *National Conference on Social Welfare Proceedings* (1913): 345–53; John Daniels, "The Social Survey: Its Reasons, Methods, and Results," *National Conference on Social Welfare Proceedings* (1910): 236–40; Edwin Bjorkman, "What Industrial Civilization May Do to Men," *World's Work* 17 (April 1909): 11479–500.

41. "Pittsburgh Women in the Metal Trades," *Survey* 20 (October 3, 1908): 34–47; Paul U. Kellogg, "The Pittsburgh Survey"; Peter Roberts, "The New Pittsburghers: Slavs and Kindred Immigrants in Pittsburgh"; Crystal Eastman, "The Temper of the Workers Under Trial"; Elizabeth Beardsley Butler, "The Working Women of Pittsburgh"; Lewis Hine, "Immigrant Types in the Steel District"; Alois B. Koukol, "The Slav's A Man for A' That"; Anna Reed, "The Jewish Immigrants of Two Pittsburgh Blocks"; Margaret F. Byington, "Homestead: A Steel Town and its People"; Paul U. Kellogg, "The Civic Responsibilities of Democracy," all in *Survey* 21 (January 2, 1909): 533–628; Margaret F. Byington, "The Mill Town Courts and their Lodgers," *Survey* 21 (January 23, 1909): 913–22; "Social Forces"; John R. Commons, "Wage Earners of Pittsburgh"; John Andrews Fitch, "The Steel Industry and the Labor Problem"; Margaret F. Byington, "Household Builded Upon Steel"; Florence Kelley, "Factory Inspection in Pittsburgh"; Elizabeth Beardsley Butler, "The Industrial Environment of Pittsburgh's Working Women," all in *Survey* 21 (March 6, 1909): 1035–1142.

42. Kellogg, "Pittsburgh Survey," 519–21.

43. See, for example, James D. McCabe, *Lights and Shadows of New York Life* (Philadelphia: National Publishing Company, 1872); George Forster, *New York by Gas-Light and Other Urban Sketches* (Berkeley: University of California Press, 1990).

44. Ernest Ingersoll, *Rand McNally & Co.'s Handy Guide to New York City* (New York: Rand McNally, 1912), 100–116.

45. "Editorial: Two Hours in the Social Cellar," *Arena* 5 (April 1892): 645–56; Campbell, *Darkness and Daylight,* ix.

46. George Allan England, *Darkness and Dawn* (Boston: Small, Maynard, 1914).

47. Muarine Greenwald, "Visualizing Pittsburgh in the 1900s: Art and Photography in the Service of Social Reform," in Greenwald and Anderson, *Pittsburgh Surveyed,* 135–36; Joseph Stella, "Discovery of America: Autobiographical Notes," *Art News* (November 1960), 41.

48. Richard Soloway, "Counting the Degenerates: The Statistics of Race Deterioration in Edwardian England," *Journal of Contemporary History* 17 (January 1982): 137–64; Charles L. Dana, "Are We Degenerating?" *Forum* 19 (June 1895): 458–66.

49. G. Frank Lydston, *The Disease of Society (The Vice and Crime Problem)* (Philadelphia: J. B. Lippincott Company, 1905), 13–20; Cesare Lombroso and Guglielmo Ferrero, *Criminal Woman, the Prostitute, and the Normal Woman,* trans. Nicole Hahn Rafter and Mary Gibson (Duke: Duke University Press, 2004). On American receptions of European theories of degeneracy, especially those of Lombroso, Rentoul, and Nordau, see Linda L. Maik, "Nordau's Degeneration: The American Controversy," *Journal of the History of Ideas* 50 (October–December 1989): 607–23; Peter D'Agostino, "Craniums, Criminals, and the 'Cursed Race': Italian Anthropology in American Racial Thought, 1861–1924," *Comparative Studies in Society and History* 44 (April 2002): 319–43.

50. Committee of Fifteen, *The Social Evil with Special Reference to Conditions Existing in the City of New York* (New York: G. P. Putnam's Sons, 1912 [1902]), v–11.

51. *Chicago's Dark Places,* 9.

52. J. W. Magruder, "Progress in Baltimore," *National Conference on Social Welfare Proceedings* (1914): 244–45; Annie W. Allen, "How to Save Girls Who Have Fallen," *Survey* 24

(April 30, 1910): 684–96; Robert A. Woods, "Banners of the New Army," *Survey* 29 (March 8, 1913): 813–14.

53. Lombroso and Ferrero, *Criminal Woman*, 182–226.

54. Committee of Fifteen, *Social Evil*, esp. 1–11, 50–97.

55. Reginald Wright Kauffman, *The House of Bondage* (Saddle River, N.J.: Gregg Press, 1968 [1910]); Theresa Malkiel, "Our Unfortunate Sisters," *Socialist Woman* 2 (November 1908); "A Capitalistic Crime Against Womanhood," *Socialist Woman* 1 (June 1907).

56. Josephine Conger Kaneko, "Old Frances," *Progressive Woman* 3 (August 1909).

57. Lee O. Harris, *The Man Who Tramps: A Story of To-Day* (Indianapolis: Douglass and Carlson, 1878); Allan Pinkerton, *Strikers, Communists, Tramps, and Detectives* (New York: G. W. Carleton, 1878); Edward E. Hale, "Report on Tramps," *Proceedings of the Conference of Charities and Corrections* (Boston: A. Williams and Co., 1877). See also Depastino, *Citizen Hobo*, 3–29.

58. W. L. Bull, "Trampery: Its Causes, Present Aspect, and Some Suggested Remedies," *National Conference on Social Welfare Proceedings* (1886): 188–206.

59. Amy Dru Stanley, *From Bondage to Contract: Wage Labor, Marriage, and the Market in the Age of Slave Emancipation* (New York: Cambridge University Press, 1998), 98–137.

60. L. L. Barbour, "Vagrancy," *National Conference on Social Welfare Proceedings* (1881): 131–38; "Tramp Act of Rhode Island," *National Conference on Social Welfare Proceedings* (1877): 128–34; Francis Wayland and F. B. Sanborn, "Tramp Laws," *National Conference on Social Welfare Proceedings* (1880): 278–81; Bull, "Trampery"; Washington Gladden, "What To Do With the Workless Man," *National Conference on Social Welfare Proceedings* (1899): 141–52; J. J. McCook, "The Tramp Problem: What It Is, and What to Do With It," *National Conference on Social Welfare Proceedings* (1895): 288–302.

61. Frank Tobias Higbie, *Indispensable Outcasts: Hobo Workers and Community in the American Midwest, 1880–1930* (Urbana: University of Illinois Press, 2003), 66–97; Theodore Waters, "Six Weeks in Beggardom," *Everybody's* (December 1904); 789–95; W. P. England, "The Lodging House," *Survey* 27 (December 2, 1911): 1313–17; "How to Tell a Hobo from a Mission Stiff," *Survey* 31 (March 21, 1914): 781; Geo. L. M'Nutt, "What is a Tramp?" *Commons* 7 (September 1902): 9–10; "The Men of the Lodging Houses," *Commons* 8 (September 1903): 1–5.

62. Robert Hunter, *Poverty* (New York: Grosset & Dunlap, 1904), 128–29.

63. Stuart A. Rice, "Vagrancy," *National Conference of Social Welfare Proceedings* (1914): 457–65.

64. Rice, "Vagrancy," 461; Alvan Francis Sanborn, *Moody's Lodging House and Other Tenement Sketches* (Boston: Copeland and Day, 1895); O. F. Lewis, "The Tramp Problem," *Annals of the American Academy of Political and Social Science* 40 (March 1912): 217–27.

65. Edmund Kelly, *The Elimination of the Tramp* (New York: G. P. Putnam's Sons, 1908); Frank Tracy Carlton, *The Industrial Situation* (New York: Fleming H. Revell Company, 1914), 116–21.

66. Alice O'Connor, *Poverty Knowledge: Social Science, Social Policy, and the Poor in Twentieth-Century U.S. History* (Princeton: Princeton University Press, 2001). Gavin Jones, *American Hungers: The Problem of Poverty in U.S. Literature, 1840–1945* (Princeton: Princeton University Press, 2008), suggests the ways that poverty knowledge inflected the literary representation of poverty and the pauper. Alice N. Lincoln, "Classification of Paupers in Almshouses," *National Conference on Social Welfare Proceedings* (1898): 184–91; George Thomas Palmer, "Why is the Pauper," *Survey* 30 (April 5, 1913): 11–15; John Lewis Gillin, *Poverty and Dependency* (New York: Century Co., 1921), 16–43; Edward Devine, *Misery and its Causes* (New York: Macmillan, 1913).

67. Rev. S. Humphreys Gurteen, *How Paupers are Made: An Address on the Prevention of Pauperism* (Chicago: R. R. Donnelley & Sons, 1883); Richard Ely, "Pauperism in the United States," *North American Review* 152 (May 1890); 345–409; Francis A. Walker, "The Causes

of Poverty," *Century* (December 1897); Richard Henry Edwards, "Poverty" (Madison, July 1909); William B. Bailey, *Modern Social Conditions* (New York: Century Co., 1906); Amos G. Warner, *American Charities*, 3rd ed. (New York: Thomas Y. Crowell Company, 1919); Warner, "The Causes of Poverty Further Considered," *American Statistical Association* 27 (September 1894): 49–68.

68. Lilian Brandt, "The Causes of Poverty," *Political Science Quarterly* 23 (December 1908): 637–51; see, for example, Richard T. Ely, "Pauperism in the United States," *North American Review* (May 1890): 395–409; Gurteen, *How Paupers are Made;* Warner, "The Causes of Poverty Further Considered," 68. For two important American works that anticipated these texts and, in their own ways, also connected degeneration and poverty, see Henry George, *Progress and Poverty: An Inquiry into the Cause of Industrial Depressions and of Increase of Want with Increase of Wealth* (New York: Robert Schalkenbach Foundation, 1960 [1879]), and Charles Loring Brace, *The Dangerous Classes of New York, and Twenty Years' Work Among Them* (New York: Wynkoop and Hallenbeck, 1872). See also Edward Devine, *Pauperism: An Analysis* (New York: New York School of Philanthropy, 1916).

69. Maurice Parmelee, *Poverty and Social Progress (New York: Macmillan, 1916),* 224–25; Devine, *Pauperism: An Analysis,* 3–4. See also Devine, *Misery and its Causes,* 241–53.

70. Devine, *Pauperism: An Analysis,* 7; Parmelee, *Poverty and Social Progress,* 223–55.

71. John R. Commons, *Races and Immigrants in America* (New York: Macmillan, 1920 [1907]), 167; Hunter, *Poverty,* 223–28.

72. London, *People of the Abyss,* 288; Ernest Crosby, "Civilization," *International Socialist Review* 1 (December 1900): 362.

73. London, *People of the Abyss,* 50.

74. Charlotte Perkins Gilman, *Women and Economics: A Study of the Economic Relation Between Men and Women as a Factor in Social Evolution* (Berkeley: University of California Press, 1998 [1898]), 171–83.

75. John J. McCook, "The Tramp Problem," *Lend A Hand* 15 (September 1895), in Social Reform Papers of John J. McCook, Antiquarian and Landmarks Society of Connecticut, Roll I, Series I: 171–73; A. O. Wright, "Vagrancy," *National Conference on Social Welfare Proceedings* (1896): 232–34; W. H. Allen, "The Vagrant: Social Parasite or Social Product," *National Conference on Social Welfare Proceedings* (1903): 379–86; Hunter, *Poverty,* 129.

76. Hunter, *Poverty,* 322–26.

77. Hastings H. Hart, "A Working Program for the Extinction of the Defective Delinquent," *Survey* 30 (May 24, 1913): 277–79; Edward T. Devine, "Protection vs. Elimination," *Survey* 31 (October 4, 1913): 35–36; Franklin Giddings, "A Moral Quarantine," *Independent* 79 (July 13, 1919): 51–52.

78. Ruth M. Alexander, *The "Girl Problem": Female Sexual Delinquency in New York, 1900–1930* (Ithaca: Cornell University Press, 1995); Mark Thomas Connelly, *The Response to Prostitution in the Progressive Era* (Chapel Hill: University of North Carolina Press, 1980); Elizabeth Alice Clement, *Love for Sale: Courting, Treating, and Prostitution in New York City, 1900–1945* (Chapel Hill: University of North Carolina Press, 2006), 45–113.

79. For examples of calls for the segregation of confirmed prostitutes, see Committee of Fifteen, *Social Evil,* 237–38; *Report of the Vice-Commission of Minneapolis, to His Honor, James C. Haynes, Mayor* (Minneapolis, 1911); *The Social Evil in Chicago: A Study of Existing Conditions with Recommendations by the Vice Commission of Chicago* (Chicago, 1911).

80. Raymond Robins, "What Constitutes a Model Municipal Lodging House," *National Conference on Social Welfare Proceedings* (1904): 155–66; Alice C. Willard, "Reinstatement of Vagrants through Municipal Lodging Houses," *National Conference on Social Welfare Proceedings* (1903): 404–11.

81. F. B. Sanborn, "Public Charities in Europe," *National Conference on Social Welfare Proceedings* (1891): 167–86; Francis G. Peabody, "The German Labor-Colonies For Tramps,"

Forum 12 (February 1892): 751–61; Henry E. Rood, "The Tramp Problem: A Remedy," *Forum* 25 (March 1898): 90–94; "Tramp Elimination," *Independent* 65 (1908): 569–70; "The Farm Colony for Tramps," *Independent* 71 (1911): 269–70; R. B. Meath, "Labour Colonies in Germany," *Nineteenth Century* (January 1891): 73–88.

82. O. F. Lewis, "Concerning Vagrancy, 1. Labor Colonies," *Charities and the Commons* 20 (September 5, 1908): 674–81; Orlando F. Lewis, "Vagrancy in the United States," *National Conference on Social Welfare Proceedings* (1907): 52–72.

83. Alexander Johnson, "The Segregation of Defectives, Report of Committee on Colonies for Segregation of Defectives," *National Conference on Social Welfare Proceedings* (1903): 245–53.

84. Alexander Cleland, "The Time to Deal with Vagrancy," *Survey* 37 (December 9, 1916): 268–69; "Tramps in Providence," *Survey* 22 (May 22, 1909): 285; Walter E. Fernald, "The Massachusetts Farm Colony for the Feeble-Minded," *National Conference on Social Welfare Proceedings* (1902): 487–90; Frank Corliss, "Vagrancy," *International Socialist Review* 11 (September 1910): 176–79; Sanny McNee, "Tramps and Decent Men," *International Socialist Review* 16 (October 1915): 201.

85. "The Farm Colony Bill," *Survey* 24 (April 30, 1910): 163; "Make Tramps Self-Supporting," *Survey* 26 (April 22, 1911): 162; "Tramp Farm Colony Assured in New York," *Survey* 26 (August 5, 1911): 633–34; Charles K. Blatchly, "A State Farm for Tramps and Vagrants," *Survey* 24 (April 9, 1910): 87–89; "Two Ways of Building Farm Colonies," *Survey* 35 (February 26, 1916): 625–26.

86. "Study of Social Conditions" in "Fifth Yearly Report of the Settlement" (December, 1896), United South End Settlements Records, Box 2, Folder: Folder: "South End House Annual Reports 1892–1910," 13; "South End House Association Ninth Annual Report" (February 1901), United South End Settlements Records, Box 2, Folder: "South End House Annual Reports 1892–1910," 8–10; "Isolating Social Contagion" in "South End House, 1910 Democracy Domesticated" (March 1910), United South End Settlements Records, Box 3, Folder: "South End House Reports (1910–11)," 12–13.

87. "Social Forces," *Survey* 22 (April 3, 1909): 1–2; "Degeneracy in Neighborhoods," in "Third Conference of the National Federation of Settlements at Pittsburgh, Pennsylvania, September 26, 27, 28, 1913," United South End Settlements Records, Box 5, Folder: "National Federation of Settlements Reports and Conference Materials (1913–1941)."

6. Dredging the Abyss

1. "Anna Richardson," *New York Times* (May 11, 1949), 29.

2. Gertrude B. Lane, preface to Anna Steese Richardson, *Better Babies and their Care* (New York: Frederick Stokes Company, 1914), x.

3. Bluford Adams, *E Pluribus Barnum: The Great Showman and the Making of U.S. Popular Culture* (Minneapolis: University of Minnesota Press, 1997), 98–117.

4. Richardson, *Better Babies and their Care;* Annette K. Vance Dorey, *Better Baby Contests: The Scientific Quest for Perfect Childhood in the Early Twentieth Century* (Jefferson, N.C.: McFarland, 1999); Marilyn Irvin Holt, *Linoleum, Better Babies, and the Modern Farm Woman, 1890–1930* (Lincoln: University of Nebraska Press, 1995).

5. Robert A. Woods and Albert J. Kennedy, *The Settlement Horizon: A National Estimate* (New York: Russell Sage, 1922), 174–75.

6. The powerful idea of degeneration gripped the American imagination but it has largely gone unrecognized by American historians with few exceptions. Daniel Pick, *Faces of Degeneration: A European Disorder, c.1848–c.1918* (Cambridge: Cambridge University Press,

1989); Sander Gilman and Edward J. Chamberlin, eds., *Degeneration: The Dark Side of Progress* (New York: Columbia University Press, 1985); Elof Axel Carlson, *The Unfit: A History of a Bad Idea* (Cold Spring Harbor, N.Y.: Cold Spring Harbor Laboratory Press, 2001); Louis S. Warren, "Buffalo Bill Meets Dracula: William F. Cody, Bram Stoker, and the Frontiers of Racial Decay," *American Historical Review* 107 (October 2002): 1124–57.

7. For an example of an otherwise superior study of the notions of civilization that largely overlooks the reform engagement with imperial culture, see Gail Bederman, *Manliness and Civilization: A Cultural History of Gender and Race in the United States, 1880–1917* (Chicago: University of Chicago Press, 1995). Matthew Jacobson, *Barbarian Virtues: The United States Encounters Foreign Peoples at Home and Abroad, 1876–1917* (New York: Hill and Wang, 2000) and Ann Laura Stoler, ed., *Haunted by Empire: Geographies of Intimacy in North American History* (Durham: Duke University Press, 2006), both point to the need to contest the divide between domestic and imperial histories. See also Kevin Gaines, "Black Americans' Racial Uplift Ideology as 'Civilizing Mission': Pauline E. Hopkins on Race and Imperialism," in *Cultures of American Imperialism*, ed. Amy Kaplan and Donald E. Pease (Durham: Duke University Press, 1993), 433–55.

8. Eileen J. Suarez, *Imposing Decency: The Politics of Sexuality and Race in Puerto Rico, 1870–1920* (Durham: Duke University Press, 1999), 53–109; Laura Wexler, *Tender Violence: Domestic Visions in an Age of U.S. Imperialism* (Chapel Hill: University of North Carolina Press, 2000); Paul A. Kramer, *The Blood of Government: Race, Empire, the United States, and the Philippines* (Chapel Hill: University of North Carolina Press, 2006); Angel Velasco Shaw and Luis Francia, eds., *Vestiges of War: The Philippine-American War and the Aftermath of an Imperial Dream, 1899–1999* (New York: New York University Press, 2002).

9. Alexander Sutherland, *The Origin and Growth of the Moral Instinct, Volumes I and II* (London: Longmans, Green, 1898), 1.

10. Sutherland, *The Origin and Growth of the Moral Instinct*, 25–120; Franklin Giddings, *A Theory of Social Causation* (New York: Macmillan, 1904), 170; Giddings, "The Laws of Evolution," *Science* 22 (August 18, 1905): 206–8.

11. Maurice Parmelee, *Poverty and Social Progress* (New York: Macmillan, 1916), 252–53; Edward Westermarck, *The Origin and Development of the Moral Ideas, Volume I* (London: Macmillan, 1924 [1906]), 386–414.

12. Sutherland, *Origin and Growth of the Moral Instinct*, 356.

13. Sutherland, *Origin and Growth of the Moral Instinct*, 400.

14. Westermarck, *Origin and Development of the Moral Ideas, Volume I*, 526–69.

15. Woods and Kennedy, *The Settlement Horizon*, 189–99; Mary Richmond, *Social Diagnosis* (New York: Russell Sage, 1917).

16. Franklin Giddings, "The Christmas Message, 1917," Franklin Giddings Papers, Columbia University, Box 4, Folder: "Printed Material."

17. Woods and Kennedy, *Settlement Horizon*, 59–60.

18. Asa Briggs, *Toynbee Hall: The First Hundred Years* (London: Routledge and Kegan Paul, 1984); Standish Meacham, *Toynbee Hall and Social Reform, 1880–1914: The Search for Community* (New Haven: Yale University Press, 1987).

19. Robert A. Woods, *English Social Movements* (New York: Charles Scribner's Sons, 1891); Woods and Kennedy, *Settlement Horizon*, 28–29.

20. "Address of the Hon. Abram S. Hewitt," "Annual Meeting of the University Settlement Society," January 27, 1898, University Settlement Society, "Report for the Year 1897," Records of the University Settlement Society of New York City, Series 3, Box 5, 64–65.

21. Woods and Kennedy, *Settlement Horizon*, 59.

22. Wm. J. Tucker, "The Andover House Association, Circular No. 1," United South End Settlements Records SW204, Social Welfare History Archives, University of Minnesota, Box 2, Folder: "South End House Annual Reports 1892–1910"; "Lincoln House Bulletin for 1897,"

United South End Settlements Records, Box 1, Folder: "Lincoln House Annual Reports, 1897–1903," 1.

23. "Third Annual Report of Hartley House" (1900), Pamphlets: Major, Neighborhood Houses and Associations Pamphlets SWP01, Social Welfare History Archives, Box 167, Folder: "Hartley House," 13; "Address of the Hon. Abram S. Hewitt and James Reynolds," "Report of the Head Worker," "Year Book of the University Settlement Society of New York, 1899," Records of the University Settlement Society of New York City, Series 3, Box 5.

24. Treasurer's Report, "Annual Meeting of the University Settlement Society" (January 27, 1898), "Report for the Year 1897."

25. Alvan F. Sanborn, "The Anatomy of a Tenement Street," *Andover House Bulletin* (Boston: Andover, 1894).

26. "Address of the Hon. Abram S. Hewitt," 65.

27. "University Settlement," signed by Henry W. Taft (1911), Records of the University Settlement Society of New York City, Series 4, Box 10.

28. "Circular No. 11 December, 1894, Third Yearly Report," United South End Settlements Records, Box 2, Folder: "South End House Annual Reports 1892–1910," 6–7.

29. Ernest Untermann, "Evolution of the Theory of Evolution," *International Socialist Review* 5 (March 1905): 513.

30. Robert Rives La Monte, "Science and Socialism," *International Socialist Review* 1 (September 1900): 160–75. For a contrast between reformist and socialist understandings of the evolution of philanthropy, see Irving Fisher, "Why I am Not a Socialist" (address at Socialist Hall, New Haven, April 5, 1914), Irving Fisher Papers, Group No. 212, Yale University, Manuscripts and Archives, Series II, Box 24, 8–9.

31. Untermann, "Evolution of the Theory of Evolution" (March 1905), 513–23; Ernest Untermann, "Evolution of the Theory of Evolution," *International Socialist Review* 6 (September 1905): 152–63.

32. Ernest Untermann, "Evolution or Revolution?" *International Socialist Review* 1 (January 1901): 406–12.

33. W. H. Miller, "A Defense of the Old Law," *International Socialist Review* 3 (November 1902): 291–95. Emphasis in original.

34. Lewis's lectures were reprinted in a book published in several editions by the leading American socialist publisher. Arthur Lewis, *Evolution: Social and Organic* (Chicago: Charles Kerr, 1908), 94–98.

35. C. A. Steere, *When Things Were Doing* (Chicago: Charles H. Kerr, 1908), 163–64.

36. Steere, *When Things Were Doing*, 215–57.

37. Woods and Kennedy, *Settlement Horizon*, 95.

38. R. [Richard] W. Gilder, "The University Settlement and Good Citizenship" (January 29, 1897), "Report for the Year 1896," Records of the University Settlement Society of New York City, Series 3, Box 5, 33.

39. "The Goodrich Social Settlement" (Cleveland, 1900), Pamphlets: Major Neighborhood Houses and Associations Pamphlets, Box 165, Folder: "Goodrich Social Settlement," 46–49.

40. "Circular No. 12, November 1, 1895," United South End Settlements Records, Box 2, Folder: "South End House Annual Reports 1892–1910," 5.

41. "Circular No. 11 December, 1894, Third Yearly Report," 1; "Circular No. 9, Dec. 9, 1893, Report for the Year 1892–93," United South End Settlements Records, Box 2, Folder: "South End House Annual Reports 1892–1910," 2.

42. "Circular No. 9, Dec. 9, 1893, Report for the Year 1892–93," 19–20; "Fifth Yearly Report of the Settlement" (December 1896), 13; "Circular No. 12, November 1, 1895," 5; "South End House Association, Eighth Annual Report" (January 1900), United South End Settlements Records, Box 2, Folder: "South End House Annual Reports 1892–1910," 7–8.

43. See, for example, Warwick Anderson, *Colonial Pathologies: American Tropical Medicine, Race, and Hygiene in the Philippines* (Durham: Duke University Press, 2006).

44. "South End House, 1910 Democracy Domesticated" (March 1910), United South End Settlements Records, Box 3, Folder: "South End House Reports 1910–1911," 12–13; "Account of Work at the University Settlement, James H. Hamilton, Headworker," *University Settlement Studies* 2, no. 1 (April 1906), 27; "Neighborhood, City, and Nation," "South End House Its 17th year of Cumulative Growth" (February 1909), United South End Settlements Records, Box 2, Folder: "South End House Annual Reports 1892–1910," 5–6.

45. B. O. Flower, *Civilization's Inferno or Studies in the Social Cellar* (Boston: Arena Publishing Company, 1893), 51–52.

46. "Neighborhood, City, and Nation," 5–6.

47. "Lincoln House Bulletin, 1900," United South End Settlements Records, Box 1, Folder: "Lincoln House Annual Reports, 1897–1903," 26–27.

48. J. H. Chase, "Child Ethics in the Street and Settlement," "Fifteenth Annual Report, 1901," Records of the University Settlement Society of New York City, Series 3, Box 4, 34–36.

49. J. Howard Moore, *The Law of Biogenesis* (Chicago: Charles H. Kerr, 1914), 69–71.

50. John W. Martin, "Social Life in the Street," "Year Book of the University Settlement Society of New York, 1899," Records of the University Settlement Society of New York City, Series 3, Box 5, 22–24.

51. Frederick A. King, "Influences of Street Life," "Year Book of the University Settlement Society of New York, 1900," Records of the University Settlement Society of New York City, Series 3, Box 5, 29–32.

52. Dorothy Ross, *G. Stanley Hall: The Psychologist as Prophet* (Chicago: University of Chicago Press, 1972); G. Stanley Hall, *Adolescence, Its Psychology and Its Relations to Physiology, Anthropology, Sociology, Sex, Crime, Religion and Education* (New York: Appleton, 1907); Hall, *Aspects of Child Life and Education* (New York: D. Appleton, 1907).

53. Lilla Estelle Appleton, *A Comparative Study of the Play Activities of Adult Savages and Civilized Children, An Investigation of the Scientific Basis of Education* (Chicago: University of Chicago Press, 1910), 4; George E. Johnson, "Games and Play," *Social Work—Twelve Monographs* (Lincoln House, 1898). On the activities of the National Recreation Association and the Playground Association of America, see "Tentative Report of the Committee on A Normal Course in Play of the Playground Association of America," "Syllabus One, Child Nature," and "Syllabus Two, The Nature and Function of Play," in "Proceedings of the Third Annual Congress of the Playground Association of America Held at Pittsburgh, PA, May 10 to 14, 1909," vol. 3 (August 1909), National Recreation Association Records SW074, Social Welfare History Archives, Box 14, Folder: "1909-NRA Congress," 111–36.

54. Moore, *Law of Biogenesis*, 61–123, and Mary Marcy, introduction to *Law of Biogenesis*, 9–13.

55. William Byron Forbush, *The Boy Problem: A Study in Social Pedagogy* (Boston: Pilgrim Press, 1902).

56. "The Boy Problem," "Getting a Vital Response," and "At Close Range with Boys" in "Third Conference of the National Federation of Settlements at Pittsburgh, Pennsylvania, September 26, 27, 28, 1913," United South End Settlements Records, Box 5, Folder: "National Federation of Settlements Reports and Conference Materials (1913–1941)."

57. "Fifteenth Annual Report of the University Settlement Society of New York, 1901," Records of the University Settlement Society of New York City, Series, 3, Box 5, 8.

58. See Maureen A. Flanagan, *Seeing with Their Hearts: Chicago Women and the Vision of the Good City, 1871–1933* (Princeton: Princeton University Press, 2002); Kathryn Kish Sklar, *Florence Kelley and the Nation's Work* (New Haven: Yale University Press, 1995); Alice O'Conner, *Poverty Knowledge: Social Science, Social Policy, and the Poor in Twentieth-Century U.S. History* (Princeton: Princeton University Press, 2001), 25–54.

59. "Fifteenth Annual Report of the University Settlement Society of New York, 1901," Records of the University Settlement Society of New York City, Series 3, Box 5, 8; King, "Influences of Street Life," 29–32.

60. Ernest Crosby, "Civilization?" *International Socialist Review* 1 (December 1900): 362.

61. Murray E. King, "Socialism and Human Nature: Do They Conflict?" *International Socialist Review* 5 (December 1904): 321–39.

62. Untermann, "Evolution or Revolution?" 410–11.

63. "Lincoln House Report for 1903," United South End Settlements Records, Box 1, Folder: "Lincoln House Annual Reports, 1897–1903," 7; Kevin P. Murphy, *Political Manhood: Red Bloods, Mollycoddles and the Politics of Progressive Era Reform* (New York: Columbia University Press, 2008), 105–124.

64. "Far-Extending Veins Struck" and "Social Science Prospecting" in "South End House, 1908" (March 1908), United South End Settlements Records, Box 2, Folder: "South End House Annual Reports 1892–1910," 9–13; "South End House, 1913" (March 1913), United South End Settlements Records, Box 2, Folder: "SEH Reports (1913–14)," 6–7.

65. "Fifth Annual Report of the College Settlements Association, From September 1, 1893 to September 1, 1894," Pamphlets: Major Neighborhood Houses and Associations Pamphlets, Box 164, Folder: "College Settlement Annual Reports, 1891–1894," 6; J. G. Phelps Stokes, "Hartley House, And Its Relations to the Social Reform Movement" (New York, 1897), Pamphlets: Major, Neighborhood Houses and Associations Pamphlets, Box 167, Folder: "Hartley House," 16.

66. "Address of Miss [Jane] Addams to the Annual Meeting," "Sixteenth Annual Report of the University Settlement Society of New York, 1902," Records of the University Settlement Society of New York City, Series 3, Box 5, 53–54; Samuel Rosensohn, "The Americanizing Influence of the Public School," 26–28, and Library Report, 28–31, 30, in "Report for the Year 1898," Records of the University Settlement Society of New York City, Series 3, Box 5; James Reynolds, "Report of the Head Worker; Moral Struggles of the Newly Arrived Immigrant," "Year Book of the University Settlement Society of New York, 1900," Records of the University Settlement Society of New York City, Series 3, Box 5, 8, 11–16; "Fifteenth Annual Report of the University Settlement Society of New York, 1901," "Year Book of the University Settlement Society of New York, 1900," Records of the University Settlement Society of New York City, Series 3, Box 5, 8.

67. August Weismann, *Essays in Heredity,* 2 vols. (Oxford: Clarendon Press, 1891); Weismann, *The Evolution Theory* (London: E. Arnold, 1904); Weismann, *The Germ-Plasm: A Theory of Heredity,* trans. W. Newton Parker and Harriet Rönnfeldt (New York: Scribner, 1893). On the persistence of Lamarckianism, see Herbert Spencer, *A Rejoinder to Professor Weismann* (London: Williams & Norgate, 1893); A. S. Packard, *Lamarck, The Founder of Evolution* (New York: Longman, Green, 1901); L. H. Bailey, "Neo-Lamarckism and Neo-Darwinism," *American Naturalist* 28 (August 1894): 661–78; *Proceedings of the First National Conference on Race Betterment, January 8–12, 1914* (Battle Creek, Mich.: Race Betterment Society, 1914), esp. 431–553. The historical work on neo-Lamarckianism is limited, especially given its historical significance. See Theodore John Greenfield, "Variation, Heredity, and Scientific Explanation in the Evolutionary Theories of Four American Neo-Lamarckians, 1867–1897" (PhD diss., University of Wisconsin, 1986); Matthew Daniel Whalen, "American Science, Society, and Civilization in the Age of Energy: An Investigation of the Relationships among Neo-Lamarckism, Social Evolutionism, and the Myth of Atlantis between 1860 and 1920" (PhD diss., University of Maryland, 1978); Peter J. Bowler, "Edward Drinker Cope and the Changing Structure of Evolutionary Theory," *Isis* 68 (June 1977): 249–65.

68. George W. Stocking Jr., *Race, Culture, and Evolution: Essays in the History of Anthropology* (Chicago: University of Chicago Press, 1982), 234–69. For important examples of the use of neo-Lamarckian theory in explaining race development, see Daniel Garrison Brinton, *The Basis of Social Relations: A Study of Ethnic Psychology* (New York: G. P. Putnam, 1902).

69. "Third Annual Report of Hartley House" (1900), 12; "Report of the Head Worker" [James B. Reynolds], "Report for the Year 1896," Records of the University Settlement Society of New York City, Series 3, Box 5, 5.

70. James B. Reynolds, "Report of the Head Worker," "Year Book of the University Settlement Society of New York, 1899," Records of the University Settlement Society of New York City, Series 3, Box 5, 11.

71. Herman H. Moellering, "Some Notes on a Weismann Lecture," *International Socialist Review* 9 (November 1908): 362–66.

72. "Report of the Head Worker," "Year Book of the University Settlement Society of New York" (1900), Records of the University Settlement Society of New York City, Series 3, Box 5.

73. "Chicago Commons, 1913–1914," Pamphlets: Major Neighborhood Houses and Associations Pamphlets, Box 165, Folder: "Chicago Commons Annual Report, 1913–1929."

74. "Second Annual Report of Hartley House" (1899), Pamphlets: Major Neighborhood Houses and Associations Pamphlets, Box 167, Folder: "Hartley House Annual Reports," 14–15.

75. Luther Halsey Gulick, president, Playground Association of America, "Play and Democracy," in "Papers of the Chicago Meeting, Playground Association of America" (June 1907), National Recreation Association Records, Box 14, Folder: "NRA-1907," 11–16.

76. *Lincoln House Monthly* (May 1898), Pamphlets: Major Neighborhood Houses and Associations Pamphlets, Box 164, Folder: "Lincoln House Monthly, 1898–1901," 37; "Lincoln House: The Annual Report of Lincoln House for the Year 1905," Pamphlets: Major Neighborhood Houses and Associations Pamphlets, Box 164, Folder: "Lincoln House Annual Reports 1905–1908," 12–15; "Lincoln House: The Annual Report of Lincoln House for the Year 1906," Pamphlets: Major Neighborhood Houses and Associations Pamphlets, Box 164, Folder: "Lincoln House Annual Reports, 1909–1915," 39; "Lincoln House A Neighborhood Club, Report for the Twenty-Fifth Year" (1912), Pamphlets: Major Neighborhood Houses and Associations Pamphlets, Box 164, Folder: "Lincoln House Monthly ca. 1898–1903," 22; *Hartley House News* (January 21, 1899): 3, and *Hartley House News* (March 28, 1900): 1, and *Hartley House News* (January 29, 1901): 1 all in Pamphlets: Major Neighborhood Houses and Associations Pamphlets, Box 167, Folder: "Hartley House News."

77. Eric Lott, *Love and Theft: Blackface Minstrelsy and the American Working Class* (New York: Oxford University Press, 1993); David R. Roediger, *Working toward Whiteness: How America's Immigrants Became White* (New York: Basic Books, 2005), 180–84.

78. "Tenth Annual Report of the East Side House" (January 1, 1902), Pamphlets: Major Neighborhood Houses and Associations Pamphlets, Box 167, Folder: "Eastside House Annual Reports," 9; Philip Joseph Deloria, *Playing Indian* (New Haven: Yale University Press, 1998).

79. "Tenth Annual Report of the East Side House," 15–16.

80. "3rd Annual Report, The Kingsley House Association" (1896), Pamphlets: Major Neighborhood Houses and Associations Pamphlets, Box 168, Folder: "Kingsley House Assn. Annual Reports," 6; "South End House 1892–1907, A Record of Fifteen Years' Work and Its Results" (March 1907), United South End Settlements Records, Box 2, Folder: "South End House Annual Reports 1892–1910," 10–11.

81. Flower, *Civilization's Inferno*, 217–21.

82. Henry S. Curtis, secretary of the Playground Association of America, "Playground Progress and Tendencies of the Year," in "Papers of the Chicago Meeting, Playground Association of America," 25–29; H. S. Braucher, "Memorandum on the Work of the Playground and Recreation Association of America," in 4th Annual Play Congress, held in Rochester, N.Y., June 7–11. (February 1912), National Recreation Association Records, Box 11, Folder: "Report of the PRAA, 1912–13;" Sarah Jo Peterson, "Voting for Play: The Democratic Potential of Progressive Era Playgrounds," *Journal of the Gilded Age and Progressive Era* 2 (April 2004): 145–75; Patricia Mooney Melvin, "Building Muscles and Civics: Folk Dancing, Ethnic Diversity, and the Playground Association of America," *American Studies* 24 (Spring 1983): 89–99.

83. Jane Addams, "Public Recreation and Social Morality," in "Papers of the Chicago Meeting, Playground Association of America," 22–24.

84. "I am the University Settlement," "Twenty-Seventh Annual Report" (1913), Records of the University Settlement Society of New York City, Series 3, Box 5.

85. J. Howard Moore, *Savage Survival in Higher Peoples* (London: Watts, 1933 [1916]), 113–15. Emphasis in original.

86. Joseph Lee, "Speech at Opening of Playground Institute in Holyoke" (1911–1912), National Recreation Association Records, Box 32, Folder: "H.S. Braucher [1911–1949]," 1; George E. Johnson, "Why Teach A Child to Play?" in "Proceedings of the Third Annual Congress of the Playground Association of America Held at Pittsburgh, PA, May 10 to 14, 1909," vol. 3 (November 1909), National Recreation Association Records, Box 14, Folder: "1909—NRA Congress," 357–65; Henry Baird Favill, M.D., president, Chicago Tuberculosis Institute, "Playgrounds in the Prevention of Tuberculosis," 33, and Joseph Lee, vice-president, Massachusetts Civic League, "Play as a School of the Citizen," 16–21, in "Papers of the Chicago Meeting, Playground Association of America"; "Hale House 24th Annual Report" (1919), United South End Settlements Records, Box 2, Folder: "Hale House Annual Report 1919," 3–5.

87. "Report on the Committee on Storytelling in the Playground" (n.d., ca. 1909), National Recreation Association Records, Box 11, Folder: "Annual Reports, 1909 and Earlier," 4–8.

88. "Life—Life—Life for All," Playground and Recreation Association of America (n.d., ca. 1912), National Recreation Association Records, Box 11, Folder: "Report of the PRAA, 1912–13," 12; "Hiram House: A Social Settlement" (n.d. ca. 1903), Pamphlets: Major Neighborhood Houses and Associations Pamphlets, Box 165, Folder: "Hiram House," 22. Emphasis in original.

89. John F. Kasson, *Houdini, Tarzan, and the Perfect Man: The White Male Body and the Challenge of Modernity in America* (New York: Hill and Wang, 2001); Clifford Putney, *Muscular Christianity: Manhood and Sports in Protestant America, 1880–1920* (Cambridge: Harvard University Press, 2001); Thomas Winter, *Making Men, Making Class: The YMCA and Workingmen, 1877–1920* (Chicago: University of Chicago Press, 2002); Richard Dyer, "The White Man's Muscles," in *The Masculinity Studies Reader,* ed. Rachel Adams and David Savran (Malden, Mass.: Blackwell, 2002), 262–73; Daniel Block, "Saving Milk through Masculinity: Public Health Officers and Pure Milk, 1880–1930," *Food and Foodways* 13 (January–June 2005): 115–34.

90. Marcellus T. Hayes, "Physical Training," *Guild Review* 1 (March 1907): 24; "The Friendly Aid, Warren Goddard House, Twelfth Annual Report" (New York, 1903), Pamphlets: Major Neighborhood Houses and Associations Pamphlets, Box 167, Folder: "The Friendly Aid Society," 53.

91. Crosby, "Civilization," 362.

92. King, "Socialism and Human Nature," 323–24. Emphasis in original.

93. Miller, "Defense of the Old Law," 294–295. Emphasis in original.

94. "1903—Modified Milk—1911," in "The Starr Centre Association" (1911), Pamphlets: Major Neighborhood Houses and Associations Pamphlets, Box 168, Folder: "Starr Centre Association," 18.

95. "Milk. Dirty! Clean!" (n.d.), Wilbur and Elsie P. Phillips Papers SW094, Social Welfare History Archives, Box 5. A film by the Edison Company, *The Man Who Learned,* presented a similar image of dirty milk leading to degeneracy. The film showed a farmer producing milk in an unhygienic way confronting the results of his milk at the bedside of a sick child. On milk campaigns, see Jennifer Koslow, "Putting It to a Vote: The Provision of Pure Milk in Progressive Era Los Angeles," *Journal of the Gilded Age and Progressive Era* 3 (April 2004): 111–44; Jacqueline H. Wolf, "'Don't Kill Your Baby': Feeding Infants in Chicago, 1903–1924," *Journal of the History of Medicine* 53 (July 1998): 219–53.

96. "Seventh Annual Report of the New York Milk Committee" (1913), 39–45, "Proceedings Conference on Milk Problems Under the Auspices of New York Milk Committee at New York City, December 2nd and 3rd, 1910," Wilbur and Elsie P. Phillips Papers, Box 5, 19.

97. "Seventh Annual Report of the New York Milk Committee," 53–59.

98. "The Better Baby Show," *Guild Journal* 2 (April 1913): 72–73; "Our Baby Contest" *Guild Journal* 3 (May 1913): 77–78; "Second Annual Baby Health Contest" *Guild Journal* 5 (November 1914): 1–2; *The Report of the Philadelphia Baby Saving Show* (Philadelphia, May 18–26, 1912).

99. *Hartley House News* (November 28, 1913): 2–3. *Hartley House News* (December 19, 1913): 4, Pamphlets: Major Neighborhood Houses and Associations Pamphlets, Box 167, Folder: "Hartley House News."

7. Of Jukes and Immigrants

1. "Charles Davenport to A.H. Estabrook" (January 8, 1911), Charles B. Davenport Papers B D27, American Philosophical Society, Series II, Folder: "Estabrook, Arthur H. #1"; Arthur H. Estabrook and Charles B. Davenport, *The Nam Family: A Study in Cacogenics,* Eugenics Records Office—Memoir No. 2 (Cold Spring Harbor, N.Y., August 1912).

2. For examples of reformist family studies, see O. C. McCulloch, "The Tribe of Ishmael: A Study in Social Degradation," *National Conference on Social Welfare Proceedings* (1888): 154–59; J. F. Wright, "Marriage Relationships Among the Tribe of Ishmael," *National Conference on Social Welfare Proceedings* (1890): 435–37; Richard Dugdale, *The Jukes: A Study in Crime, Pauperism, Disease and Heredity* (New York: G. P. Putnam's Sons, 1877). Both the Jukes and the Ishmaelites would be updated by eugenicists to reflect changing ideas about heredity. Elof Axel Carlson, *The Unfit: A History of a Bad Idea* (Cold Spring Harbor, N.Y.: Cold Spring Harbor Laboratory Press, 2001), 161–278. Recent scholars have described the global reach and intersections of eugenics. They have also critiqued the notion of a cohesive eugenics movement. Nancy Ordover, *American Eugenics: Race, Queer Anatomy, and the Science of Nationalism* (Minneapolis: University of Minnesota Press, 2003); Alexandra Minna Stern, *Eugenic Nation: Faults and Frontiers of Better Breeding in Modern America* (Berkeley: University of California Press, 2005); Nancy Stepan, *The Hour of Eugenics: Race, Gender, and Nation in Latin America* (Ithaca: Cornell University Press, 1991). This chapter explores the engagement of eugenicists with the politics of reform and socialism. These eugenicists were grouped around the Eugenics Records Office and claimed positions as national leaders of eugenics.

3. "Report of the Head Worker," "Twenty-Seventh Annual Report" (1913), Records of the University Settlement Society of New York City—Jacob S. Eisinger Collection, Wisconsin Historical Society, Series III, Box 5, 9–27; "The Better Baby Show," *Guild Journal* 2 (April 1913): 72.

4. "After Citizenship Papers—What?" (paper read by Robbins Gilman before the Minnesota Sate Council of Americanization, May 15, 1925), Records of the University Settlement Society of New York City, Series 4, Box 10, 4.

5. "A Scientific Basis for the Treatment of Problems of Criminology and Penology," *Proceedings of the National Conference of Charities and Corrections* (St. Louis, May 19–26, 1910): 81–87, in Maurice Parmelee Papers MS 1744, Yale University, Manuscripts and Archives, Series III, Box 3, Folder 2; "Report of the Head Worker," "Twenty-Sixth Annual Report" (1912), Records of the University Settlement Society of New York City, Series 3, Box 5, 4; Edwin Grant Conklin, *Heredity and Environment in the Development of Men* (Princeton: Princeton University Press, 1915), 91, 342–48; Thomas Nixon Carver, *Essays in Social Justice* (Harvard: Harvard University Press, 1915), 159–65; Samuel J. Holmes, *The Trend of the Race, A Study of the Present Tendencies in the Biological Development of Civilized Mankind* (New York: Harcourt, Brace and Company, 1921), 13–26, 126–33; Helen H. Gardner, "Heredity: Is Acquired

Character or Condition Transmittable?" *Arena* 9 (May 1894): 769–76; "Heredity vs. Environment," in Ira S. Wile, "Health and the Nations," *Survey* 29 (November 2, 1912): 150; "Health," *Survey* 29 (November 23, 1912): 240; Amey B. Eaton, "The Eugenics Movement," *Survey* 29 (November 23, 1912): 242–44. See also Franklin Giddings's concerns about the application of ideas of germ plasm and changing ideas about the way to eliminate the unfit. Giddings, "Lecture Notes" (October 12–23, 1925), Franklin Giddings Papers, Columbia University, Box 2, Folder "Lecture Notes, 1925."

6. Alice Hamilton, "Heredity and Responsibility," *Survey* 29 (March 22, 1913): 865–66; Edwin Conklin, "Heredity and Responsibility," *Science* 37 (January 10, 1913): 46–54.

7. Edward T. Devine, "Protection vs. Elimination," *Survey* 31 (October 4, 1913): 35–36.

8. Carl Kelsey response to D. Collin Wells, "Social Darwinism," *American Journal of Sociology* 12 (March 1907): 711.

9. Simon Patten, "The Reorganization of Social Work," *Survey* 30 (July 5, 1913): 468–72.

10. Herbert William Conn, *Social Heredity and Social Evolution* (New York: Abingdon Press, 1914), 4–43; Conn, "Social Heredity," *Independent* 56 (January 21, 1904): 143–46; Scott Nearing, *Poverty and Riches: A Study of the Industrial Regime* (Philadelphia: John C. Winston, 1916), 157; John R. Commons, "Natural Selection, Social Selection, and Heredity," *Arena* 18 (July, 1897): 90–97. Nearing was using a phrase coined by Lester Frank Ward in his *Applied Sociology* (Boston: Ginn, 1906), 281.

11. Stephen Smith, "The Basic Principles of Race Betterment," in *Proceedings of the First National Conference on Race Betterment* (Battle Creek, Mich.: Race Betterment Foundation, 1914), 11.

12. Conklin, *Heredity and Environment*, 320–21; Edwin Grant Conklin, "The Mechanism of Heredity," *Science* 27 (January 17, 1908): 89–99; F. Adams Woods, "Separating Heredity from Environment," *American Breeders Magazine* 2 (1911): 194–95; Ophelia L. Amigh, "Alcoholism as a Cause of Degeneracy," *National Conference on Social Welfare Proceedings* (1901): 282–83; Kathy J. Cooke, "The Limits of Heredity: Nature and Nurture in American Eugenics before 1915," *Journal of the History of Biology* 31 (Summer 1998): 263–78; Mark Haller, *Eugenics: Hereditarian Attitudes in American Thought* (New Brunswick, N.J.: Rutgers University Press, 1963); Gregg Mitman, *The State of Nature: Ecology, Community, and American Social Thought, 1900–1950* (Chicago: University of Chicago Press, 1992); Hamilton Cravens, *The Triumph of Evolution: American Scientists and the Heredity-Environment Controversy, 1900–1941* (Philadelphia: University of Pennsylvania Press, 1978); Marouf A. Hasian Jr., *The Rhetoric of Eugenics in Anglo-American Thought* (Athens: University of Georgia Press, 1996).

13. Herman Whittaker, "Weismannism and Its Relation to Socialism," *International Socialist Review* 1 (March 1901): 513–23.

14. Arthur Lewis, *Evolution: Social and Organic* (Chicago: Charles Kerr, 1908); J. A. A. Watson, "The Inheritance of Acquired Characteristics," *International Socialist Review* 18 (July 1917): 33–35; Eva Trew, "Sex Sterilization," *International Socialist Review* 13 (May 1913): 814–17, and 14 (July 1913): 29–31.

15. Smith, "The Basic Principles of Race Betterment," 5–22.

16. *Proceedings of the First National Conference on Race Betterment* (1914), 1.

17. Jacob Riis, "The Bad Boy," in *Proceedings of the First National Conference on Race Betterment* (1914), 241–50; "The Boy Problem" in "Third Conference of the National Federation of Settlements at Pittsburgh, Pennsylvania, September 26, 27, 28, 1913," United South End Settlements Records SW204, Social Welfare History Archives, University of Minnesota, Box 5, Folder: "National Federation of Settlements Reports and Conference Materials (1913–1941)."

18. Maynard M. Metcalf, "Relation of Eugenics and Euthenics to Race Betterment," and Herbert Adolphus Miller, "The Psychological Limits of Eugenics," in *Proceedings of the First*

National Conference on Race Betterment (1914), 456–64; "Eugenics, Euthenics, and Eudemics, Lecture Delivered before Federation for Child Study, January 30, 1913," Lester Frank Ward Papers, Brown University, Box 62, Folder 3. On euthenics, see Ellen H. Richards, *Euthenics: The Science of Controllable Environment; A Plea for Better Living Conditions as a First Step Toward Higher Human Efficiency* (Boston: Whitcomb and Borrows, 1910); Leon J. Cole, "The Relation of Eugenics to Euthenics," *Popular Science Monthly* 81 (November 1912): 475–82; Lester F. Ward, "Eugenics, Euthenics, and Eudemics," *American Journal of Sociology* 18 (May 1913): 737–754; Sarah Stage, "Ellen Richards and the Social Significance of the Home Economics Movement," in *Rethinking Home Economics: Women and the History of a Profession,* ed. Sarah Stage and Virginia B. Vincenti (Ithaca: Cornell University Press, 1997), 17–33; Cooke, "Limits of Heredity," 270–73.

19. Charles B. Davenport, "Relative Effects of Heredity and Environment," in *Proceedings of the First National Conference on Race Betterment,* 471–72; Daniel J. Kevles, *In the Name of Eugenics: Genetics and the Uses of Human Heredity* (Cambridge: Harvard University Press, 1995), 54–55. The Eugenics Record Office had the extraordinary endowment of $300,000. Garland E. Allen, "The Eugenics Record Office at Cold Spring Harbor, 1910–1940: An Essay in Institutional History," *Osiris* 2 (1986): 225–64.

20. Irving Fisher, "The Importance of Hygiene for Eugenics," in *Proceedings of the First National Conference on Race Betterment* (1914), 473.

21. John J. Stevenson, "The Social Problem," *Popular Science Monthly* 78 (March 1911): 258–67; Franklin H. Giddings, "Public Charity and Private Vigilance," *Popular Science Monthly* 55 (August 1899): 433–38; Harriet A. Townsend, "Phases of Practical Philanthropy," *Popular Science Monthly* 55 (August 1899): 534–42; Henry C. Potter, "The Help that Harms," *Popular Science Monthly* 55 (October 1899): 721–32; Donald K. Pickens, *Eugenics and the Progressives* (Nashville: Vanderbilt University Press, 1968). See also George W. Stocking Jr., "Lamarckianism in American Social Science: 1890–1915," *Journal of the History of Ideas* 23 (1962): 239–56.

22. Leon Cole, "The Relation of Philanthropy and Medicine to Race Betterment," in *Proceedings of the First National Conference on Race Betterment* (1914), 494–508.

23. Walter F. Martin, "Report of the Contest," in *Proceedings of the First National Conference on Race Betterment* (1914), 620–24.

24. "Permanent Family Archives," *Eugenical News* 3 (February 1918): 16; "The Burden of the Feeble-Minded," Franklin B. Kirkbride, "The Right to be Well-Born" and Eleanor Hope Johnson, "Feeble-Minded as City Dwellers," and Charles B. Davenport, "The Nams: The Feeble-Minded as Country Dwellers," all in *Survey* 27 (March 2, 1912): 1837, 1838–39, 1840–43, 1844–45; Simon Patten, "The Reorganization of Social Work," *Survey* 30 (July 5, 1913): 468–72; Edward T. Devine, "Protection vs. Elimination," *Survey* 31 (October 4, 1913): 35–36; L. Stern, "Heredity and Environment in the Study of the Bilder Clan," *National Conference on Social Welfare Proceedings* (1922): 179; C. B. Davenport, "Social Work and Eugenics," *National Conference on Social Welfare Proceedings* (1912): 280.

25. "Fifth Meeting of the Board of Directors of the American Eugenics Society" (January 25, 1928), Davenport Papers, Series I, Folder "American Eugenics Society—Folder 2," 4.

26. H. H. Laughlin, "The Relation of Eugenics to Other Sciences," reprinted from the *Eugenics Review* (July 1919): 1–12.

27. Charles B. Davenport, "Heredity and Eugenics" (1920), Davenport Papers, Series I, Folder: "Davenport, Charles B. Lecture: Heredity and Eugenics," 3.

28. C. B. Davenport, "Field Work Indispensable to the Care of Defectives by State," *National Conference on Social Welfare Proceedings* (1915): 312.

29. "The Progress of Eugenics in America" (1912), Davenport Papers, Series I, Folder: "Davenport, Charles B. Lecture: The Progress of Eugenics in America"; "Foreword," *Eugenical News* 1 (January 1916): 1; "Field-Workers' Reports," *Eugenical News* 1 (May 1916): 9–10.

30. "A Strenuous Life," *Eugenical News* 1 (March 1916): 17–18; H. H. Laughlin (superintendent), *Eugenics Record Office: Report No. 1* (Cold Spring Harbor, N.Y.: Eugenics Record Office, 1913), 12–14; H. H. Laughlin, "Report on the Organization and the First Eight Months' Work of the Eugenics Record Office," *American Breeders Magazine* 2, no. 2 (1911): 107–12.

31. "Qualities Desired in a Eugenical Field Worker [list]" (1921), Harry H. Laughlin Papers, Truman State University, Pickler Memorial Library, Box C-2-4, Folder 9; "Second Field Workers' Conference" (June 20–21, 1913), Eugenics Records Office Papers Ms. Coll. 77, American Philosophical Society, Series II, Folder: "Eugenics Records Office—Field Workers' Conference, #3," 28–29.

32. "Timber Rats" (1917), Eugenics Records Office Papers, Series VII, Box 1, Folder: "Devitt, S.C. #1."

33. "A.H. Estabrook to Charles Davenport, May 15, 1912," Davenport Papers, Series II, Folder: "Estabrook, Arthur H. #1"; "A.H. Estabrook to Charles Davenport, November 18, 1924," Davenport Papers, Series II, Folder: "Estabrook, Arthur H. #2"; "Ivan E. McDougle to Arthur Estabrook, April 13, 1923," Arthur Estabrook Papers, M. E. Grenander Department of Special Collections and Archives, University at Albany, State University of New York, Box 1, Folder 2.

34. "Directions for the Guidance of Field Workers" (n.d.), Eugenics Records Office Papers, Series II, Folder: "Eugenics Records Office—Field Workers."

35. "Field Worker's Routes, F.H. Danielson's" (October–December, 1910), Davenport Papers, Series II, Folder: "Eugenics Records Office—Field Workers."

36. "Training Course for Field Workers" (1921), Laughlin Papers, Box C-2-6, Folder 17.

37. Charles Davenport et al., "The Family-History Book," *Eugenics Record Office, Bulletin No. 7* (Cold Spring Harbor, September 1912), 46–51. There is no evidence that these two Davenports were related. "Memorandum for Meeting of the Board of Scientific Directors" (1914), Davenport Papers, Series II, Folder: "Eugenics Records Office—Board of Scientific Directors, Minutes #1"; "Udell, Z.E., #36," "Thayer, Ethel H., #48," and "Letchworth Village, #95," Eugenics Records Office Papers, Series VII, Box 1: "Meeting of the Field Workers" (1915).

38. Charles Davenport et al., "The Family-History Book," 7, 46–51; Charles B. Davenport, "The Nams: The Feeble-Minded as Country Dwellers," *Survey* 27 (March 2, 1912): 1844–45.

39. Laughlin, "Acquired or Inherited?" (1912–13), Davenport Papers, Series II, Folder: "Eugenics Records Office—Play."

40. Mary Storer Kostir, "The Ohio Board of Administration, The Family of Sam Sixty," Publication No. 8 (January 1916), Eugenics Records Office Papers, Series I, Box 36, Folder A: 3134 #4, 74. Emphasis in original.

41. Davenport and Estabrook, *Nam Family*, 76–84.

42. "Family-Tree Folder" (1921–22), Eugenics Records Office Papers, Series I, Box I, Folder A: 01#6; "How to Prepare a Family Pedigree" (n.d.), Eugenics Records Office Papers, Series I, Box I, Folder A: 01#6, Folder A: 01#10.

43. Leland Griggs, "The Inheritance of Acquired Characters," *Popular Science Monthly* 82 (January 1913): 46–52; Frederick Adams Wood, "Heredity and the Hall of Fame," *Popular Science Monthly* 82 (May 1913): 445–52; Charles B. Davenport, "Heredity, Culpability, Praiseworthiness, Punishment, and Reward," *Popular Science Monthly* 83 (July 1913): 33–39; Edward L. Thorndike, "Eugenics: With Special Reference to Intellect and Character," *Popular Science Monthly* 83 (August 1913): 125–38; Woods, "The Racial Origin of Successful Americans," *Popular Science Monthly* 84 (May 1914): 397–402.

44. "Huntington Family, Form V Individual Record Sheet" (1929); "Ellsworth Huntington to Members of the Committee on Family Records" (May 27, 1929), Laughlin Papers, Box C-2-2, Folder 10.

45. "The Family and the Nation" (Minneapolis, November 7, 1912), Davenport Papers, Series I, Folder: "Davenport, Charles B. Lecture: The Family and the Nation," 1–4.

46. "Fitter Family Examinations" (1925), Eugenics Records Office Papers, Series VI, Box 1, Folder: "Fitter Families Examinations—Arkansas #1"; "Dr. Otto C. Gaser to Tracy E. Tuthill" (March 21, 1917), and Eugenics Department of the Kansas Free Fair, "Fitter Families for Future Firesides" (1920–24), Eugenics Records Office Papers, Series VI, Box 4, Folder: "Selected List of 500 Best Family Records."

47. Estabrook and Davenport, *Nam Family,* 84–85.

48. "Arthur Estabrook to Miss Ethel R. Evans, February 17, 1917," Estabrook Papers, Box 1, Folder 1.

49. Charles B. Davenport, *Nomadism or the Wandering Impulse, with Special Reference to Heredity* (Washington, D.C.: Carnegie Institute, 1915); Davenport, "Feebly Inhibited," *Eugenical News* 1 (February 1916): 6; "Tramp and Poet," *Eugenical News* 2 (June 1917): 42; William H. Davies, "The Autobiography of a Super-tramp" (New York: A. A. Knopf, 1917).

50. Elizabeth S. Kite, "Unto the Third Generation," *Survey* 28 (September 28, 1912): 789–91.

51. Henry H. Goddard, *The Kallikak Family: A Study in the Heredity of Feeble-Mindedness* (New York: Macmillan, 1912); "A Tale of Two Families: When the Kallikaks Moved to Harrisburg," *Survey* 38 (April 7, 1917): 23–24.

52. Davenport and Estabrook, *Nam Family,* 1–67. Harry H. Laughlin, *Eugenical Sterilization: 1926: Historical, Legal, and Statistical Review of Eugenical Sterilization in the United States* (New Haven: American Eugenics Society, 1926), 1–3, 10–65; Wendy Kline, *Building a Better Race: Gender, Sexuality, and Eugenics from the Turn of the Century to the Baby Boom* (Berkeley: University of California Press, 2001), 32–94.

53. Harry H. Laughlin, "The Socially Inadequate: How Shall We Designate and Sort Them?" *American Journal of Sociology* 27 (July 1921), 54–70; Franklin H. Giddings, "Seven Devils," *Independent* 94 (September 13, 1919): 356–57.

54. Davenport, "Heredity and Eugenics" (1920), 9.

55. "A Century of Immigration," *Eugenical News* 5 (January 1920): 6–7; "Non-Fecundity of the Fit," *Eugenical News* 5 (August 1920): 60; "Metal and Dross in the 'Melting Pot'," *Eugenical News* 8 (April 1923): 32; "The Trend of the Science of Eugenics," *Eugenical News* 32 (July 11, 1914): 388.

56. Davenport and Estabrook, *Nam Family,* 73–74.

57. Great Britain, *Parliamentary Papers,* Cd. 2210, *Report of the Interdepartmental Committee on Physical Deterioration* (1904); George F. Shee, "The Deterioration in National Physique," *Nineteenth Century* 53 (1903): 798–801; "Physical Deterioration," *Man* 5 (1905): 83–84; Richard Soloway, "Counting the Degenerates: The Statistics of Race Deterioration in Edwardian England," *Journal of Contemporary History* 17 (January 1982): 137–64.

58. "A.H. Estabrook to Charles Davenport" (September 19, 1919), Davenport Papers, Series I, Folder: "Estabrook, Arthur H. #2."

59. "Research on Psychological Records of the Army" (August 25, 1920), Robert M. Yerkes Papers, Group No. 569, Yale University, Manuscripts and Archives, Folder 1443, 1–6; "Examiner's Guide for Psychological Examining in the Army" (1918), Yerkes Papers, Folder 1766; "Army Mental Tests: Methods, Typical Results and Practical Applications" (1918), Yerkes Papers, Folder 1768; Daniel J. Kevles, "Testing the Army's Intelligence: Psychologists and the Military in World War I," *Journal of American History* 55 (December 1968): 565–81; Stephen Jay Gould, *The Mismeasure of Man* (New York: W. W. Norton, 1996), 222–63; John Carson, "Army Alpha, Army Brass, and the Search for Army Intelligence," *Isis* 84 (June 1993): 278–309.

60. Lt. Colonel A. G. Love and Major Davenport (Sanitary Corps), "A Comparison of White and Colored Troops in Respect to the Incidence of Disease," *Proceedings of the National Academy of Sciences* 5 (March 1919): 58–67; Lt. Colonel A. G. Love and Major Davenport (Sanitary Corps), "Defects Found in Drafted Men," reprinted from *Scientific Monthly*

(January 1920), in Charles Davenport Papers, Cold Spring Harbor Laboratory, 1–141; Albert G. Love and Charles B. Davenport, *Defects Found in Drafted Men: Statistical Information Compiled from the Draft Records Showing the Physical Condition of the Men Registered and Examined in Pursuance of the Requirements of the Selective-Service Act* (Washington, D.C.: GPO, 1920); Albert Gallatin Love, *Physical Examination of the First Million Draft Recruits: Methods and Results* (Washington, D.C.: GPO, 1919); "Physique of Drafted Men," *Survey* 42 (March 27, 1920): 807–8; Horace B. English, "Is American Feeble-Minded?" *Survey* 44 (October 15, 1922): 79–81.

61. Robert de C. Ward, "Some Aspects of Immigration to the Untied States in Relation to the Future American Race," *Eugenics Review* 7, no. 4 (1916): 263–82.

62. E. K. Sprague, "Medical Inspection of Immigrants," *Survey* 30 (June 21, 1913): 420–22.

63. "Harry Laughlin to Charles B. Davenport" (April 14, 1921), Davenport Papers, Series II, Folder: Folder: "Laughlin, H.H.—1921 #1."

64. "Statement of Harry Laughlin," *Biological Aspects of Immigration, Hearings before the Committee on Immigration and Naturalization* (Washington, D.C: GPO, 1921), 4–5.

65. *Analysis of the Metal and Dross in America's Modern Melting Pot in Hearings before the Committee on Immigration and Naturalization* (Washington, D.C.: GPO, 1923), 727–62.

66. Mae Ngai, *Impossible Subjects: Illegal Aliens and the Making of Modern America* (Princeton: Princeton University Press, 2004), 37. Ngai focuses especially on the process of defining and enforcing quotas as a process of racial construction. For alternative explanations, see John Higham, *Strangers in the Land: Patterns of American Nativism, 1860–1925* (New Brunswick, N.J.: Rutgers University Press, 1998), 300–316; Desmond King, *The Liberty of Strangers: Making the American Nation* (New York: Oxford University Press, 2005), 63–82.

67. H. S. Jennings, "'Undesirable Aliens': A Biologists Examination of the Evidence Before Congress," *Survey* 51 (December 15, 1923): 309–12, 364; "Taking the Queue Out of Quota, An Interview with W.W. Husband, Commissioner-General of Immigration," *Survey* 51 (March 15, 1924): 666–69.

68. "H.H. Laughlin to Charles Davenport" (November 22, 1923), Davenport Papers, Series II, Folder: "Laughlin, H.H.—1923 #2"; Laughlin, "Experimental Family Histories" (1924), Davenport Papers, Series II, Folder: "Laughlin, H.H.—1924."

8. Following the Monkey

1. "Following the Monkey" (n.d., ca. 1925), Ben Lewis Reitman Papers, File List 426, University of Illinois—Chicago, Series: Writings, Folder: 11, 4–5.

2. The classic history of the IWW is Melvyn Dubofsky, *We Shall Be All: A History of the Industrial Workers of the World* (Chicago: Quadrangle Books, 1969). Frank Tobias Higbie, *Indispensable Outcasts: Hobo Workers and Community in the American Midwest, 1880–1930* (Urbana: University of Illinois Press, 2003), 134–72, examines the intersection between tramping and IWW politics. Tim Cresswell, *The Tramp in America* (London: Reaktion Books, 2001), 70–86, discusses Reitman and his obsession with tramp taxonomies. See also Roger Bruns, *The Damndest Radical: The Life and World of Ben Reitman* (Urbana: University of Illinois Press, 1987).

3. The correspondence between McCook and Aspinwall is included in Social Reform Papers of John J. McCook, Antiquarian and Landmarks Society of Connecticut, Microfilm Roll 12; John J. McCook, "Leaves from the Diary of a Tramp," *Independent* 53 (November 21, 1901): 2760–67; (December 5, 1901): 2880–88; (December 19, 1901): 3009–13; 54 (January 2, 1902): 23–28; (January 16, 1902): 154–60; (February 6, 1902): 332–37; (March 13,

1902): 620–24; (April 10, 1902): 873–74; (June 26, 1902): 1539–44; Cresswell, *The Tramp in America*, 181–95.

4. Todd Depastino, *Citizen Hobo: How a Century of Homelessness Shaped America* (Chicago: University of Chicago Press, 2003), 49–58; Kenneth L. Kusmer, *Down and Out, on the Road: The Homeless in American History* (New York: Oxford University Press, 2002).

5. Josiah Flynt Willard, *Tramping with Tramps: Studies and Sketches of Vagabond Life* (New York: Century Co., 1907 [1899]); Willard, *My Life* (New York: Outing Publishing Company, 1908).

6. Jack London, *The Road* (New Brunswick, N.J.: Rutgers University Press, 2006 [1907]); London, *Jack London on the Road: The Tramp Diary and Other Hobo Writing*, ed. Richard E. Etulain (Logan: Utah State University Press, 1979); Jonathan Auerbach, *Male Call: Becoming Jack London* (Durham: Duke University Press, 1996).

7. London, *The Road*, 124–29.

8. London, *The Road*, 133. Emphasis in original.

9. London, *The Road*, 83–88.

10. London, *The Road*, 61–63, 132.

11. Willard, *Tramping with Tramps*, 381–98.

12. London, *The Road*, 123–24, 101–6.

13. Reitman, "Brotherhood Welfare Association" (1908), Reitman Papers, Series: Organizations, Folder: "Brotherhood Welfare Association."

14. London, *The Road*, 132.

15. Nels Anderson, *The American Hobo* (Leiden: E. J. Brill, 1975 [1923]), 160–61.

16. Nels Anderson, *The Hobo: The Sociology of the Homeless Man* (Chicago: University of Chicago Press, 1923); Dean Stiff [Nels Anderson], *The Milk and Honey Route: A Handbook for Hobos* (New York: Vanguard Press, 1930).

17. Reitman, "Untitled Notes" (ca. 1908), Reitman Papers, Series: Organizations, Folder: Brotherhood Welfare Association.

18. Reitman, "Factors that Make Men Social Outcasts and How to Overcome Them" (n.d.), Reitman Papers, Supplement V.

19. "Reitman to 'Little Ben' Reitman" (May 6, 1918), Reitman Papers, Series: Topics, "Outcast Narratives" (ca. 1918–1919), Reitman Papers, Series: Writings, Folder 9.

20. "The Tramp. What Brought Him. And What Will Remove Him," *Appeal to Reason* (December 4, 1897); "Tramp, Tramp, Tramp," *Appeal to Reason* (January 25, 1895).

21. "Evolution of the Hobo" (n.d.), Reitman Papers, Series: Correspondence, 240.

22. Reitman, "Following the Monkey," 25–26; Reitman, "The American Tramp," (1908), Reitman Papers, Series: Organizations, Folder: "Brotherhood Welfare Association."

23. "Answers to Questions by Cartoonist Walker," *Appeal to Reason* (October 2, 1904).

24. "Jim and James, No. 10—The End," *Appeal to Reason* (September 2, 1905).

25. Thorstein Veblen, *The Theory of the Leisure Class* (New York: Macmillan, 1899), chap. 10. For a typical hesitantly admiring review, see John Cummings, "The Theory of the Leisure Class," *Journal of Political Economy* 7 (September 1899): 425–55. Lester Frank Ward would defend the book, claiming "the trouble with this book is that it contains too much truth." See Lester F. Ward, "The Theory of the Leisure Class," *American Journal of Sociology* 5 (May 1900): 829–37. See also Clare Virginia Bay, "Thorstein Veblen and the Rhetoric of Authority," *American Quarterly* 46 (June 1994): 139–73.

26. Harry Kemp, "The Lure of the Tramp," *Independent* 70 (June 8, 1911): 1271.

27. William Hard, *The Women of Tomorrow* (New York: Baker & Taylor, 1911).

28. Reitman, "Following the Monkey," 190–98.

29. Robert Rives La Monte, "A Forgotten 'Tramp,'" *Industrial Union Bulletin* (April 13, 1907).

30. London, *The Road*, 33.

31. "The Free-Footed Rebel," *Solidarity* (September 6, 1913); J. H. Walsh, "Shall We Die Starving or Shall We Die Fighting?" *Industrial Union Bulletin* (July 25, 1908); Walsh, "I.W.W. 'Red Special' Overall Brigade," *Industrial Union Bulletin* (September 19, 1908).

32. *Solidarity* (November 21, 1914).

33. "Evolution of the Hobo"; "Seizing a Train," *Solidarity* (February 25, 1911).

34. Jack London, *War of the Classes* (New York: Macmillan, 1905), 267–78. Emphasis in original.

35. "My Friend, Mr. Block," *Solidarity* (March 15, 1913).

36. "Will Dyson's Cartoons," *International Socialist Review* 16 (December 1915): 336.

37. "The Jungles in California," *Industrial Worker* (May 6, 1909); "The Social Evil," *Industrial Worker* (April 8, 1909); Louis Duchez, "Stories from Real Life," *Solidarity* (August 27, 1910).

38. "Sanny M'Nee on Tramps and Decent Men," *International Socialist Review* 16 (October 1915): 201; *Solidarity* (November 21, 1914).

39. London, *The Road*, 15–31; Anderson, *American Hobo*, 96.

40. "You Need Industrial Unionism" (1922), IWW Papers, Walter P. Reuther Library and Archive, Box 44, Folder, 35.

41. "The 'Simple Life' in the Jungles," *Industrial Worker* (March 18, 1909). Gerald Ronning, "Jackpine Savages: Discourses of Conquest in the 1916 Mesabi Iron Range Strike," *Labor History* 44 (August 2003): 359–82, examines the subversive use of notions of savagery and civilization in IWW politics.

42. Kemp, "Lure of the Tramp," 1270. This reading of tramp masculinity as a subversion of chivalrous manhood presents a proletarian counterpart to the gendered transition noted by Gail Bederman, *Manliness and Civilization: A Cultural History of Gender and Race in the United States, 1880–1917* (Chicago: University of Chicago Press, 1995). See also Elliott Gorn, *The Manly Art: Bare-Knuckle Prize Fighting in America* (Ithaca: Cornell University Press, 1986), and Howard P. Chudacoff, *The Age of the Bachelor: Creating an American Subculture* (Princeton: Princeton University Press, 1999). The expression of masculinity through sexual relations with other men—and the absence of women—confirms the shifting constructions of sexuality and gender noted by George Chauncey, *Gay New York: Gender, Urban Culture, and the Making of the Gay Male World* (New York: Basic Books, 1994).

43. Reitman, "Following the Monkey," 61–86.

44. "The 'Simple Life' in the Jungles," "Out in the Jungles," *Industrial Worker* (April 1, 1909); "Notes Along the N.P.," *Industrial Worker* (April 22, 1909).

45. "The Jungles for Workers," "Banquet Room for Loafers" *Industrial Worker* (April 29, 1909).

46. Walsh, "I.W.W. 'Red Special.'"

47. Anderson, *The Hobo*, 144.

48. London, *The Road*, 115, 133.

49. Reitman, "Homosexuality" (n.d.), Reitman Papers, Supplement IV, 11–35. For discussions of manhood and sexuality among hoboes and tramps, see Higbie, *Indispensable Outcasts*, 173–200; Depastino, *Citizen Hobo*, 81–91.

50. William Wellman (director), *Beggars of Life* (Paramount Famous Lasky Corp., 1928), two reels, eighty-three minutes; Jim Tully, *Beggars of Life* (Garden City, N.Y.: Garden City Publishing Co., 1924).

51. "Ladylike Men," *Solidarity* (December 14, 1909); "A (Bohn) Head," *Industrial Worker* (July 27, 1911).

52. "The Social Evil"; "Freedom," *Solidarity* (May 14, 1910); Elizabeth Gurley Flynn, "The IWW Call to Women," *Solidarity* (July 31, 1915); "Prostitution and Wage Slavery," *Solidarity* (September 23, 1916).

53. David Starr Jordan, *War and Waste: A Series of Discussions of War and War Accessories* (Garden City, N.Y.: Doubleday, Page, 1914), 6–7; Jordan, *Imperial Democracy* (New York: D. Appleton, 1899).

54. Jordan, *War and Waste*, 25–31.

55. David Starr Jordan, "The Eugenics of War," *Eugenics Review* 5 (April 1913–January 1914): 197–213.

56. Jordan, *War and Waste*, 173–75.

57. Alice Hamilton, "Race Suicide," *Survey* 35 (January 1, 1916): 407–8.

58. Seth K. Humphrey, *Mankind: Racial Values and the Racial Prospect* (New York: Charles Scribner's Sons, 1917), 132; David Starr Jordan, "War and Genetic Values," *Journal of Heredity* 10 (May 1919): 223–25.

59. "The Aftermath of War," *Eugenical News* 1 (March 1916): 15; Robert de C. Ward, "Some Aspects of Immigration to the United States in Relation to the Future American Race," *Eugenics Review* VII (April 1915–January 1916): 263–82.

60. "Heart of a Jewish Child," *Eugenical News* 4 (August 1919): 61; Rose Cohen, *Out of the Shadow: A Russian Jewish Girlhood on the Lower East Side* (Ithaca: Cornell University Press, 1995 [1918]).

61. "Lenin, the Dictator," *Eugenical News* 4 (December 1919): 93; A. R. Williams, *Lenin: The Man and his Work* (New York: Scott & Seltzer, 1919); Thomas Jefferson Downing, "A Possible Factor of Degeneracy," *New York Medical Journal* 108 (July 20, 1918): 103–5; Frederick Adams Woods, "The Racial Limitation of Bolshevism," *Journal of Heredity* 10 (April, 1919): 188–190; Franklin Giddings, "Bolsheviki Must Go," *Independent* 98 (January 18, 1919): 88.

62. Lothrop Stoddard, *The Revolt against Civilization: The Menace of the Under Man* (New York: Charles Scribner's Sons, 1922), 1–19. Emphasis in original. See also Matthew Pratt Guterl, *The Color of Race in America, 1900–1940* (Cambridge: Harvard University Press, 2001), 100–153; Gary Gerstle, *American Crucible: Race and Nation in the Twentieth Century* (Princeton: Princeton University Press, 2001), 81–127.

63. Stoddard, *Revolt against Civilization*, 32–33. Emphasis in original.

64. Stoddard, *Revolt against Civilization*, 120–25.

65. Basil M. Manly, "Less Than Third of Men Are Fit for U.S. Army," *Solidarity* (February 24, 1917).

66. Stoddard, *Revolt against Civilization*, 125–225; Victor Yarros, "Bolshevism: Its Rise, Decline, and—Fall?" *International Journal of Ethics* 30 (April 1920): 267–83; Edward Ross, *The Russian Soviet Republic* (New York: Century Co., 1923); Philip Ainsworth Means, *Racial Factors in Democracy* (Boston: Marshall Jones, 1918), 141–84.

67. Stoddard, *Revolt against Civilization*, 177–236; "A New Basis for History," *New York Times* (July 11, 1920): 82; Charles Conant Josey, *Race and National Solidarity* (New York: Charles Scribner's Sons, 1923).

68. "Davenport to Lothrop Stoddard, July 31, 1920"; "Lothrop Stoddard to Charles Davenport, May 3, 1922"; "Charles Davenport to Lothrop Stoddard, May 10, 1922," Charles B. Davenport Papers B D27, American Philosophical Society, Series I, Folder: "Stoddard, Lothrop"; "The Color-Races," *Eugenical News* 5 (September 1920): 70–71.

69. Lothrop Stoddard, *The Rising Tide of Color* (New York: Charles Scribner's Sons, 1922 [1920]), 196–97.

70. Stoddard, *Rising Tide*, 13.

71. "The Chinamen is Coming," *Solidarity* (February 11, 1911).

72. Stoddard, *Rising Tide*, 10–53; Irving Fisher, "Impending Problems of Eugenics," *Scientific Monthly* 13 (September 1921): 218–31.

73. Stoddard, *Rising Tide*, 11, 77–86, 130–38; Abdullah Achmet, "Seen Through Mohammedan Spectacles," *Forum* 53 (October 1914): 484–97.

74. "Harding Says Negro Must Have Equality in Political Life," *New York Times* (October 27, 1921).

75. F. Scott Fitzgerald, *The Great Gatsby* (New York: Charles Scribner's Sons, 1925), 19.

76. *New York Times* (May 30, 1920).

77. For a transcript of Harding's speech, see "Harding Says Negro Must Have Equality in Political Life."

78. "Negro Efficiency," *Eugenical News* 1 (November 1916): 79.

79. Clayton Sedgwick Cooper, *The Modernizing of the Orient* (New York: McBride, 1914), 5.

80. Stoddard, *Rising Tide,* 241–97.

81. Stoddard, *Rising Tide,* 295–97; John Carter, "Little Brown Men Carry Britain's Flag," *New York Times* (August 30, 1925).

82. Stoddard, *Rising Tide,* 266–67, 297–98; Stoddard, *Re-Forging America* (New York: Charles Scribner's Sons, 1927), 256–57.

83. Stoddard, *Rising Tide,* 230–34, 270–71; B. L. Putnam Weale, *The Conflict of Color: The Threatened Upheaval throughout the World* (Miami: Mnemosyne Publishing Co., 1969 [1910]), 98–99; Michael Adas, *Machines as the Measure of Man: Science, Technology, and Ideologies of Western Dominance* (Ithaca: Cornell University Press, 1989), 361–65.

84. James Weldon Johnson, "The Changing Status of Negro Labor," *National Conference on Social Welfare Proceedings* (1918): 383–88.

85. "Reasons Why Negroes Go North," *Survey* 38 (June 2, 1917): 226–27; Oscar Leonard, "Welcoming Southern Negroes: East St. Louis and Detroit—A Contrast," *Survey* 38 (July 14, 1917): 331–35; George Edmund Haynes, "Negroes Move North," *Survey* 40 (May 4, 1918 and January 4, 1919): 115–22, 455–461.

86. Charles S. Johnson, "Substitution of Negro Labor for European Immigrant Labor," *National Conference on Social Welfare Proceedings* (1926): 317–25; T. Arnold Hill, "Recent Developments in the Problem of Negro Labor," *National Conference on Social Welfare Proceedings* (1921): 321–25; George E. Haynes, "Negro Labor and the New Order," *National Conference on Social Welfare Proceedings* (1919): 531–38.

87. Blanton Fortson, "Northward to Extinction," and William Pickens, "Migrating to Fuller Life," *Forum* 72 (November 1924): 593–607. For two important readings of the social and cultural reactions to African American industrial workers and the African American interaction with modernity, see Kimberly L. Phillips, *Alabama North: African-American Migrants, Community, and Working-Class Activism in Cleveland, 1915–1945* (Urbana: University of Illinois Press, 1999); Joel Dinnerstein, *Swinging the Machine: Modernity, Technology, and African American Culture between the World Wars* (Amherst: University of Massachusetts Press, 2003).

88. Edward Eggleston, *The Ultimate Solution of the American Negro Problem* (Boston: R.G. Badger, 1913); Raymond Pearl, "The Vitality of the Peoples of America," *American Journal of Hygiene* 1 (September–November 1921): 592–674.

89. Jerome Dowd, *The Negro in American Life* (New York: Century Co., 1926), 525; S. J. Holmes, *Studies in Evolution and Eugenics* (New York: Harcourt, Brace and Company, 1923), 229–55.

90. *Analysis of the Metal and Dross in America's Modern Melting Pot in Hearings Before the Committee on Immigration and Naturalization* (Washington, D.C.: GPO, 1923); "Shall We All Be Mulattoes?" *Literary Digest* (March 7, 1925): 23–24, in Arthur Estabrook Papers, University at Albany, State University of New York, M. E. Grenander Department of Special Collections and Archives, Box 1, Folder 32.

91. Charles B. Davenport, "The Immigration Policy of the United States" (n.d., ca. 1926), Davenport Papers, Series I, Folder: "Davenport, Charles B. Lecture: Immigration Policy of the United States"; "Do Races Differ in Mental Capacity?" (ca. 1924), Davenport Papers, Series I, Folder: "Davenport, Charles B. Lecture: Do Races Differ in Mental Capacity?"

92. "Charles B. Davenport to Dr. R.J. Terry, January 4, 1927" and "R.J. Terry to Charles B. Davenport, February 9, 1927," "Chas. B. Davenport to the Chairman, Division of Anthropology and Psychology, National Research Council, June 27, 1927," "Chas. B. Davenport to R.J. Terry, March 4, 1929," "Committee on the American Negro, Proposals for the Organization of Investigations on the American Negro" (1929), and "National Research Council, Division of Anthropology and Psychology Committee on the Negro, Preliminary Report" (1929), all in Davenport Papers, Series I, Folder: "Committee on the American Negro."

93. "The New Immigration Bill," *Eugenical News* 9 (January 1924): 1; "Report of the Committee on Selective Immigration of the Eugenics Committee of the United States of America," *Eugenical News* 9 (February 1924): 21–24; Arthur H. Estabrook, "Presidential Address: Blood Seeks Environment," *Eugenical News* 11 (August 1926): 106–14.

94. Stoddard, *Re-Forging America*, 191–93.

95. *"American History In Terms of Human Migration" Extracts from Hearings before the Committee on Immigration and Naturalization House of Representatives* (Washington, D.C.: GPO, 1928), 1–15.

96. Henry Pratt Fairchild, *The Melting-Pot Mistake* (Boston: Little, Brown, 1926).

97. Stoddard, *Re-Forging America*, 215–16.

98. Stoddard, *Re-Forging America*, 255–325; Wilbur C. Abbott, *The New Barbarians* (Boston: Little, Brown, 1925).

99. Stoddard, *Re-Forging America*, 13.

9. Failing of Art and Science: The Abyss in a New Era

1. Alexandra Minna Stern, *Eugenic Nation: Faults and Frontiers of Better Breeding in Modern America* (Berkeley: University of California Press, 2005).

2. Arthur H. Estabrook and Ivan E. McDougle, *Mongrel Virginians: The Win Tribe* (Baltimore: Williams and Wilkins, 1926); "Mongrel Virginians," *Annals of the American Academy of Political and Social Science* 126 (July 1926): 165–66.

3. Simeon Strunsky, "About Books, More or Less: Nordics and Applied Science," *New York Times* (October 26, 1924).

4. "Debate on Menace to Western World," *New York Times* (January 10, 1926); Carleton P. Barnes, "*Re-Forging America* by Lothrop Stoddard," *Economic Geography* 4 (April 1928): 211–12; "*Lonely America* by Lothrop Stoddard," *Journal of Negro History* 18 (January 1933): 85–86.

5. Lothrop Stoddard, *Into the Darkness* (New York: Duell, Sloan, & Pearce, 1940).

6. "Plot Denied In Berlin," *New York Times* (October 11, 1933).

7. Edward Alsworth Ross, "The Post-War Intellectual Climate," *American Sociological Review* 10 (October 1945): 648–50.

8. Elizabeth Faue, *Community of Suffering and Struggle: Women, Men, and the Labor Movement in Minneapolis, 1915–1945* (Chapel Hill: University of North Carolina Press, 1991); Barbara Melosh, *Engendering Culture: Manhood and Womanhood in New Deal Public Art and Theater* (Washington, D.C.: Smithsonian Institution Press, 1991).

9. Waldemar Kaempffert, "The Week in Science: Can the White Race Survive?" *New York Times* (June 9, 1935).

10. *The Race Concept: Results of an Inquiry* (Paris: UNESCO, 1952), 5; Michelle Brattain, "Race, Racism, and Antiracism: UNESCO and the Politics of Presenting Science to the Postwar Public," *American Historical Review* 112 (December 2007): 1386–1413.

11. The most obvious example would be Richard J. Herrnstein and Charles Murray, *The Bell Curve: The Reshaping of American Life by Difference in Intelligence* (New York: Free Press,

1994), and the response in Stephen Jay Gould, *The Mismeasure of Man* (New York: W. W. Norton, 1996). Brattain also points to the current focus on DNA testing and genomic research.

12. "The Rising Tide," *New York Times* (April 3, 1960).

13. "A Warning from the Past" (January 2000), http://www.amren.com/ar/2000/01.

14. Samuel Huntington, "The Clash of Civilizations," *Foreign Affairs* 72 (Summer 1993): 22–49, and Huntington, *The Clash of Civilizations and the Remaking of World Order* (Cambridge: Harvard University Press, 1996); Robert G. Lee, "Brown Is the New Yellow: The Yellow Peril in an Age of Terror," in *Race, Nation, and Empire in American History*, ed. James T. Campbell, Matthew Pratt Guterl, and Robert G. Lee (Chapel Hill: University of North Carolina Press, 2007): 335–51; Lothrop Stoddard, "Pan-Turanism," *American Political Science Review* 11 (February 1917): 12–23.

15. Patrick J. Buchanan, *The Death of the West: How Dying Populations and Immigrant Invasions Imperil Our Country and Civilization* (New York: Thomas Dunne Books/St. Martin's Press, 2002).

16. Donald McGranahan, "Development Indicators and Development Models," *Journal of Development Studies* 8, no. 3(1972): 91–102. See also Uma Kothari, ed., *A Radical History of Development Studies: Individuals, Institutions, and Ideologies* (London: Zed Books, 2005).

17. William M. Adler, *Mollie's Job: A Story of Life and Work on the Global Assembly Line* (New York: Scribner, 2000); Jefferson Cowie, *Capital Moves: RCA's Seventy-Year Quest for Cheap Labor* (Ithaca: Cornell University Press, 1999).

BIBLIOGRAPHIC ESSAY

This book depends on an array of sources including archival records of leading individuals (such as Edward Ross or Charlotte Perkins Gilman) and important institutions (such as the Eugenics Records office or the University Settlement Society). It makes ample use of the avalanche of printed books, pulp fiction, pamphlets, magazines, and journals that reacted in different ways to American industrial evolution—and all the problems associated with it. It engages with the far-ranging historical literature on the Gilded Age and the Progressive Era, including works on labor, progressive reform, socialist and radical politics, and empire, African American, cultural, and immigration histories. This analysis connects often discrete histories. Sometimes, it clarifies or extends the arguments accepted by historians; at other times, it offers starkly different conclusions.

The historical literature touched upon and used in this text is vast. It is most useful to highlight a few key texts and their arguments as well to suggest a set of new typographies to group and link the diverse histories of this period.

Industrial and Social Evolution

Most histories of American industry tend to place the "r" back in industrial evolution. Although they have elucidated the experience of industrial workers, the larger intellectual and cultural history of industry is less clear. For the last several decades, scholars of the history of labor, industry, and the working class have been indebted to the landmark scholarship of David Montgomery, *Workers' Control in America: Studies in the History of Work, Technology, and Labor Struggles* (New York: Cambridge University Press, 1981), and Herbert Gutman, *Work, Culture, and Society in Industrializing*

America: Essays in American Working-Class and Social History (New York: Vintage Books, 1977). Much of American labor and industrial history begins in the early republic; see Jonathan Prude, *The Coming of Industrial Order: Town and Factory Life in Rural Massachusetts, 1810–1860* (Amherst: University of Massachusetts Press, 1999 [1983]).

T. Jackson Lears, *No Place of Grace: Anti-Modernism and the Transformation of American Culture, 1880–1970* (Chicago: University of Chicago Press, 1994), Jeffrey Sklansky, *The Soul's Economy: Market Society and Selfhood in American Thought, 1820–1920* (Chapel Hill: University of North Carolina Press, 2002), Amy Dru Stanley, *From Bondage to Contract: Wage Labor, Marriage, and the Market in the Age of Slave Emancipation* (New York: Cambridge University Press, 1998), and Alexander Trachtenberg, *The Incorporation of America: Culture and Society in the Gilded Age,* 25th anniversary ed. (New York: Hill and Wang, 2007), note both the anxiety about and central place of industry in turn-of-the-twentieth-century American cultural and intellectual history. Michael Adas, *Machines as the Measure of Men: Science, Technology, and Ideologies of Western Dominance* (Ithaca: Cornell University Press, 1989), is notable for its global frame.

The rise of evolutionary thought in the United States is examined by Ronald L. Numbers, *Darwinism Comes to America* (Cambridge: Harvard University Press, 1998); Richard Hofstadter, *Social Darwinism in American Thought* (Philadelphia: University of Pennsylvania Press, 1944); Gregg Mitman, *The State of Nature: Ecology, Community, and American Social Thought, 1900–1950* (Chicago: University of Chicago Press, 1992); and Cynthia Russett, *Darwin in America: The Intellectual Response, 1865–1912* (New York: Freeman, 1976). Daniel T. Rodgers, *Atlantic Crossings: Social Politics in a Progressive Age* (Cambridge: Harvard University Press, Belknap Press, 1998), describes the exchange of progressive ideas between Europe and North America.

For histories of American social science that note the significance of evolutionary thought and the cross-fertilization between biology and the social sciences, see Dorothy Ross, *The Origins of American Social Science* (New York: Cambridge University Press, 1991); George W. Stocking Jr., *Race, Culture, and Evolution: Essays in the History of Anthropology* (Chicago: University of Chicago Press, 1982 [1968]); and George W. Stocking Jr., *Victorian Anthropology* (New York: Free Press, 1987).

Explorations

The turn of the twentieth century is best understood as a time of interconnected movements of ideas, labor, armies, and goods. It is useful to think about these issues in common. Matthew Frye Jacobson, *Barbarian Virtues: The United States Encounters Foreign Peoples at Home and Abroad, 1876–1917* (New York: Hill and Wang, 2000), hints at connections between histories of trade, immigration, and empire. Such connections produced interchangeable ways of evaluating supposedly primitive peoples at home and abroad. For significant works that suggest (in different contexts) a cross-fertilization between imperial travel narratives, urban tourism, exploration, and social reform investigation, see Mary Louise Pratt, *Imperial Eyes: Travel Writing and Transculturation* (New York: Routledge, 1992); Seth Koven, *Slumming: Sexual and Social Politics in Victorian London* (Princeton: Princeton University Press, 2004); Angela M. Blake, *How New York Became American, 1890–1924* (Baltimore: Johns Hopkins University Press, 2006); and Catherine Cocks, *Doing the Town: The Rise of Urban Tourism in the United States, 1850–1915* (Berkeley: University of California Press, 2001).

Among works that examine the confluence of class, empire, and the domestic sphere in shaping interconnections between the domestic and the imperial are Robert Rydell, *All the World's a Fair: Visions of Empire at American International Expositions, 1876–1916* (Chicago: University of Chicago Press, 1984), and James T. Campbell, Matthew Pratt Guterl, and Robert G. Lee, eds., *Race, Nation, and Empire in American History* (Chapel Hill: University of North Carolina Press, 2007). Amy Kaplan and Donald E. Pease, eds., *Cultures of United States Imperialism* (Durham: Duke University Press, 1993); Paul A. Kramer, *The Blood of Government: Race, Empire, the United States, and the Philippines* (Chapel Hill: University of North Carolina Press, 2006); Ann Laura Stoler, ed., *Haunted by Empire: Geographies of Intimacy in North American History* (Durham: Duke University Press, 2006); and Warwick Anderson, *Colonial Pathologies: American Tropical Medicine, Race, and Hygiene in the Philippines* (Durham: Duke University Press, 2006), highlight the need to question histories of empire as purely issues of expansion. Rather, the experience of empire reshaped American understandings of race, gender, sexuality, and class. It transformed understandings of industry and immigration as well.

Historians of immigration have come to focus especially on the role of nativism and the regulation and restriction of immigration in the shaping of the nation. See John Higham, *Strangers in the Land: Patterns of American Nativism, 1860–1925* (New Brunswick, N.J.: Rutgers University Press, 2002); Erika Lee, *At America's Gates: Chinese Immigration during the Exclusion Era, 1882–1943* (Chapel Hill: University of North Carolina Press, 2003); Mae Ngai, *Impossible Subjects: Illegal Aliens and the Making of Modern America* (Princeton: Princeton University Press, 2004); Amy L. Fairchild, *Science at the Borders: Immigrant Medical Inspection and the Shaping of the Modern Industrial Labor Force* (Baltimore: Johns Hopkins University Press, 2003); Desmond King, *The Liberty of Strangers: Making the American Nation* (New York: Oxford University Press, 2005); Evelyn Nakano Glenn, *Unequal Freedom: How Race and Gender Shaped American Citizenship and Gender* (Cambridge: Harvard University Press, 2002); and Eiichiro Azuma, *Between Two Empires: Race, History, and Transnationalism in Japanese America* (New York: Oxford University Press, 2005). Taken together, these works suggest the importance of locating debates about immigration in an imperial context.

The analysis in this book highlights the need to examine African American workers and their relationship to the industrial economy. For two books that begin this examination, see Joel Dinnerstein, *Swinging the Machine: Modernity, Technology, and African-American Culture between the World Wars* (Amherst: University of Massachusetts Press, 2003), and William P. Jones, *The Tribe of Black Ulysses: African American Lumber Workers in the Jim Crow South* (Urbana: University of Illinois Press, 2005).

Visions of Civilization and Savagery

As Gail Bederman, *Manliness and Civilization: A Cultural History of Gender and Race in the United States, 1880–1917* (Chicago: University of Chicago Press, 1995), has demonstrated, ideas about civilization were omnipresent in both the imperial and domestic imagination. It is important to note that the fixation with civilization and its fragility were dependent on a fascination with the primitive and the notion that the present of one race might help explain the past of another. The place of the primitive in American culture is best represented by Adam Kuper, *The Invention of Primitive Society: Transformations of an Illusion* (New York: Routledge, 1988), and Patrick Brantlinger, *Dark Vanishings: Discourse on the Extinction of Primitive Races, 1800–1930* (Ithaca: Cornell University Press, 2003).

The significance of ideas about savagery and its persistence at the heart of civilization can help formulate a history of poverty that includes the study of the experience of poverty as well as its larger cultural contexts. Two important texts that examine the place of poverty in literature or intellectual life are Keith Gandal, *The Virtues of the Vicious: Jacob Riis, Stephen Crane, and the Spectacle of the Slum* (New York: Oxford University Press, 1997), and Alice O'Connor, *Poverty Knowledge: Social Science, Social Policy, and the Poor in Twentieth-Century U.S. History* (Princeton: Princeton University Press, 2002).

There is a longstanding historical literature about prostitution and a more recent literature about tramping. These histories are, in fact, connected by larger cultural concerns about savagery and civilization. See Ruth M. Alexander, *The "Girl Problem": Female Sexual Delinquency in New York, 1900–1930* (Ithaca: Cornell University Press, 1995); Mark Thomas Connelly, *The Response to Prostitution in the Progressive Era* (Chapel Hill: University of North Carolina Press, 1980); Elizabeth Alice Clement, *Love for Sale: Courting, Treating, and Prostitution in New York City, 1900–1945* (Chapel Hill: University of North Carolina Press, 2006); Frank Tobias Higbie, *Indispensable Outcasts: Hobo Workers and Community in the American Midwest, 1880–1930* (Urbana: University of Illinois Press, 2003); and Todd Depastino, *Citizen Hobo: How a Century of Homelessness Shaped America* (Chicago: University of Chicago Press, 2003),

There are numerous superb social and intellectual histories of progressivism and reform politics. See Mina Julia Carson, *Settlement Folk: Social Thought and the American Settlement Movement, 1885–1930* (Chicago: University of Chicago Press, 1990); Kathryn Kish Sklar, *Florence Kelley and the Nation's Work* (New Haven: Yale University Press, 1995); and Alan Dawley, *Changing the World: American Progressives in War and Revolution* (Princeton: Princeton University Press, 2003). Louise Michele Newman, *White Women's Rights: The Racial Origins of Feminism in the United States* (Oxford: Oxford University Press, 1999), and Mark Pittenger, *American Socialists and Evolutionary Thought, 1870–1920* (Madison: University of Wisconsin Press, 1993), highlight the engagement of feminists, reformers, and socialists with theories of social evolution. They describe parallel socialist and reform efforts to preserve civilization—even at the cost of the annihilation of the unfit.

The Problem of Degeneration

Fears of degeneration run through the political thought of socialists, reformers, and eugenicists. Yet most studies focus either globally or on Europe. See Daniel Pick, *Faces of Degeneration: A European Disorder, c.1848–c.1918* (Cambridge: Cambridge University Press, 1989), and Sander Gilman and Edward J. Chamberlin, eds, *Degeneration: The Dark Side of Progress* (New York: Columbia University Press, 1985). Elof Axel Carlson, *The Unfit: A History of a Bad Idea* (Cold Spring Harbor, N.Y.: Cold Spring Harbor Laboratory Press, 2001), is one of the few books to highlight widespread American fears of the survival of the unfit.

American eugenics, by contrast, has received a great deal of recent attention. Daniel J. Kevles, *In the Name of Eugenics: Genetics and the Uses of Human Heredity* (Cambridge: Harvard University Press, 1998), has become the classic in the field. Alexandra Minna Stern, *Eugenic Nation: Faults and Frontiers of Better Breeding in Modern America* (Berkeley: University of California Press, 2005), helps extend the history of eugenics both in terms of chronology and geographic focus. The relationship of reform and eugenics was first examined by Donald K. Pickens, *Eugenics and the Progressives* (Nashville: Vanderbilt University Press, 1968).

The study of whiteness has reshaped American history and forced historians to think about changing constructions of race and the relationship of categories of race and ethnicity: Matthew Frye Jacobson, *Whiteness of a Different Color: European Immigrants and the Alchemy of Race* (Cambridge: Harvard University Press, 1998); David R. Roediger, *Working toward Whiteness: How America's Immigrants Became White; The Strange Journey from Ellis Island to the Suburbs* (New York: Basic Books, 2005); and Matthew Pratt Guterl, *The Color of Race in America, 1900–1940* (Cambridge: Harvard University Press, 2001). It is useful as well to think about whiteness as a constructed category linked to ideas of civilization and the achievement of industry.

Historians know less about the relationship of class, race, and the construction of gender. Among works that examine the confluence of class, empire, and the domestic in shaping gender ideals, see Thomas Winter, *Making Men, Making Class: The YMCA and Workingmen, 1877–1920* (Chicago: University of Chicago Press, 2002), and John F. Kasson, *Houdini, Tarzan, and the Perfect Man: The White Male Body and the Challenge of Modernity in America* (New York: Hill and Wang, 2001). Notions of fitness and unfitness can help historians examine changing ideas about masculinity and manhood.

INDEX

Note: Page numbers in *italics* indicate figures.

Adams, Henry Carter, 17
Addams, Jane, 166, 178, 183. *See also* Hull
  House
Adler, Felix, 123, 124
Africa, 4, 36, 85, 235
African Americans: birthrate among, 243; civil
  rights movement and, 252; Communist
  Party and, 246; eugenicists and, 201, 216,
  249–52; immigrant laborers and, 71; as
  industrial workers, 6, 84–91, *90*, 237, 240;
  migration of, 13, 213, 239–44; segregation
  of, 88, 89, 157–60, 171, 245, 250; slavery of,
  46–47, 55, 84–86; socialist views of, 88–89;
  tramps and, 232; in unions, 242; vaudeville
  shows and, 171, 181–82; for workers in
  Philippines, 88, 89
Aguinaldo, Emilio, 55
Albania, 36–37
alcoholism, 2, 193, 197, 200, 201, 208; pauperism and, 209; tramps and, 219, 224
Alexander, Ruth, 158
Allen, William H., 156
American Economic Association (AEA),
  18–19, 33, 87, 89, 148
American Federation of Labor (AFL), 214
American Indians, 23–27, 229, 235; Ely on,
  23, 30–31; Huntington on, 58–59; industri-
  alization and, 87; migrations of, 31
"American race," 2, 70–71, 74–77, 80–81, 86,
  102, 126
anarchists, 22, 80, 170, 214. *See also* radicals
Anderson, Eric, 89
Anderson, Nels, 216, 219–20, 227, 230, 231
Anderson, Warwick, 63

Andover House, 168, 172
Andover University, 167
anthropology, 4–7, 252; ethnography and, 1,
  26–29, 81–82; imperial, 164–65. *See also*
  primitivism
anti-Communism, 213, 234–38, 245–46, 248.
  See also socialism
Appleton, Lilla Estelle, 175
Arabs, 31, 239
Armenians, 81
Asian immigrants, 55–57, 235, 237, 253;
  birthrates of, 127; as industrial workers, 85,
  87–88, 90–97, 240; steamship companies
  and, 80; in United Kingdom, 240; U.S.
  restrictions on, 56, 244–46
Aspinwall, William, 217
assimilationism, 94, 173, 241, 244–45, 249
Australia, 21, 48, 112, 118
Aztecs, 24

baby contests, 161–62, 186, 188–90, 195, 196;
  at Iowa State Fair, 207; at Race Betterment
  Conference, 198, 201. *See also* children
Baguio, Philippines, 49–52, *50*
Bahamas, 47
Baker, Frank, 178
Bandelier, Adolf, 109
Bankoff, Greg, 48
barbarism. *See* civilization/savagery binary
Barbour, L. L., 154
Barker, L. F., 55
Barnes, Carleton P., 250
Barnum, P. T., 135, 162
Battle Creek Sanitarium, 198, 201